SpringerWienNewYork

Viktor I. Korobov • Valery F. Ochkov

Chemical Kinetics with Mathcad and Maple

SpringerWienNewYork

Ph.D. Viktor I. Korobov
Dnipropetrovsk National University
Department of Physical and Inorganic Chemistry
Gagarin Avenue 72
49050 Dnipropetrovsk
Ukraine
korvik58@mail.ru
http://twt.mpei.ac.ru/TTHB/korobov.html

Professor Valery F. Ochkov
Moscow Power Engineering Institute (TU)
Krasnokazarmennaya st. 14
Moscow
Russia
ochkov@twt.mpei.ac.ru
http://twt.mpei.ac.ru/ochkov/v_ochkov.htm

Printed in Germany

SpringerWienNewYork is a part of Springer Science+Business Media
springer.at

Cover design: WMXDesign GmbH, Heidelberg, Germany

Typesetting: SPi, Pondicherry, India

Printed on acid-free and chlorine-free bleached paper
SPIN: 80029760

Library of Congress Control Number: 2011928800

ISBN 978-3-7091-0530-6 e-ISBN 978-3-7091-0531-3
DOI 10.1007/978-3-7091-0531-3
SpringerWienNewYork

Preface

Chemical kinetics is one of the parts of physical chemistry with the most developed mathematical description. Studying basics of chemical kinetics and successful practical application of knowledge obtained demand proficiency in mathematical formalization of certain problems on kinetics and making rather sophisticated calculations. In this respect, it is difficult or sometimes even impossible to make considerable part of such calculation without using a computer. With a mass of literature on chemical kinetics the problems of practical computing the kinetics are not actually discussed. For this reason the authors consider useful to state basics of the formal kinetics of chemical reactions and approaches to two main kinetic problems, direct and inverse, in terms of up-to-date mathematical packages Maple and Mathcad.

Why did the authors choose these packages?

The history of using computers for scientific and technical calculation can be conveniently divided into three stages:

- Work with absolute codes
- Programming using high-level languages
- Using mathematical packages such as Mathcad, Maple, Matlab,[1] Mathematica, MuPAD, Derive, Statistica, etc.

There are no clear boundaries between listed stages (*information technologies*). Working in mathematical program one may insert an Excel table,[2] as the need arose, or some user functions written in C language, which codes contain fragments of assembler. Besides, absolute codes are still using in calculators, which are of great utility in scientific and technical calculations. It is better to consider here not isolated stages of computer technique development but a range of workbenches that expand and interweave. This tendency results in sharp decrease of time required for creating calculation methods and mathematical models, leads to refusal of a

[1]Matlab is more likely a special programming language rather than a mathematical package.

[2]The list of packages does not include Excel tables that are still the most popular application for computing. However, Excel just as Matlab holds the half-way position between programming languages and mathematical packages.

programmer as an additional link between a researcher and a computer, to openness of calculations, enable us to see not only result, but all formulas in traditional notation and also all intermediate data reinforced by plots and diagrams. It is openness, clearness of Mathcad calculations that makes the package attractive calculation and effective *educational* tool enabling us to use it as visualization of the basics of chemical kinetics.

Maple is justly considered to be the best package of the symbolic mathematics, particularly for analytical solution of differential equations. With this, the problems on chemical kinetics are often resolved into this task.

Mathcad and Maple were developed as program applications *alternative* to traditional programming languages. Sometimes a student or even a professional cannot solve a chemical problem because a certain step transforms it from chemistry into informatics demanding deep knowledge of programming languages. However, as a rule a chemist has no such knowledge (and does not have to). Mathcad and Maple enable us to solve a wide range of scientific, engineering, technical and training problems without using traditional programming.

A reader getting acquainted with the book content may form an opinion that the authors gave a slant towards Mathcad package and its "server development", Mathcad Application/Calculation Server. The case is that Mathcad was initially developed as a tool for *numerical* calculations. In fact, numerical calculations lie at the center of the book. At the same time, chemical kinetics also requires *analytical*, symbolic calculations. If symbolic tools of Mathcad become insufficient for solving a particular problem we decided to take advantages of Maple – an acknowledged leader among systems of computer mathematics designed for analytical calculations.

Mathcad and Maple possess some properties allowing them to be popular both among "non-programmers" and even among aces of programming. The point is that work with these packages accelerates several times (in order) statement and solving a problem. The similar situation occurred during conversion from absolute codes to high-level programming languages (FORTRAN, Pascal, BASIC, etc.).

Even if a user knows programming languages quite well it is often appear helpful to use Mathcad on a stage of formulation and debugging of a mathematical model. One of the authors leads a team of programmers and engineers that has developed and successfully markets WaterSteamPro program package (http://www.wsp.ru) designed for calculations of thermal physical properties of the heat carriers at power stations. The final version of the package was written and compiled on Visual C++, but this project was hardly implemented without previous analysis of its formulae and algorithms in Mathcad, which possesses convenient visualization tools for numbers and formulae.

It should be also noted that in distinction from Excel Mathcad enables us to create calculations open for studying and further improvements.

"There is no rose without a thorn". The main limitation of the mathematical packages was that, as a rule, they couldn't generate executable files (exe files), which could be launched without the original program. In particular, this prevented a progressive phenomenon - dividing those sitting in front of a screen into *users* and *developers*. Usually, specialists working with mathematical programs, with

Mathcad, kept "subsistence production": developed calculations for their own or for the a few colleagues, who can work with Mathcad. They could give their results only to those who had installed corresponding package. However, this person may not buy a ready-made file but try to create required file by his own. Actually, we consider now small calculations, which creating and checking out require time is comparable with time for searching, installing, and learning the same ready – made version. Although, rather bulky calculations find difficulty in opening the way to the market: the personal calculation can be improved or enlarged at any time but in case of somebody else's calculation it is not. Here we can draw an analogy with another "internal Mathcad" problem. Sometimes it is easier to create a user function than find its completed version in the wilds of built-in Mathcad functions.

If a user was not acquainted with Mathcad package and did not have it in a computer it was possible to give him (her) a file only with a significant load, on condition that he (she) would install corresponding Mathcad version and would learn at least the basics of the program. Often it required to upgrade both operating system and hardware, or even to buy a new computer. It was also necessary to learn Mathcad.

Mathsoft Engineering and Education Inc., which was bought by PTC (http://www.ptc.com), a new owner of Mathcad package, in 2006, took actions to improve the situation. Firstly, they try to launch a free and shorten version, Mathcad Explorer, together with the eighth Mathcad release. Mathcad Explorer enabled us to open Mathcad files and calculate by them but not edit and save the documents on disks. The program itself could be downloaded from the internet free of charge. Secondly, the company actively developed tools for publication Mathcad worksheets on local networks and on the internet for studying but not for calculating. One of the main consumers of the mathematical programs is education branch in which the *way* to result, studying of the calculation methods, is more important than the result itself. In particular, Mathcad 2001i version, in which the letter "i" meant interactive, was designed for this.

However, all these solutions had no distant future. As was noted above, Mathcad Explorer, a rather bulky program, should be downloaded from the network and installed into a computer. In this case for solving intricate problems it is better to install Mathcad itself, which is recently possible to download from the network for prior charge, rather than a shorten version. On the other hand, it is desirable not only to view Mathcad worksheets, or rather their html or MathML copies (casts), opened on the network but also to transform them, change initial data and view (print, save) a new result. The solution of this problem, almost complete rather than partial, turned out to be possible with the help of internet again.

Mathcad Application Server (MAS) was put into operation in 2003 (it was renamed into *Mathcad Calculation Server – MA/CS* in 2006) enabling us to run Mathcad worksheets and call them *remotely* via the internet.

MA/CS technology allows us to solve the following problems:

- There is no need to install required version of Mathcad, or it's shorten version Mathcad Explorer (see above), find somewhere, check executable files for viruses and run them. We only have to connect computers to the Internet, access MA/CS server using Internet Explorer (version 5.5 and higher) or any browser installed in the palmtop computers or smart phones. This looks as if we work with Mathcad worksheet: we can change source data; make calculation and get a result (save and print). Calculation method (formulas in traditional notation but not in the program form; this feature makes Mathcad widespread) and intermediate data can be visible or hidden partially or completely.

- New methods of calculation, open for studying, become available instantly to all surfers of the Internet, or staff of a corporation, or researchers of a university. We should only give corresponding addresses to users. Moreover, information about suchlike calculations can appear in the databases of different search engines (yandex, google).
- Any error, misprint, imperfection and assumption in a calculation noted by an author or users can be corrected quickly. It is also easy to upgrade and extend calculations.
- The MA/CS technology does not exclude tradition capability to download Mathcad websheets from a server for their upgrading or extending. We only should make a corresponding reference in a document. There are two ways of using mcd-files. We can transfer them only for calculations on a working station having Mathcad installed and lock documents with passwords. Another way is to give them freely or sell them for work without limitations.
- The MAS technology allows us to cut down expenses for mathematical software for a corporation or a university. There is no need to install Mathcad to every computer for routine calculations, to equip all computer classrooms. Mathcad package is required now only for those who develop methods of calculations. The others can use corporation (university or open to public) MA/CS.

It was MA/CS technology that used by the authors to develop educational project on chemical kinetics: http://twt.mpei.ac.ru/TTHB/ChemKin.html. It is possible to get access to Mathcad web sheets collection by this reference and make basic kinetic calculations in remote access mode. For this purpose a user does to have to install Mathcad on the computer.

This book was published in Russian in 2009. Due to the internet project on chemical kinetics noted above almost all its pages have been translated into English, and the book became familiar to a large number of chemists all over the word. It becomes necessary to translate the book into English and supplement it with new data that have been done by the authors.

The authors would like to acknowledge their colleagues and former students who assisted in publication of the book:

Anna Grynova, Australian National University.
Natalia Yurchenko, Dnipropetrovsk National University.
Julia Chudova, Moscow Power Engineering Institute.
Alexander Zhurakovski, Oxford University.
Sasha Gurke, Knovel Corp.

Dnipropetrovsk, Ukraine — Viktor I. Korobov
Moscow, Russia — Valery F. Ochkov

Contents

Chapter 1
Formally-Kinetic Description of One- and Two-Step Reactions

1.1 Main Concepts of Chemical Kinetics

We must accept that in order to describe the chemical system it is urgent for us to know the exact way it follows during the transformation of the reagents into the products of the reaction. Knowledge of that kind gives us a possibility to command chemical transformation deliberately. In other words, we need to know the mechanism of the chemical transformation. Time evolution of the transition of the reactionary system from the unconfigured state (parent materials) to the finite state (products of the reaction) is of great importance too, because it is information of how fast the reaction goes. Chemical kinetics is a self-contained branch of chemical knowledge, which investigates the mechanisms of the reactions and the patterns of their passing in time, and which gives us the answers to questions from above.

Chemical kinetics together with chemical thermodynamics forms two corner stones of the chemical knowledge. However, classic thermodynamic approach to the description of the chemical systems is based on the consideration of the unconfigured and finite states of the system exceptionally, with the absolute abstracting from any assumptions about the methods (ways) of the transition of the system from the unconfigured to the finite state. Thermodynamics can define whether the system is in equilibrium state. If it is not, than thermodynamics states that the system would certainly pass into the equilibrium state, because the factors for such transition exist. Still, it is impossible to predict, what the dynamics of such transition would be, that is in what time the equilibrium state will come, in terms of classic thermodynamic approaches. Such problems are not in interest of thermodynamics, and the time coordinate is absolutely extraneous to the thermodynamics approach. This is the distinction of kind between thermodynamic and kinetic methods of the description of the chemical systems.

The mainframe notion of the chemical kinetics is *the rate of the reaction.* Reaction rate is defined as the change in the quantity of the reagent in time unit, referred to the unit of the reactionary space. The concept of reactionary space differs depending on the nature of the reaction. In the homogeneous system the

V.I. Korobov and V.F. Ochkov, *Chemical Kinetics with Mathcad and Maple*,
DOI 10.1007/978-3-7091-0531-3_1, © Springer-Verlag/Wien 2011

reaction is carried out in the whole volume, whereas in the heterogeneous system – at the phase interface. In mathematic terms:

$$r = \pm \frac{dn}{Vdt} \text{(homogeneous reaction)},$$

$$r = \pm \frac{dn}{Sdt} \text{(heterogeneous reaction)}.$$

Derivative's side here formally shows if the current substance is accumulated or consumed during the reaction. If the volume of the system in the homogeneous reaction is constant (closed system), then $dn/V = dC$, and therefore, the rate is interrelated to the change of the molar concentration of the reagent in time:

$$r = \pm \frac{dC}{dt}.$$

The change of the concentrations of the substances is different due to the different stoichiometry of the interaction between them; therefore more exact expression for the rate is as following:

$$r = \nu_i^{-1} \frac{dC_i}{dt},$$

where ν_i is a stoichiometric coefficient of the i participant of the reaction. For example, for reaction:

$$aA + bB \rightarrow cC + dD,$$

$$r = -\frac{1}{a}\frac{dC_A(t)}{dt} = -\frac{1}{b}\frac{dC_B(t)}{dt} = \frac{1}{c}\frac{dC_C(t)}{dt} = \frac{1}{d}\frac{dC_D(t)}{dt}.$$

It is considered that the reaction rate is a positive magnitude; therefore the stoichiometric coefficients of parent substances are taken with negative side.

The mathematic basis for the quantitative description of the reaction is the fundamental postulate of the chemical kinetics – *the law of mass action*. In kinetic formulation this law expresses the proportionality between the rate and the concentrations of the reagents:

$$r = k \prod_i c_i^{n_i}.$$

Here k – is a *rate constant*. It is a major kinetic parameter, which formally expresses the value of the rate when the concentrations of the reagents equal to one. The rate constant does not depend on the concentrations of the substances and on time, but for most reactions it depends on the temperature. The exponent of concentrations n is called *a reaction order for i substance*. To understand this notion we need to define *simple* and *complex* reaction.

It is assumed in the formal kinetics that if the transition of the unconfigured reagents into the products is not accompanied by the formation of intermediates of any kind, i.e., goes in one stage, and then such reaction is simple or *elementary*. For example, if it is known, that reaction

$$A + 2B \rightarrow \text{Products},$$

is simple, then the equation for the law of mass action for it would be:

$$r = kC_A(t)^1 C_B(t)^2.$$

In this case the reaction orders for each substance are exactly equal to their stoichiometric coefficients. In this case one says, that reaction has first order for substance A and second order for substance B. The sum of the reaction orders for each substance gives a common reaction order. It is obvious then, that given example is a reaction of third order.

If the process of chemical transformation goes in more than one stage, than such reaction is *complex*. Generally for complex reaction there is inconsistency between stoichiometric and kinetic equations. In the equation for the law of mass action for the complex reaction the exponents of the concentrations are some numbers, defined experimentally, and in most cases are not equal to the stoichiometric coefficients.

There are *direct* and *inverse* problems of chemical kinetics.

Starting point for solution of the direct problem of chemical kinetics is a kinetic scheme of the reaction, which reflects assumed *mechanism* of the chemical transformation. The *mechanism* in terms of formal kinetics is a certain totality of the elementary stages (elementary reactions), through which a transformation of the unconfigured substances into the finite products goes. Furthermore a *mathematic model* of the reaction is formed on the basis of the postulate scheme. From the definition of the reaction rate as a time derivative of the concentration of the reagent it follows, that for the N participants of a multi-stage reaction its mathematic model is a set of N differential equations, with each of them describing the rate of expense or accumulation of each participant of the reaction. Time dependences of concentrations, obtained in the result of the equations set solution, are so called *kinetic curves*. Analytic solution gives a set of equations of kinetic curves in the integrated form, numerical – a set of concentrations of the substances in certain moments of time.

In the *inverse* problem of chemical kinetics kinetic parameters of the reaction (reaction orders for reagents, rate constants) are calculated using experimental data. The goal of the inverse problem is to reconstruct of the kinetic scheme of the transformations, i.e., to define the mechanism of the reaction.

1.2 Kinetics of Simple Reactions

Let us consider a direct kinetic problem for simple reactions in the closed exothermic system (the volume and the temperature are constant). If we assume the correspondence between kinetic and stoichiometric equations, the scheme of the simple reaction with sole reagent going in one stage could be written as:

$$nA \xrightarrow{k} \text{Products},$$

where n is a reaction order, in this case has a value equal to the stoichiometric coefficient. We can mark out the cases of *mono-*, *bi-* and *three-molecular* reactions with only one reagent depending on the value of n. The mathematic model of such reactions could be expressed by the differential equation

$$\frac{dC_A(t)}{dt} = -kC_A(t)^n. \tag{1.1}$$

with an initial condition, which corresponds to the concentration of reagent A at the moment of the start of the reaction ($t = 0$):

$$C_A(0) = C_{A_0}.$$

Concentration C_{A_0} is called *an initial concentration*, and the values $C_A(t)$ in each moment of time – *the current concentrations.* An analytic solution of the direct kinetic problem is a definition of the functional connection between current concentration and time.

In (1.1) variables are separated, therefore its solution could be accomplished in MathCAD (Fig. 1.1). Prior to the interpretation the results of the solution we need to examine document in Fig. 1.1 in detail. In the strict sense MathCAD does not have on-board sources for the analytic solution of the differential equations, therefore given solution is obtained in a little artificial way. Firstly, the variables were preliminarily separated, and the equation was represented in the form of equality, whose both parts were completely prepared for the integration. Secondly, both parts of the equation were written in such a way, that the names of the integration variables differed from the names of the variables, used as the limits of integration. However, we have obtained a solution of the direct kinetic problem, which allows writing a time-dependence of the reagent's current concentration:

$$k = \frac{1}{(n-1)t}\left(\frac{1}{{C_A}^{n-1}} - \frac{1}{{C_{A_0}}^{n-1}}\right). \tag{1.2}$$

Evidently concentration of the reagent decreases in time differently depending on the reaction order. Thus, if the reaction order is formally conferred to the values of 0, 2 or 3, we will obtain:

$$\frac{d}{dt}C_A(t) = -k \cdot C_A(t)^n \qquad C_A(0) = C_{A0}$$

$$\int_{C_{A0}}^{C_A} \frac{1}{xx^n}\,dxx = -k \cdot \int_0^t 1\,dyy \text{ solve}, k \rightarrow \frac{-\left(-C_A \cdot C_{A0}^{\ n} + C_{A0} \cdot C_A^{\ n}\right)}{t \cdot C_A^{\ n} \cdot C_{A0}^{\ n} \cdot (-1+n)}$$

$$k = \frac{-\left(-C_A \cdot C_{A0}^{\ n} + C_{A0} \cdot C_A^{\ n}\right)}{t \cdot C_A^{\ n} \cdot C_{A0}^{\ n} \cdot (-1+n)} \left|\begin{array}{l}\text{substitute}, n = 0 \\ \text{solve}, C_A\end{array}\right. \rightarrow -k \cdot t + C_{A0}$$

$$k = \frac{-\left(-C_A \cdot C_{A0}^{\ n} + C_{A0} \cdot C_A^{\ n}\right)}{t \cdot C_A^{\ n} \cdot C_{A0}^{\ n} \cdot (-1+n)} \left|\begin{array}{l}\text{substitute}, n = 2 \\ \text{solve}, C_A\end{array}\right. \rightarrow \frac{C_{A0}}{k \cdot t \cdot C_{A0} + 1}$$

$$k = \frac{-\left(-C_A \cdot C_{A0}^{\ n} + C_{A0} \cdot C_A^{\ n}\right)}{t \cdot C_A^{\ n} \cdot C_{A0}^{\ n} \cdot (-1+n)} \left|\begin{array}{l}\text{substitute}, n = 3 \\ \text{solve}, C_A\end{array}\right. \rightarrow \begin{bmatrix} \dfrac{1}{\left(2 \cdot k \cdot t \cdot C_{A0}^{\ 2} + 1\right)^{\frac{1}{2}}} \cdot C_{A0} \\ \dfrac{-1}{\left(2 \cdot k \cdot t \cdot C_{A0}^{\ 2} + 1\right)^{\frac{1}{2}}} \cdot C_{A0} \end{bmatrix}$$

Fig. 1.1 Analytic solution of the direct kinetic problem for the simple reaction by the means of Mathcad package

$$C_A(t) = C_{A_0} - kt \text{ (zero order)}, \tag{1.3}$$

$$C_A(t) = \frac{C_{A_0}}{1 + kC_{A_0}t} \text{ (second order)}, \tag{1.4}$$

$$C_A(t) = \frac{C_{A_0}}{\sqrt{1 + 2kC_{A_0}{}^2 t}} \text{ (third order)}. \tag{1.5}$$

Apparently an equation in the form (1.2) is inapplicable to the first-order reaction, since when n = 1 it contains uncertainty of a type 0/0. However, uncertainty could be expanded due to l'Hopital's rule. Getting of the integrated form of the kinetic equation by differentiation with respect to the variable n of the numerator and the denominator for the (1.2) is shown in Fig. 1.2.

Thus in the first-order reaction current concentration decreases in time by the exponential law:

$$C_A(t) = C_{A_0}e^{-kt}. \tag{1.6}$$

Obtained dependences (1.3)–(1.6) are also called *the equations of kinetic curves.* Kinetic curves themselves are properly represented with graphs. Thus kinetic

$$k = \frac{1}{(n-1)\cdot t}\cdot\left(\frac{1}{C_A{}^{n-1}} - \frac{1}{C_{A0}{}^{n-1}}\right)$$

$$k = \frac{\frac{d}{dn}\,\mathrm{numer}\left[\frac{1}{(n-1)\cdot t}\cdot\left(\frac{1}{C_A{}^{n-1}} - \frac{1}{C_{A0}{}^{n-1}}\right)\right]}{\frac{d}{dn}\,\mathrm{denom}\left[\frac{1}{(n-1)\cdot t}\cdot\left(\frac{1}{C_A{}^{n-1}} - \frac{1}{C_{A0}{}^{n-1}}\right)\right]} \text{ substitute}, n = 1 \rightarrow k = \frac{\ln(C_{A0}) - \ln(C_A)}{t}$$

$$k = \frac{\ln(C_{A0}) - \ln(C_A)}{t} \text{ solve}, C_{A0} \rightarrow \exp(k\cdot t)\cdot C_A$$

Fig. 1.2 The derivation of the kinetic equation of the first-order reaction

$$C_{A\_0order}(C_{A0}, k, t) := C_{A0} - k\cdot t \qquad C_{A\_1order}(C_{A0}, k, t) := C_{A0}\cdot e^{-k\cdot t}$$

$$C_{A\_2order}(C_{A0}, k, t) := \frac{C_{A0}}{1 + k\cdot C_{A0}\cdot t} \qquad C_{A\_3order}(C_{A0}, k, t) := \frac{C_{A0}}{\sqrt{1 + k\cdot C_{A0}{}^2\cdot t}}$$

$$k := 1 \qquad C_{A0} := 1$$

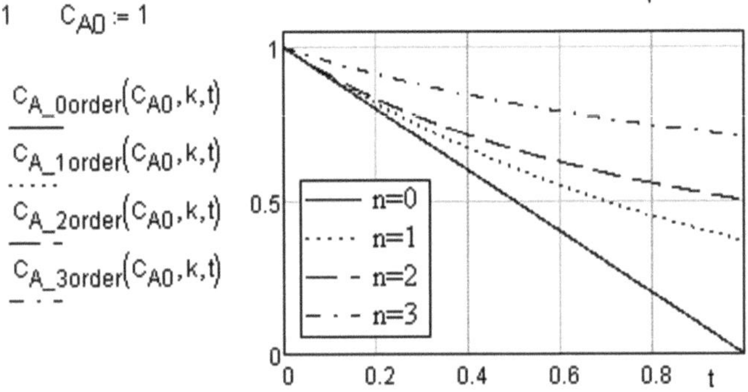

Fig. 1.3 Kinetic curves of the reagent in the elementary reactions of various orders

curves for the reagent in hypothetic reactions of different orders and same values of rate constant and initial concentrations of the reagent are shown in Fig. 1.3. As seen with the increase of the order the decrease in reagent's concentration in time becomes less intensive.

Examined case of simple reaction with sole reagent can be extended to some reaction with few reagents. For example, let the kinetic scheme of the reaction is

$$A + B \xrightarrow{k} \text{Products.}$$

If initial concentrations of the reagents A and B are equal, i.e., $C_{A_0} = C_{B_0}$, then due to the stoichiometry of the reaction:

$$C_A(t) = C_{A_0} - x(t),$$

$$C_B(t) = C_{B_0} - x(t).$$

Same quantities of both reagents, equal to x moles, would have reacted in the unit of volume by the moment of time t. Hence, $C_A(t) = C_B(t)$ and

$$\frac{dC_A(t)}{dt} = -kC_A(t)C_B(t) = -kC_A(t)^2.$$

Consequently, time-dependence of the reagent's A concentration is described by (1.4). In the same manner it could be shown that (1.5) is true for the description of the simple reaction's kinetics

$$A + B + C \xrightarrow{k} \text{Products},$$

when the initial concentrations of all three reagents are equal.

Other versions of the transformations with the participation of several reagents are also possible, and for them obtained kinetic equations are true too. Let us assume, that there is an interactions by the scheme $A + B \rightarrow \mathit{Products}$, but reagent B is taken in such excess comparing to reagent A before the start of the reaction, that the change of its concentration during the reaction could be neglected and we can consider $C_B(t) = \text{const}$. Then

$$\frac{dC_A(t)}{dt} = -kC_A(t)C_B(t) \approx k'C_A(t).$$

In this case constant k' include practically unchangeable in time concentration of the substance B and is called an *effective rate constant* unlike *true rate constant* k. The change of reagent's A concentration corresponds to the patterns of the first-order reaction (1.6). However in this case it is said that reaction has a *pseudo-first* order.

Another important characteristic of the simple reaction is a *half-life time* $\tau_{1/2}$ – time from the moment of the beginning of the reaction, during which half of the initial quantity of the substance reacts:

$$C_A(t)_{t=\tau_{1/2}} = C_{A_0}/2.$$

It is to determine the connection between the half-life time and the initial reagent's concentration on the basis of the integrated forms of the kinetic equations of various orders (Fig. 1.4).

Due to Fig. 1.4, the character of this connection changes in principle with the change of the reaction order. Thus, half-life time in the zero-order reaction is directly proportional to the reagent's initial concentration. Half-life time in the first-order reaction is defined only by the value of the rate constant and does not depend on C_{A_0}. $\tau_{1/2}$ in the second-order reaction is inversely proportional to the initial concentration of the reagent, and in the third-order reaction – to the square of the reagent's initial concentration. These kinds are used in practice to define the order of the investigated reaction by the experimental data.

$$C_A = C_{A0} - k \cdot t \quad \left| \begin{array}{l} \text{substitute}, t = \tau_{1\_2} \\ \text{substitute}, C_A = \frac{1}{2} \cdot C_{A0} \rightarrow \frac{1}{2} \cdot \frac{C_{A0}}{k} \\ \text{solve}, \tau_{1\_2} \end{array} \right.$$

$$C_A = C_{A0} \cdot e^{-k \cdot t} \quad \left| \begin{array}{l} \text{substitute}, t = \tau_{1\_2} \\ \text{substitute}, C_A = \frac{1}{2} \cdot C_{A0} \rightarrow \frac{\ln(2)}{k} \\ \text{solve}, \tau_{1\_2} \end{array} \right.$$

$$C_A = \frac{C_{A0}}{1 + k \cdot C_{A0} \cdot t} \quad \left| \begin{array}{l} \text{substitute}, t = \tau_{1\_2} \\ \text{substitute}, C_A = \frac{1}{2} \cdot C_{A0} \rightarrow \frac{1}{k \cdot C_{A0}} \\ \text{solve}, \tau_{1\_2} \end{array} \right.$$

$$C_A = \frac{C_{A0}}{\sqrt{1 + 2 \cdot k \cdot C_{A0}^2 \cdot t}} \quad \left| \begin{array}{l} \text{substitute}, t = \tau_{1\_2} \\ \text{substitute}, C_A = \frac{1}{2} \cdot C_{A0} \rightarrow \frac{3}{2 \cdot k \cdot C_{A0}^2} \\ \text{solve}, \tau_{1\_2} \end{array} \right.$$

Fig. 1.4 Half-life times for the reactions of various orders

Kinetic equations of the reactions of various orders are often represented in the *linear form*. Thus, (1.6) for the first-order reaction looks as following after taking the logarithm:

$$\ln C_A(t) = \ln C_{A_0} - kt. \tag{1.7}$$

Kinetic equations for the second- and third-order reactions could be expressed as following:

$$\frac{1}{C_A(t)} = \frac{1}{C_{A_0}} + kt \text{ (second order)}, \tag{1.8}$$

$$\frac{1}{2C_A(t)^2} = \frac{1}{2C_{A_0}(t)^2} + kt \text{ (third order)}. \tag{1.9}$$

It is follows from (1.7)–(1.9), that for the reaction of each order *linearize coordinates* exist. These are the coordinates, in which kinetic curves could be represented in the form of straight line. Thus, kinetic curve of the reagent in the first-order reaction is linearized in the coordinates $ln\ C_A$ from t. For the second- and third-order reactions linearize coordinates are $1/C_A$ from t and $1/C_A{}^2$ from t correspondingly. In the zero-order reaction, as it follows from (1.3), time-dependence of the reagent's current concentration is linear. Model kinetic curves for the reactions of various orders and their anamorphosises in corresponding coordinates are given in Fig. 1.5. There is a very important condition: the slopes of the obtained straight lines are defined by the value of the rate constant. This fact gives us an opportunity to define the rate constant on the basis of the experimental kinetic data (see Chap. 4).

For the second-order reaction with two reagents that have different initial concentration C_{A_0} and C_{B_0}, mathematical model is:

$$\frac{dx(t)}{dt} = k(C_{A_0} - x(t))(C_{B_0} - x(t)),$$

where $x(t)$ is quantity of the reagent, that has had reacted by the moment of time t (initial condition – $x(0) = 0$). The solution of the direct kinetic problem could be expressed as

$$x(t) = C_{A_0} C_{B_0} \frac{e^{\left(C_{A_0} - C_{B_0}\right)kt} - 1}{C_{A_0} e^{\left(C_{A_0} - C_{B_0}\right)kt} - C_{B_0}}.$$

$k := 0.1 \quad C_{A0} := 1 \quad C_{A\_0o}(t) := (C_{A0} - k \cdot t) \quad C_{A\_1o}(t) := C_{A0} \cdot e^{-k \cdot t}$

$C_{A\_2o}(t) := \dfrac{C_{A0}}{1 + k \cdot C_{A0} \cdot t} \quad C_{A\_3o}(t) := \dfrac{C_{A0}}{\sqrt{1 + 2 \cdot k \cdot C_{A0}{}^2 \cdot t}}$

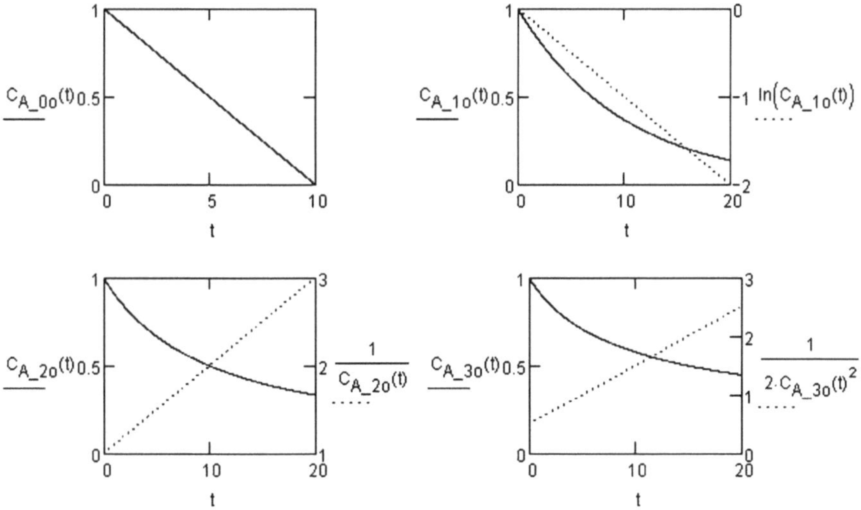

Fig. 1.5 Model kinetic curves for the reactions of various orders and their anamorphosises in linearize coordinates

At that the kinetic curves of the separate reagents are defined by the ratios:

$$C_A(t) = C_{A_0} - x(t),$$

$$C_B(t) = C_{B_0} - x(t).$$

The integrated form of the kinetic equation for this case could be also represented in the form, which indicates the possibility of the linearization of the kinetic data:

$$kt = \frac{1}{C_{A_0} - C_{B_0}} \ln \frac{C_{B_0}[C_{A_0} - x(t)]}{C_{A_0}[C_{B_0} - x(t)]}.$$

Getting corresponding equations and the calculation of the kinetic curves by the means of Mathcad is shown in Fig. 1.6.

Now we discuss the questions of the kinetics of the reactions of various orders, and in many respects we consider order as a formal value, and do not use the specific examples of chemical transformations. Essentially, this is the very peculiarity of the formal-kinetic approach to the description of the reactions. Reaction

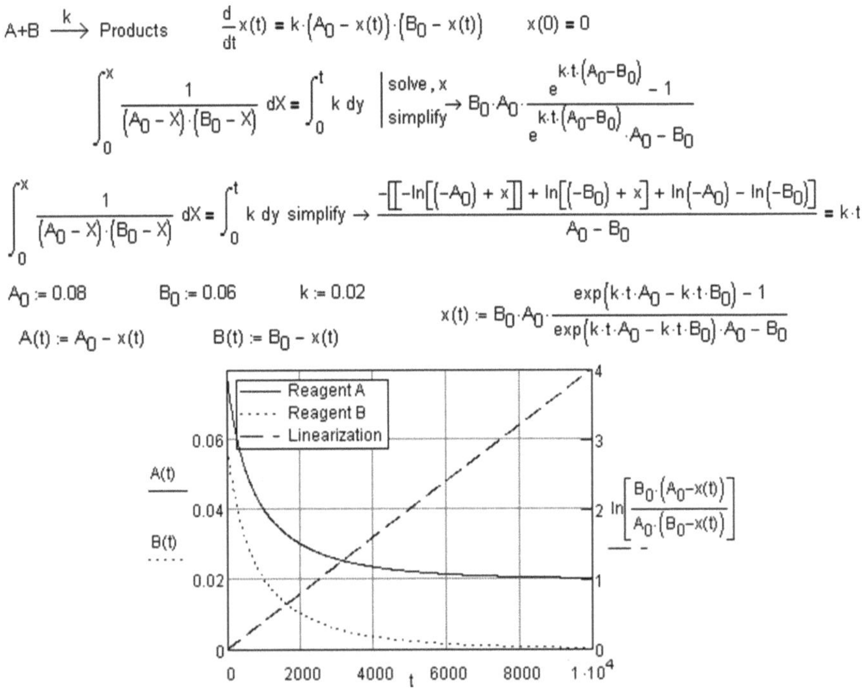

Fig. 1.6 The solution of the direct kinetic problem for the second-order reaction in case of inequality of the reagents' initial concentrations (on-line calculation http://twt.mpei.ac.ru/MCS/Worksheets/Chem/ChemKin-1-06-MCS.xmcd)

order is a value, which could not be calculated theoretically for the specific reaction, it could only be defined on the basis of data, obtained from the chemical experiment. Practice proves that the majority of the reactions have first or second order. Third-order reactions are extremely uncommon. The conception of the collision of two reacting particles in the reactionary medium is a very convenient visual metaphor of the chemical interaction. If we imagine such collision as an elementary act, leading to the appearance of the products of the reaction, then it becomes obvious, that the probability of two particles meeting each other at some point of the medium is much higher, then the probability of the collision between three particles. Because of this reason there is much more second-order reactions, then third-order reactions. And we do not even take in an account the possibility of the reactions of higher orders.

From the other side, the knowledge of the individual reaction order says nothing about its mechanism. For example, if we experimentally define the first kinetic order of the reaction, it does not necessarily mean, that this reaction is simple. Experimentally defined order can be a pseudo-order or it can indicate, that the investigated reaction is complex, and the behavior of the system is defined by some limitative stage, which has the same order as the one defined experimentally. We can state unambiguously, that the presence of the fractional or negative order of the reaction is the evidence of its complex mechanism. Some reactions have zero order. This value of the order is typical either for complex or for simple reactions that follow special mechanism, which provides such energetic conditions of the interactions between reactionary particles, in which the rate of the reaction does not depend on the concentration.

1.3 Reactions, Which Include Two Elementary Steps

A complex reaction includes more than one elementary stage. Formal-kinetic description of the complex reactions is based upon the *principle of the independence of the reactions' passing*. The main point of this principle is that if some reaction is a separate stage of a complex chemical transformation, then it goes under the same kinetic rules, as if the other stages were absent. A consequent of that principle is used in mathematic analysis: if there are several elementary stages with the participation of the same substance, then the change of its concentration is an algebraic sum of the rates of all those stages, multiplied by the stoichiometric coefficient of this substance in each stage. In this case a stoichiometric coefficient is taken with the positive sign, if in this particular stage the substance is formed, and with the negative sign, if it is expended. Let us illustrate the essence of this principle with the example of the interaction between the nitric (III) oxide and chlorine due to the overall reaction:

$$2NO + Cl_2 \rightarrow 2NOCl.$$

Reductive mechanism of this reaction could be expressed with the order of two elementary stages:

$$NO + Cl_2 \xrightarrow{k_1} NOCl_2,$$

$$NOCl_2 + NO \xrightarrow{k_2} 2NOCl.$$

Full mathematic model of the process is a set of differential equations, which describe the change of the concentration of each reaction participant in time. Let us draw attention to the fact, that initial reagent NO and intermediate product $NOCl_2$ participate in both stages of the process. Due to the principle of the independence of reactions' passing

$$\frac{dC_{NO}(t)}{dt} = (-1)r_1 + (-1)r_2 = -r_1 - r_2 = -k_1 C_{NO}(t)C_{Cl_2}(t) - k_2 C_{NOCl_2}(t)C_{NO}(t),$$

$$\frac{dC_{NOCl_2}(t)}{dt} = (+1)r_1 + (-1)r_2 = r_1 - r_2 = k_1 C_{NO}(t)C_{Cl_2}(t) - k_2 C_{NOCl_2}(t)C_{NO}(t).$$

Let us examine the regularities of some complex reactions, consisting of two elementary first-order stages. Such reactions could be expressed with following transformation schemes:

$$A \underset{k_2}{\overset{k_1}{\rightleftarrows}} B(\text{reversible reaction}),$$

$$A \xrightarrow{k_1} B \xrightarrow{k_2} C(\text{successive reaction}),$$

$$A \underset{k_2}{\overset{k_1}{\rightrightarrows}} {}^{B}_{C}(\text{parallel reaction}).$$

1.3.1 Reversible (Two-Way) Reactions

In reversible reactions the transformation of reagent into product is complicated by simultaneous reverse conversion. Due to the principle of independence of the elementary stages passing the rate of the reversible reaction is defined by the difference between rates of direct and reverse stages. Example for reaction

$$A \underset{k_2}{\overset{k_1}{\rightleftarrows}} B,$$

in which forward and reverse stages have first kinetic order

$$r(t) = \overrightarrow{r}(t) - \overleftarrow{r}(t) = k_1 C_A(t) - k_2 C_B(t).$$

It is possible that the term "reversible" reactions is not quite well turned in this particular case of chemical transformations, because from the position of thermodynamics all reaction are reversible without any exclusion. Here we talk about reversibility in formal-kinetic sense, e.g., it is implied that the rate constants of both stages have commensurable values, and we cannot neglect the rate of any of those stages.

Mathematic analysis of this kinetic scheme becomes more convenient, when we use a helper function $x(t)$ – the quantity of reagent A in volume unit that have had reacted by the moment of time t. In this case mathematic model of the reaction could be expressed by the differential equation

$$\frac{dx(t)}{dt} = k_1[C_{A_0} - x(t)] - k_2[C_{B_0} + x(t)]. \tag{1.10}$$

Here variables are easily separated, therefore, the function $x(t)$ could be defined by the means of MathCAD (Fig. 1.7). For initial condition $x(0) = 0$ we get

$$x(t) = \frac{k_1 C_{A_0} + k_2 C_{B_0}}{k_1 + k_2}\left[1 - e^{-(k_1+k_2)t}\right].$$

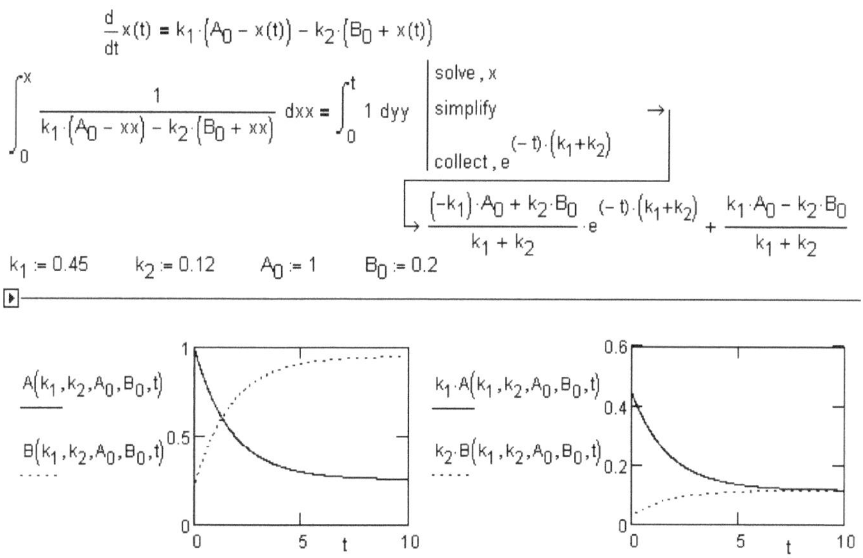

Fig. 1.7 Solution of the direct problem for the first-order two-way reaction (on-line calculation http://twt.mpei.ac.ru/MCS/Worksheets/Chem/ChemKin-1-07-MCS.xmcd)

Current concentrations of the reagents are:

$$C_A(t) = C_{A_0} - x(t),$$

$$C_B(t) = C_{B_0} + x(t).$$

Trend of kinetic curves for each participant of the reaction is modeled in Fig. 1.7. It is seen, that in some time the concentrations of the substances become time-undependable, which corresponds to the equilibrium condition. Time-dependences of the rate of each stage are also shown here, and that illustrates the dynamic character of the chemical equilibrium: reaching equilibrium condition does not mean the end of the reaction. In equilibrium condition the total rate of the reaction is equal to zero because of the equality of the rates of direct and reverse reaction, which have quiet definite values.

Often kinetic equation of two-way reaction is written in another form. By expanding (1.10) we get:

$$\frac{dx(t)}{dt} = k_1 C_{A_0} - k_2 C_{B_0} - (k_1 + k_2)x(t).$$

Assume $\overline{x}$ is the quantity of substance A that have had reacted by the moment of equilibrium establishment, then

$$\frac{d\overline{x}}{dt} = k_1 C_{A_0} - k_2 C_{B_0} - (k_1 + k_2)\overline{x} = 0.$$

It follows that:

$$k_1 C_{A_0} - k_2 C_{B_0} = (k_1 + k_2)\overline{x}.$$

Therefore, initial differential equation could be written as:

$$\frac{dx(t)}{dt} = (k_1 + k_2)\overline{x} - (k_1 + k_2)x(t) = (k_1 + k_2)[\overline{x} - x(t)],$$

and its solution for $x(0) = 0$ would look like

$$(k_1 + k_2)\,t = \ln\frac{\overline{x}}{\overline{x} - x(t)}.$$

This correlation is used for the solution of inverse kinetic problem. However, for such solution it is necessary to have specific information about equilibrium condition (values of equilibrium constant or equilibrium concentrations).

In the *second-order reversible reaction* at least one of the stages must have second kinetic order. For example, reversible bimolecular transformation is a reaction of that kind:

$$A + B \underset{k_2}{\overset{k_1}{\rightleftharpoons}} C + D,$$

to which following mathematic model corresponds

$$\frac{dx(t)}{dt} = k_1[C_{A_0} - x(t)]\ [C_{B_0} - x(t)] - k_2[C_{C_0} + x(t)]\ [C_{D_0} + x(t)].$$

Solution of this equation in Mathcad is shown in Fig. 1.8. Solution's result could be substantially simplified by using symbols: $k_1 - k_2 = \beta$; $k_1 C_{A_0} C_{B_0} - k_2 C_{C_0} C_{D_0} = \delta$; $k_1(C_{A_0} + C_{B_0}) + k_2(C_{C_0} + C_{D_0}) = \gamma$; and also $\sqrt{4\beta\delta - \gamma^2} = \lambda$. Then, as it follows from Fig. 1.8, kinetic equation of the reaction is

$$x(t) = 2tg\left(\frac{\lambda}{2}t\right) \frac{\delta}{\lambda + \gamma \cdot tg\left(\frac{\lambda}{2}t\right)}. \tag{1.11}$$

Though the presence of trigonometric functions in kinetic equation looks pretty unusual, obtained expression is quiet suitable for the calculations. Thus, kinetic curves for all participants of the reaction, calculated with (1.11) for arbitrarily chosen values of constants k_1, k_2 and initial concentrations of the substances, are shown in Fig. 1.8.

In total eight types of kinetic schemes for two-way second-order reactions exist. All of them are given in Table 1.1. We can say, that (1.11) is true for any of those eight types зa reactions, but the parameters are the different combinations of rate constants and initial concentrations of the reagents in each case.

1.3.2 Consecutive Reactions

Till now we have used only symbolic resources of Mathcad system to solve differential equations. We again accentuate, that all kinetic equations from above are the equations with separatable variables. User need to separate those variables without assistance to obtain a solution. And getting final solution is in fact an integration of both parts of the obtained equalities.

We now will separately discuss the possibilities of Mathcad's symbolic transformations. Symbolic transformations in Mathcad became possible after authors had instilled the core of Maple V R4 package symbolic operations in the program. But, symbolic core was instilled in a slightly topped version, apparently in order not to overload the package – only the simplest symbolic constructions could be calculated. Unfortunately, functions present in Maple to solve differential equations were not included in the list of functions, available for work in MathCAD.

$$\frac{d}{dt}x = k_1\cdot(A_0 - x)\cdot(B_0 - x) - k_2\cdot(C_0 + x)\cdot(D_0 + x)$$

$$k_1\cdot(A_0 - x)\cdot(B_0 - x) - k_2\cdot(C_0 + x)\cdot(D_0 + x) \;\Big|\; \begin{array}{l}\text{expand}\\ \text{collect}, x\end{array} \rightarrow$$

$$\rightarrow (k_1 - k_2)\cdot x^2 + [(-k_1)\cdot A_0 - k_1\cdot B_0 - k_2\cdot C_0 - k_2\cdot D_0]\cdot x + k_1\cdot A_0\cdot B_0 - k_2\cdot C_0\cdot D_0$$

$$\beta = k_1 - k_2 \qquad \delta = k_1\cdot A_0\cdot B_0 - k_2\cdot C_0\cdot D_0$$

$$\gamma = -(-k_1\cdot A_0 - k_1\cdot B_0 - k_2\cdot C_0 - k_2\cdot D_0) \;\Bigg|\; \begin{array}{l}\text{expand}\\ \text{collect}, k_1\\ \text{collect}, k_2\end{array} \rightarrow \gamma = (C_0 + D_0)\cdot k_2 + (A_0 + B_0)\cdot k_1$$

$$\int_0^x \frac{1}{\beta\cdot xx^2 - \gamma\cdot xx + \delta}\,dxx = \int_0^t 1\,dyy \;\Bigg|\; \begin{array}{l}\text{solve}, x\\ \text{substitute}, (4\cdot\delta\cdot\beta - \gamma^2)^{\frac{1}{2}} = \lambda\\ \text{simplify}\end{array} \rightarrow 2\cdot\tan\left(\frac{1}{2}\cdot t\cdot\lambda\right)\cdot\frac{\delta}{\lambda + \tan\left(\frac{1}{2}\cdot t\cdot\lambda\right)\cdot\gamma}$$

$$A_0 := 0.06 \quad B_0 := 0.05 \quad C_0 := 0.04 \quad D_0 := 0.03 \quad k_1 := 0.5 \quad k_2 := 0.25$$

$$\delta := k_1\cdot A_0\cdot B_0 - k_2\cdot C_0\cdot D_0 \quad \gamma := (C_0 + D_0)\cdot k_2 + (A_0 + B_0)\cdot k_1 \quad \beta := k_1 - k_2$$

$$\lambda := (4\cdot\delta\cdot\beta - \gamma^2)^{\frac{1}{2}} \qquad x(t) := 2\cdot\tan\left(\frac{1}{2}\cdot t\cdot\lambda\right)\cdot\frac{\delta}{\lambda + \tan\left(\frac{1}{2}\cdot t\cdot\lambda\right)\cdot\gamma}$$

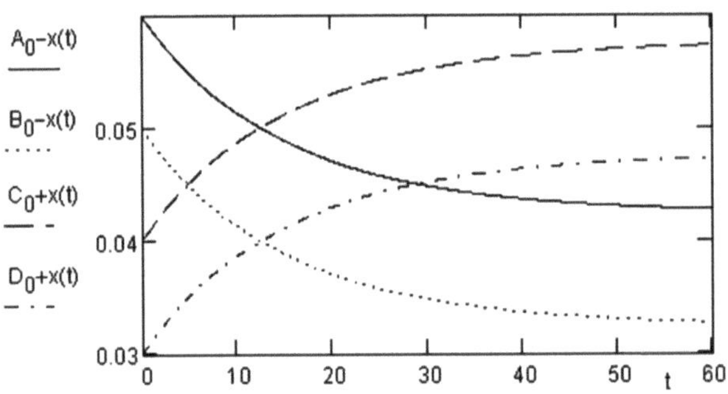

Fig. 1.8 Solution of direct kinetic problem for second-order reversible reaction $A + B \Leftrightarrow C + D$ (on-line calculation http://twt.mpei.ac.ru/MCS/Worksheets/Chem/ChemKin-1-08-MCS.xmcd)

Therefore we need to accept, that many kinetic problems could not be solved analytically in this package. And it is often very important to know analytic expressions for the time-dependences of current concentrations in particular.

In this case Maple system opens wide facilities for the analysis of mathematic models of complex reactions, primarily owing to the built-in function dsolve for solution of ordinary differential equations and sets.

Table 1.1 Parameters β, γ, δ in (1.12)

Reaction	β	γ	δ
$A + B \leftrightarrow C + D$	$k_1 - k_2$	$k_1(C_{A_0} + C_{B_0}) + k_2(C_{C_0} + C_{D_0})$	$k_1 C_{A_0} C_{B_0} - k_2 C_{C_0} C_{D_0}$
$A + B \leftrightarrow 2C$	$k_1 - 4k_2$	$k_1(C_{A_0} + C_{B_0}) + 4k_2 C_{C_0}$	$k_1 C_{A_0} C_{B_0} - k_2 C_{C_0}^2$
$2A \leftrightarrow C + D$	$4k_1 - k_2$	$k_1 C_{A_0} + k_2(C_{C_0} + C_{D_0})$	$k_1 C_{A_0}^2 - k_2 C_{C_0} C_{D_0}$
$2A \leftrightarrow 2C$	$k_1 - k_2$	$2(k_1 C_{A_0} + k_2 C_{C_0})$	$k_1 C_{A_0}^2 - k_2 C_{C_0}^2$
$A + B \leftrightarrow C$	k_1	$k_1(C_{A_0} + C_{B_0}) + k_2$	$k_1 C_{A_0} C_{B_0} - k_2 C_{C_0}$
$2A \leftrightarrow C$	$4k_1$	$4k_1 C_{A_0} + k_2$	$k_1 C_{A_0}^2 - k_2 C_{C_0}$
$A \leftrightarrow C + D$	$-k_2$	$k_1 + k_2(C_{C_0} + C_{D_0})$	$k_1 C_{A_0} - k_2 C_{C_0} C_{D_0}$
$A \leftrightarrow 2C$	$-4k_2$	$k_1 + 4k_2 C_{C_0}$	$k_1 C_{A_0} - k_2 C_{C_0}^2$

We need to apply dsolve in following format in order to find partial solution of differential equation:

```
dsolve({ODE, IC}, y(x))
```

where `IC` is an entry condition for equation `ODE`, which we need to be solved with respect to function `y(x)`.

If a set of ordinary differential equations `sysODE` is being solved with respect to a multitude of unknown functions `funcs` for given multitude of initial conditions `ICs`, then we use next format:

```
dsolve({sysODE, ICs}, {funcs})
```

Let us not forget, that entry conditions in kinetic problem are the initial values of concentrations of the reagents in the moment of time $t = 0$, and time-dependences of the same reagents' concentrations are the desired functions.

Now we will directly move to the description of successive reactions. Reactions are called *consecutive* (successive), when the products, formed in previous stages, are the reagents for the next stages.

The simplest example of a successive reaction is a formation of final product P from reagent A through the stage of intermediate product B formation:

$$A \xrightarrow{k_1} B \xrightarrow{k_2} P.$$

If each stage is a first-order reaction, then mathematic model for the given kinetic scheme, is a set of differential equations:

$$\frac{dC_A(t)}{dt} = -k_1 C_A(t),$$

$$\frac{dC_B(t)}{dt} = k_1 C_A(t) - k_2 C_B(t),$$

$$\frac{dC_P(t)}{dt} = k_2 C_B(t).$$

If only reagent A was present in reactionary medium before reaction started, then we have a set of initial conditions: $C_A(0) = C_{A_0}$, $C_B(0) = C_p(0) = 0$.

Differential equation, describing time-expense of initial reagent, is completely congruent in its form with equation of first-order simple reaction, therefore

$$C_A(t) = C_{A_0} e^{-k_1 t}. \tag{1.12}$$

We get an expression below by substituting (1.12) in the differential equation for intermediate B:

$$\frac{dC_B(t)}{dt} = k_1 C_{A_0} e^{-k_1 t} - k_2 C_B(t).$$

Unfortunately, obtained equation could not be solved by means of Mathcad symbolic core "directly". We definitely can differentiate it with respect to variable t and get linear inhomogeneous differential equation of second-order with respect to derivative. After that we can use the methodic of getting partial solution, given nearly in every handbook of differential equations. However it will mainly be a "hand" work, and not a computational calculation. Symbolic resources of Maple allow finding the solution of its equation directly and getting analytic expression for time-dependence of intermediate's concentration (Fig. 1.9):

$$C_B(t) = C_{A_0} \frac{k_1}{k_2 - k_1} \left(e^{-k_1 t} - e^{-k_2 t} \right). \tag{1.13}$$

Finally, applying material balance equation

$$C_{A_0} = C_A(t) + C_B(t) + C_C(t),$$

```
> restart;
> eq:=diff(C[B](t),t)=k[1]*C[A0]*exp(-k[1]*t)-k[2]*C[B](t);
```

$$eq := \frac{d}{dt} C_B(t) = k_1\, C_{A0}\, \mathrm{e}^{(-k_1\, t)} - k_2\, C_B(t)$$

```
> collect(dsolve({eq,C[B](0)=0},C[B](t)),C[A0]);
```

$$C_B(t) = \left(\frac{k_1\, \mathrm{e}^{(-t\,(k_1 - k_2))}}{-k_1 + k_2} + \frac{k_1}{k_1 - k_2} \right) \mathrm{e}^{(-k_2\, t)}\, C_{A0}$$

Fig. 1.9 Derivation of kinetic curve equation for intermediate product in successive first-order reaction with the help of `dsolve` function it is easy to obtain the equation of kinetic curve for the final product P:

$$C_P(t) = C_{A_0}\left(1 - \frac{k_2}{k_2 - k_1}e^{-k_1 t} + \frac{k_1}{k_2 - k_1}e^{-k_2 t}\right). \tag{1.14}$$

Kinetic curves of successive reaction participants, calculated with formulas (1.12)–(1.14), are shown in Fig. 1.10. As we see, initial reagent is expensed due to traditional patterns of first-order simple reaction: its concentration exponentially decreases in time. Kinetic curve of intermediate product has a characteristic maximum. The location of the maximum point abscissa could be found from the condition of the equality of derivative $dC_B(t)/dt$ to zero:

$$t_{\max} = \frac{\ln(k_2/k_1)}{k_2 - k_1}. \tag{1.15}$$

In the moment of time $t = t_{\max}$ intermediate's concentration reaches its maximum value, defined by expression:

$$C_{\max_B} = C_{A_0}\frac{k_1}{k_2 - k_1}\left[\left(\frac{k_2}{k_1}\right)^{-\frac{k_1}{k_2-k_1}} - \left(\frac{k_2}{k_1}\right)^{-\frac{k_2}{k_2-k_1}}\right].$$

The curve of accumulation of the final product also has an original form. First the rate of P accumulation is little, but it grows with the formation of intermediate. Further intermediate is exhausted, and the rate of final product's formation decreases again. In such a way there is a bend on the kinetic curve of substance P, and its abscissa could be found by equating the second derivative $d^2C_p(t)/dt$ with zero. It could be easily seen, that bend point abscissa corresponds to abscissa of maximum point at intermediate's curve (Fig. 1.10).

Let us draw attention to the fact, that trend of curve for final product is different in principle, then in simple reactions, considered above. The presence of bend on the curve could already be considered as some criterion, denoting the fact, that product's formation is not accompanied by previous stages in this case.

Thus, product's accumulation on initial stage goes with considerably little rate. It might happen in practice, that resulting quantities of the product are so small, that they could not be detected with present chemical-analytic methods before certain point. In this case we discuss *breakdown time of the reaction*.

Obtained analytic formulas give a possibility to build correspondent graphs, showing the degree of influence of reaction's kinetic parameters values on the location of critical points on some kinetic curves (Fig. 1.11).

We can see, that coordinates of maximum point on the curve for intermediate depend both on absolute values of rates and on their ratio (Fig. 1.11, (1.15)). It also follows from Fig. 1.11, that maximum point abscissa shifts to the smaller values of time with the increase of ratio k_2/k_1, and at the same time maximum concentration

$$C_A(C_{A0},k_1,t) := C_{A0}\cdot e^{-k_1\cdot t} \qquad C_B(C_{A0},k_1,k_2,t) := \frac{C_{A0}\cdot k_1}{k_2-k_1}\cdot\left(e^{-k_1\cdot t}-e^{-k_2\cdot t}\right)$$

$$C_P(C_{A0},k_1,k_2,t) := C_{A0}\cdot\left(1-\frac{k_2}{k_2-k_1}\cdot e^{-k_1\cdot t}+\frac{k_1}{k_2-k_1}\cdot e^{-k_2\cdot t}\right)$$

$$t_{max}(k_1,k_2) := \frac{d}{dt}C_B(C_{A0},k_1,k_2,t) = 0 \text{ solve},t \rightarrow \frac{-\ln\left(\frac{k_1}{k_2}\right)}{k_2-k_1}$$

$$t_{change}(k_1,k_2) := \frac{d^2}{dt^2}C_P(C_{A0},k_1,k_2,t) = 0 \text{ solve},t \rightarrow \frac{-\ln\left(\frac{k_1}{k_2}\right)}{k_2-k_1}$$

$$C_B\left(C_{A0},k_1,k_2,\frac{\ln\left(\frac{k_2}{k_1}\right)}{k_2-k_1}\right) \text{ simplify} \rightarrow -k_1\cdot C_{A0}\cdot\frac{\left[-\left(\frac{k_2}{k_1}\right)^{\frac{-k_1}{k_2-k_1}}\right]+\left(\frac{k_2}{k_1}\right)^{\frac{-k_2}{k_2-k_1}}}{k_2-k_1}$$

$$C_P\left(C_{A0},k_1,k_2,\frac{\ln\left(\frac{k_2}{k_1}\right)}{k_2-k_1}\right) \text{ simplify} \rightarrow -C_{A0}\cdot\frac{(-k_2)+k_1+k_2\cdot\left(\frac{k_2}{k_1}\right)^{\frac{-k_1}{k_2-k_1}}-k_1\cdot\left(\frac{k_2}{k_1}\right)^{\frac{-k_2}{k_2-k_1}}}{k_2-k_1}$$

$$C_{A0} := 1 \quad k_1 := \frac{1}{2} \quad k_2 := \frac{1}{3} \qquad t_{maxB} := t_{max}(k_1,k_2)$$

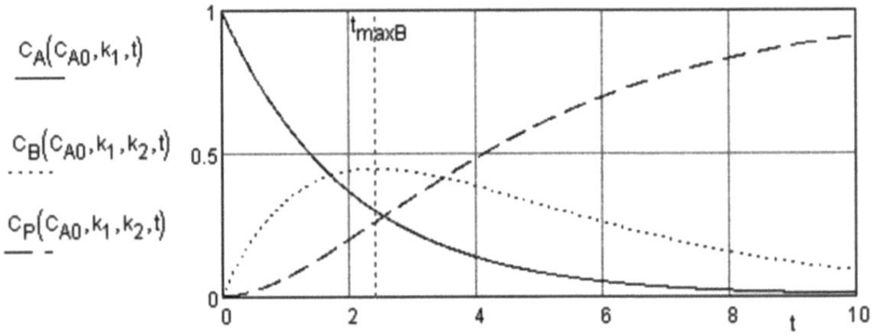

Fig. 1.10 Kinetic curves of the participant of successive first-order reaction (on-line calculation http://twt.mpei.ac.ru/MCS/Worksheets/Chem/ChemKin-1-10-MCS.xmcd)

of substance B decreases. From the other side, the bigger the constants ratio is the more intensive product's formation on the initial stage of the reaction goes. Bend point's abscissa also shifts to the home.

Direct problem of successive first-order reaction with initial concentrations $C_A(0) = C_{A_0},\ C_B(0) = C_{B_0},\ C_P(0) = C_{P_0}$, is of interest too. Analytic solution for this case could be obtained in Maple by using dsolve command for a set of differential equations (Fig. 1.12). As we see, equations of kinetic curves for intermediate and final product of the reaction could be written as:

$$C_B(C_{A0}, k_1, k_2, t) := \frac{C_{A0} \cdot k_1}{k_2 - k_1} \cdot \left(e^{-k_1 \cdot t} - e^{-k_2 \cdot t}\right)$$

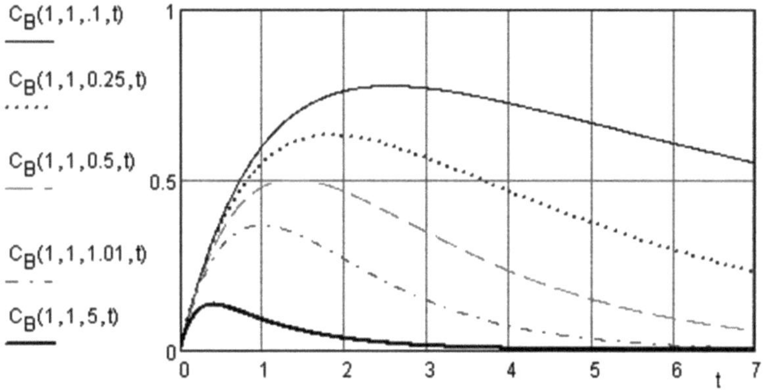

$$C_P(C_{A0}, k_1, k_2, t) := C_{A0} \cdot \left(1 - \frac{k_2}{k_2 - k_1} \cdot e^{-k_1 \cdot t} + \frac{k_1}{k_2 - k_1} \cdot e^{-k_2 \cdot t}\right)$$

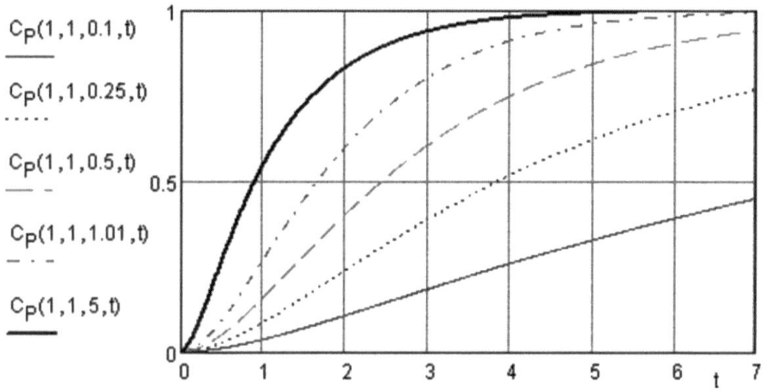

Fig. 1.11 Trend of kinetic curves for intermediate and final product of the successive reactions with different ratio k_2/k_1

$$C_B(t) = C_{B_0} e^{-k_2 t} + C_{A_0} \frac{k_1}{k_2 - k_1} \left(e^{-k_1 t} - e^{-k_2 t}\right),$$

$$C_P(t) = C_{A_0} + C_{B_0} + C_{C_0} + \frac{C_{A_0} + C_{B_0}}{k_2 - k_1} \left(k_1 e^{-k_2 t} - k_2 e^{-k_1 t}\right).$$

At that abscissa of critical points at those curves will depend not only on the constants ratio but also on initial concentrations $C_{B_0}$and C_{C_0} and:

$$t_{cr} = \frac{1}{k_2 - k_1} \ln \left[\frac{k_2}{k_1} \left(1 + \frac{C_{B_0}}{C_{A_0}} - \frac{k_2}{k_1} \frac{C_{B_0}}{C_{A_0}}\right)\right].$$

```
> restart;
> SE:=D(A)(t)=-k[1]*A(t),D(B)(t)=k[1]*A(t)-k[2]*B(t),D(P)(t)=k[2]*B(t);
```

$$SE := \mathrm{D}(A)(t) = -k_1\,A(t),\ \mathrm{D}(B)(t) = k_1\,A(t) - k_2\,B(t),\ \mathrm{D}(P)(t) = k_2\,B(t)$$

```
> InC:=A(0)=C[A0],B(0)=C[B0],P(0)=C[P0];
```

$$InC := A(0) = C_{A0},\ B(0) = C_{B0},\ P(0) = C_{P0}$$

```
> S:=dsolve({SE,InC},{A(t),B(t),P(t)}): assign(S):
> A(t),B(t);
```

$$C_{A0}\,\mathbf{e}^{(-k_1\,t)},\ \left(\frac{k_1\,C_{A0}\,\mathbf{e}^{(-k_1\,t+k_2\,t)}}{-k_1+k_2}+\frac{k_1\,C_{A0}+C_{B0}\,k_1-C_{B0}\,k_2}{k_1-k_2}\right)\mathbf{e}^{(-k_2\,t)}$$

```
> simplify(P(t));
```

$$-\frac{1}{k_1-k_2}\Big(k_1\,\mathbf{e}^{(-k_2\,t)}\,C_{B0}-k_1\,C_{A0}-C_{P0}\,k_1-C_{B0}\,k_1+k_1\,\mathbf{e}^{(-k_2\,t)}\,C_{A0}-\mathbf{e}^{(-k_2\,t)}\,C_{B0}\,k_2+C_{A0}\,k_2$$
$$-\,\mathbf{e}^{(-k_1\,t)}\,C_{A0}\,k_2+C_{P0}\,k_2+C_{B0}\,k_2\Big)$$

```
> t[max]:=solve(diff(B(t),t)=0,t);t[change]:=solve(diff(P(t),t$2)=0,t);
```

$$t_{max} := -\frac{\ln\left(\dfrac{k_2\,(k_1\,C_{A0}+C_{B0}\,k_1-C_{B0}\,k_2)}{k_1^{\,2}\,C_{A0}}\right)}{k_1-k_2}$$

$$t_{change} := -\frac{\ln\left(\dfrac{k_2\,(k_1\,C_{A0}+C_{B0}\,k_1-C_{B0}\,k_2)}{k_1^{\,2}\,C_{A0}}\right)}{k_1-k_2}$$

Fig. 1.12 Derivation of kinetic equations for successive reactions in Maple

It follows from the last expression that intermediate's concentration passes through the maximum only if next condition is fulfilled:

$$C_{A_0} + C_{B_0} > \frac{k_2}{k_1} C_{B_0}.$$

The same condition is true for the bend point presence on the final product's curve. Time-dependence of concentration for reagent A is expressed with (1.12).

We can also use special visually oriented interface elements, created by authors on the basis of bump pack Maplets, to solve direct kinetic problem, starting from Maple 9. Let us demonstrate some facilities of this package. Particularly, command

```
dsolve[interactive]({sysODE,ICs})
```

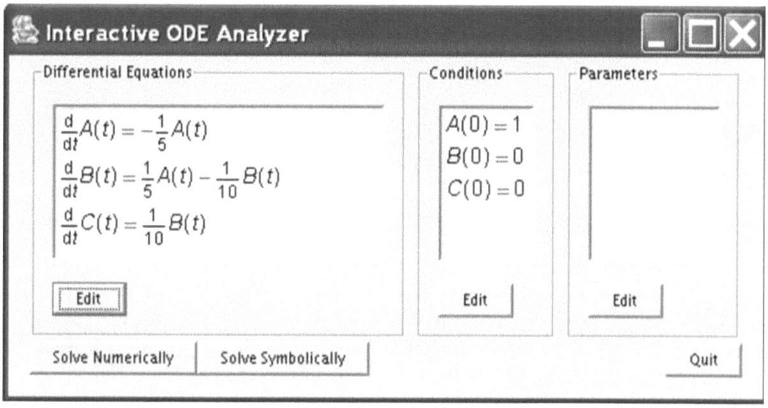

Fig. 1.13 Dialog box interactive ODE analyzer

activates a window `Interactive ODE Analyzer` (Fig. 1.13), in which information about current system and its given entry conditions is reflected. This information could be changed at user's will (`Edit` button) directly in this window, and not in a command line. Further we can choose symbolic (`Solve Analytically`) calculations mode. This would lead to the appearance of corresponding results output window. Thus, results of the analytic solution of the direct kinetic problem for successive first-order reaction with the visualization of kinetic curves are given in Fig. 1.14 (window interface resources `solve`, `plot`, `plot options` are used).

Even in the presence of two elementary stages kinetic equations of successive reaction become noticeably more complicated, if at least one of them passes due to the patterns of second-order reaction. Mathematic analysis of such mechanisms with the object of symbolic solution of direct kinetic problem in Mathcad is difficult, therefore in this case it is appropriate to use Maple's analytic facilities. It is still possible to get integrated forms of equations for some kinetic schemes in Maple.

Thus, let us examine a successive reaction:

$$A \xrightarrow{k_1} B;\ B + C \xrightarrow{k_2} P.$$

Here intermediate B, formed on the first stage, interacts with reagent C on the second stage. Second stage has a second order. Mathematic model for this reaction is:

$$\frac{dC_A(t)}{dt} = -k_1 C_A(t),$$

$$\frac{dC_B(t)}{dt} = k_1 C_A(t) - k_2 C_B(t) C_C(t),$$

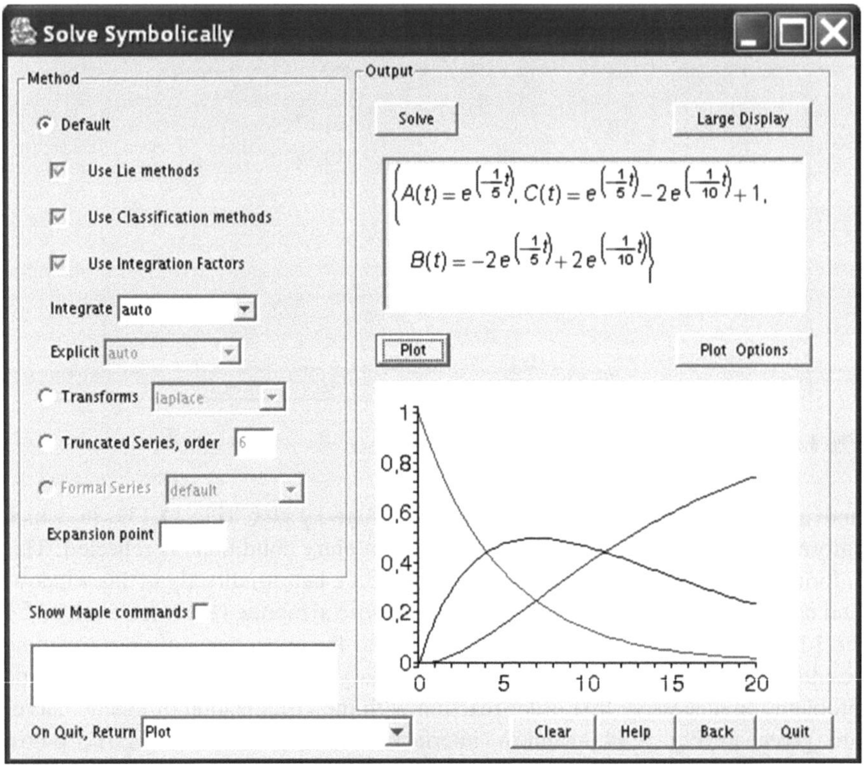

Fig. 1.14 Solution of kinetic problem with the help of visually oriented elements of Maple system

$$\frac{dC_C(t)}{dt} = -k_2 C_B(t) C_C(t),$$

$$\frac{dC_P(t)}{dt} = k_2 C_B(t) C_C(t).$$

Initial conditions for this set of ODE: $C_A(0) = C_{A_0}$, $C_C(0) = C_{C_0}$, $C_B(0) = C_P(0) = 0$. Time-dependence of reagent's A concentration is obvious:

$$C_A(t) = A_0 e^{-k_1 t}.$$

Maple gives following result for substances B and C:

$$C_B(t) = \frac{\left[(1 - e^{-k_1 t}) k_2 C_0 \alpha(t) - e^{\frac{k_1^2 t - k_2 A_0}{k_1}} + \beta\right] A_0 - k_2 {C_0}^2 \alpha(t) + \left[e^{k_2 \frac{-A_0 e^{-k_1 t} - k_1 (A_0 - C_0) t}{k_1}} - \beta\right] C_0}{k_2 C_0 \alpha(t) + \beta},$$

$$C_C(t) = \frac{e^{-\frac{k_2}{k_1}A_0 e^{-k_1 t} - k_2(A_0 - C_0)t}}{k_2\alpha(t) - \frac{\beta}{C_0}},$$

where $\beta = e^{-\frac{k_2}{k_1}A_0}$. Function α(t) is expressed by an integral with an auxiliary variable u. Independent variable t in this case is an upper integration limit:

$$\alpha(t) = \int_0^t e^{-\frac{k_2}{k_1}\left[k_1(A_0 - C_0)u + A_0 e^{-k_1 u}\right]} du.$$

Such integral could be calculated only numerically.

Equation of final product's time-dependence could be derived from material balance ratio:

$$P(t) = C_0 - C(t).$$

Given formulas allow to show trend of participants' kinetic curves with graphs. Corresponding curves, calculated for arbitrary values of rate constants and reagents' initial concentrations, are given in Fig. 1.15.

As it seen in Fig. 1.15, common trend of kinetic curves is similar to the behavior of substances in successive first-order reaction except for reagent C, participating in the second stage. On this curve there also is a characteristic bend, consistent with time-change of intermediate's B formation and expense rates.

Sufficiently complicated equations of kinetic curves for intermediate and final product are obtained during analysis of the scheme of successive transformation

$$A \xrightarrow{k_1} B\,,\ 2B \xrightarrow{k_2} P.$$

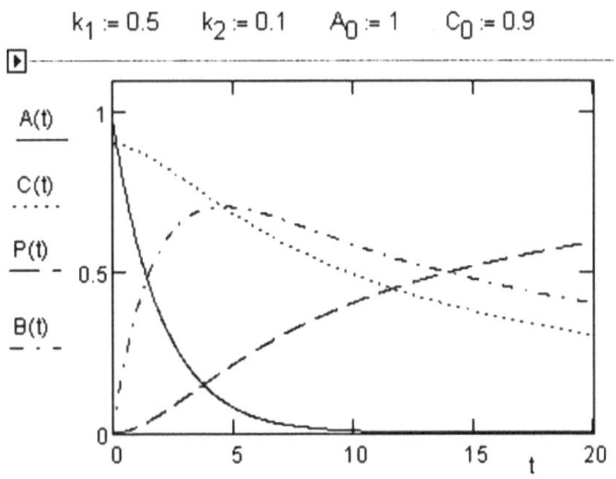

Fig. 1.15 Kinetic curves of the participants of successive second-order reaction (on-line calculation http://twt.mpei.ac.ru/MCS/Worksheets/Chem/ChemKin-1-15-MCS.xmcd)

If we assume, that kinetics of this reaction is similar to written stoichiometry, then

$$\frac{dC_A(t)}{dt} = -k_1 C_A(t),$$

$$\frac{dC_B(t)}{dt} = k_1 C_A(t) - 2k_2 C_B(t)^2,$$

$$\frac{dC_P(t)}{dt} = k_2 C_B(t).$$

It is remarkable, that function `dsolve` cannot solve this set analytically, if it is written in the working paper as it is given above. However, if we exclude first equation and put obvious expression for initial reagent into the second one

$$C_A(t) = C_{A_0} e^{-k_1 t},$$

then in that case Maple will find a solution. Let us write obtained result due to Maple's syntax:

$$C_B(t) = \frac{1}{2}\sqrt{2C_{A_0}\frac{k_1}{k_2}} \times \frac{BesselI(1,\alpha)BesselK\left(1,\alpha e^{-\frac{1}{2}k_1 t}\right) - BesselI\left(1,\alpha e^{-\frac{1}{2}k_1 t}\right)BesselK(1,\alpha)}{BesselI(1,\alpha)BesselK\left(0,\alpha e^{-\frac{1}{2}k_1 t}\right) + BesselI\left(0,\alpha e^{-\frac{1}{2}k_1 t}\right)BesselK(1)} e^{-\frac{1}{2}k_1 t},$$

$$C_p(t) = \frac{1}{2}C_{A_0}k_1 \times \int_0^t \frac{\left[BesselI(1,\alpha)BesselK\left(1,\alpha e^{-\frac{1}{2}k_1 u}\right) - BesselI\left(1,\alpha e^{-\frac{1}{2}k_1 u}\right)BesselK(1,\alpha)\right]^2}{\left[BesselI(1,\alpha)BesselK\left(0,\alpha e^{-\frac{1}{2}k_1 u}\right) + BesselI\left(0,\alpha e^{-\frac{1}{2}k_1 u}\right)BesselK(1,\alpha)\right]^2} e^{-k_1 u} du.$$

Here $\alpha = 2\sqrt{2C_{A_0}\frac{k_2}{k_1}}$.

As we see, special Bessel functions are included in the equations. Despite an outward difficult look of those equations they are completely usable to calculate kinetic curves of substances B and C. Bessel functions' values for given parameters' values could be computed with the help of mathematic packages. Let us pint, that Maple's built-in functions `BesselI(0,x)`, `BesselI(1,x)`, `BesselK(0,x)`, `BesselK(1,x)` have their Mathcad-analogs in the form of built-in functions `I0(x)`, `I1(x)`, `K0(x)`, `K1(x)` correspondingly.

Let is also give equations of kinetic curves for reaction, in which intermediate is formed as a result of two different substances interaction:

$$A + B \xrightarrow{k_1} C,$$

$$C \xrightarrow{k_2} P.$$

If $C_A(0) = C_{A_0}$, $C_{B_0}(0) = C_{B_0}$, $C_C(0) = C_P(0) = 0$, then for this kinetic scheme

$$C_A(t) = \frac{C_{A_0}\delta e^{k_1\delta t}}{C_{A_0}e^{k_1\delta t} - C_{B_0}},$$

$$C_B(t) = \frac{C_{B_0}\delta}{C_{A_0}e^{k_1\delta t} - C_{B_0}},$$

$$C_C(t) = C_{A_0}C_{B_0}\delta^2 k_1 \int_0^t \frac{e^{[k_2+k_1\delta]u}}{[C_{A_0}e^{k_1\delta u} - C_{B_0}]^2} du\, e^{-k_2 t},$$

$$C_P(t) = C_{A_0}C_{B_0}\delta^2 k_1 k_2 \int_0^t \int_0^u \frac{e^{[k_2+k_1\delta]u}}{[C_{A_0}e^{k_1\delta u} - C_{B_0}]^2} du\, e^{-k_2 u} du,$$

where $\delta = C_{A_0} - C_{B_0}$.

1.3.3 Parallel Reactions

If reagent participates in several elementary stages simultaneously, then such reaction is called *parallel*. Parallel reactions are probably the most common in practice, especially in organic synthesis. Kinetic scheme of parallel reaction with two elementary stages of first order and two final products:

$$A - \begin{matrix} \xrightarrow{k_1} B \\ \xrightarrow{k_2} C \end{matrix},$$

has its mathematic model in the form of set of ordinary differential equations

$$\frac{dC_A(t)}{dt} = -k_1 C_A(t) - k_2 C_A(t) = -(k_1 + k_2)C_A(t),$$

$$\frac{dC_B(t)}{dt} = k_1 C_A(t),$$

$$\frac{dC_C(t)}{dt} = k_2 C_A(t),$$

with set of initial conditions $C_A(0) = C_{A_0},\ C_B(0) = C_C(0) = 0$, (in the absence of products before reaction begins). First equation in mathematic model is per se a kinetic equation for irreversible first-order reaction, in which constant is expressed by sum of rate constants of parallel stages $k_1 + k_2$. Therefore it is obvious, that

$$C_A(t) = C_{A_0} e^{-(k_1+k_2)t}.$$

Substitution this ration into second and third equations of the set allows to obtain integrated forms of kinetic correlations for reaction's products (Fig. 1.16):

$$C_B(t) = \frac{k_1}{k_1 + k_2} C_{A_0}\left[1 - e^{-(k_1+k_2)t}\right],$$

$$C_C(t) = \frac{k_2}{k_1 + k_2} C_{A_0}\left[1 - e^{-(k_1+k_2)t}\right],$$

and

$$\frac{C_B(t)}{C_C(t)} = \frac{k_1}{k_2}.$$

If, for example, substance B is a target product, then we can define its yield without awaiting for the end of the process.

This conclusion, however, should not be extended on the parallel reactions, in which products are already present in reactionary mixture at zero time, which corresponds to the set of entry conditions $C_A(0) = C_{A_0},\ \ C_B(0) = C_{B_0}$, $C_C(0) = C_{C_0}$. In this case a technique of *excluding time as an independent variable* could be properly applied to find integrated forms of kinetic equations.

It follows from initial ODE set, that

$$\frac{dC_B(t)}{dt} \Big/ \frac{dC_C(t)}{dt} = \frac{k_1 C_A(t)}{k_2 C_A(t)},$$

or

$$\frac{dC_B(C_C)}{dC_C} = \frac{k_1}{k_2}.$$

After separating variables we get:

$$\int\limits_{C_{B_0}}^{C_B} dC_B = \frac{k_1}{k_2} \int\limits_{C_{C_0}}^{C_C} dC_C.$$

$$\frac{d}{dt}C_A(t) = -(k_1 + k_2)\cdot C_A(t) \qquad \frac{d}{dt}C_B(t) = k_1\cdot C_A(t) \qquad \frac{d}{dt}C_C(t) = k_2\cdot C_A(t)$$

$$C_A(0) = C_{A0} \qquad C_B(0) = 0 \qquad C_C(0) = 0$$

$$\int_{C_{A0}}^{C} \frac{1}{xx}\,dxx = -(k_1 + k_2)\cdot\int_0^t 1\,dyy \text{ solve}, C \rightarrow e^{(-t)\cdot k_1 - t\cdot k_2}\cdot C_{A0}$$

$$\frac{d}{dt}C_B(t) = k_1\cdot C_A(t) = k_1\cdot C_{A0}\cdot e^{-(k_1+k_2)\cdot t} \qquad \frac{d}{dt}C_C(t) = k_2\cdot C_A(t) = k_2\cdot C_{A0}\cdot e^{-(k_1+k_2)\cdot t}$$

$$\int_0^{C_B} 1\,dxx = \int_0^t k_1\cdot C_{A0}\cdot e^{-[(k_1+k_2)\cdot yy]}\,dyy \text{ simplify} \rightarrow C_B = (-k_1)\cdot C_{A0}\cdot\frac{e^{(-t)\cdot(k_1+k_2)} - 1}{k_1 + k_2}$$

$$\int_0^{C_C} 1\,dxx = \int_0^t k_2\cdot C_{A0}\cdot e^{-(k_1+k_2)\cdot yy}\,dyy \text{ simplify} \rightarrow C_C = (-k_2)\cdot C_{A0}\cdot\frac{e^{(-t)\cdot(k_1+k_2)} - 1}{k_1 + k_2}$$

$$C_A(C_{A0}, k_1, k_2, t) := C_{A0}\cdot e^{-(k_1+k_2)\cdot t} \qquad C_B(C_{A0}, k_1, k_2, t) := \frac{k_1\cdot C_{A0}}{k_1 + k_2}\cdot\left[1 - e^{-(k_1+k_2)\cdot t}\right]$$

$$\frac{C_B(C_{A0}, k_1, k_2, t)}{C_C(C_{A0}, k_1, k_2, t)} \rightarrow \frac{k_1}{k_2} \qquad C_C(C_{A0}, k_1, k_2, t) := \frac{k_2\cdot C_{A0}}{k_1 + k_2}\cdot\left[1 - e^{-(k_1+k_2)\cdot t}\right]$$

$$k_1 := 0.81 \qquad k_2 := 0.27 \qquad C_{A0} := 1$$

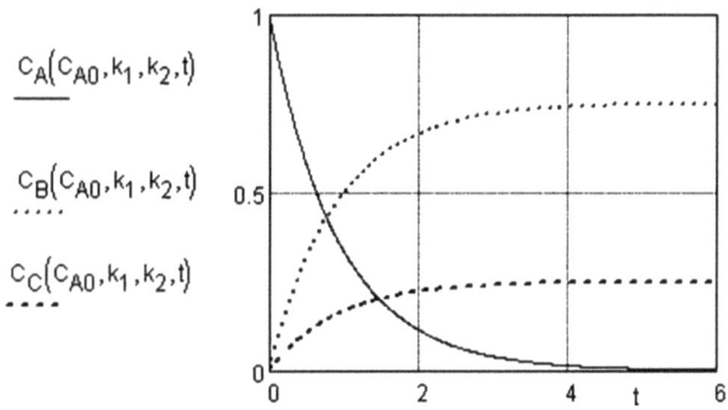

Fig. 1.16 Kinetic equations and corresponding kinetic curves for the participants of parallel first-order reaction (on-line calculation http://twt.mpei.ac.ru/MCS/Worksheets/Chem/ChemKin-1-16-MCS.xmcd)

Hence, we can express current concentration of product B through current concentration of substance C:

$$C_B(t) = C_{B_0} + \frac{k_1}{k_2}(C_C(t) - C_0).$$

In this case ratio of products' concentrations is not a constant and changes in time.

Integrated forms of kinetic equations for parallel reactions of second-order are quiet bulky in most cases. Let us confine to the examination of one of practically important examples. For example, it is well-known, that methyl halogenides CH_3X hydrolysis in alkaline solutions proceeds due to following mechanism:

$$CH_3X + H_2O \xrightarrow{k_1} CH_3OH + H^+ + X,$$

$$CH_3X + OH^- \xrightarrow{k_2} CH_3OH.$$

Here molecules of solvent and hydroxyl ions are in a way compete with each other in interacting with the molecules of organic substrate. At that the first stage could be considered as a first-order (or more exactly pseudo first) reaction on account of great solvent's excess. Second stage is an elementary bimolecular reaction. If initial concentrations of methyl halogenide and OH^- ions are equal to a and b correspondingly, then according to *Melvin-Hughes* kinetics of reaction is described by differential equation:

$$\frac{dx(t)}{dt} = k_1(a - x(t)) + k_2(a - x(t))(b - x(t)),$$

Its solution with respect to the fractional conversion of the initial reagent $x(t)/a$ is given in a Maple worksheet in Fig. 1.17.

Given solution could be additionally simplified by substitutions:

$$\alpha = k_1 + k_2(b - a),$$

$$\beta = \frac{k_2 a}{k_1 + k_2 b}.$$

```
> restart;
> DE_Moelwin_Hughes:=diff(x(t),t)=k[1]*(a-x(t))+k[2]*(a-x(t))*(b-x(t));
```

$$DE\_Moelwin\_Hughes := \frac{d}{dt}x(t) = k_1\,(a - x(t)) + k_2\,(a - x(t))\,(b - x(t))$$

```
> InValues:=x(0)=0;
```

$$InValues := x(0) = 0$$

```
> Sol:=simplify(dsolve({DE_Moelwin_Hughes,InValues},x(t))):
> assign(Sol);x(t)/a;
```

$$\frac{(k_1 + k_2\,b)\left(e^{(t\,(-k_1 - k_2\,b + k_2\,a))} - 1\right)}{-k_1 - k_2\,b + k_2\,a\,e^{(t\,(-k_1 - k_2\,b + k_2\,a))}}$$

Fig. 1.17 For the derivation of kinetic equation for methyl halogenide hydrolysis process

As a result we get a simple expression:

$$\frac{x(t)}{a} = \frac{e^{\alpha t} - 1}{e^{\alpha t} - \beta}.$$

1.3.4 Simplest Self-Catalyzed Reaction

A reaction, speeded up by one of its products, is called *self-catalyzed.* Classic example of self-catalysis is a reaction of acetone iodination in acid medium, described by total scheme

$$CH_3COCH_3 + I_2 \rightarrow CH_3COCH_2I + HI.$$

Reaction consists of two stages. On its first stage ketone transforms into enol (slowed stage):

$$CH_3COCH_3 + H_3O^+ \xrightarrow{k} CH_3COHCH_3 + H_2O.$$

Further enol reacts with iodide to form final products. As reaction passes, hydronium content increases, which speeds up the reaction. Thus, rate of the reaction is defined by rate of acetone enolization:

$$dC_A/dt = kC_AC_{H_3O^+},$$

where C_A is current concentration of acetone, k is a rate constant of rate-determining stage. If initial concentrations of acetone and hydronium are marked as C_{A_0} and C_{B_0}, then kinetic equations of self-catalyzed reaction could be written in the next form

$$\frac{dx(t)}{dt} = k(C_{A_0} - x(t))(C_{B_0} + x(t)),$$

with entry condition $x(0) = 0$. Solution of direct problem for self-catalyzed reaction, performed in Mathcad, is shown in Fig. 1.18. Peculiar forms of kinetic curves for reagent and intermediate with bends on each are of interest. Thus, equation of time-dependence of substance B concentration with expression for $x(t)$ (Fig. 1.18) being taken into account could be written as

$$C_B(t) = C_{B_0} + x(t) = \frac{C_{A_0} + C_{B_0}}{1 + \frac{A_0}{B_0} e^{-k(C_{A_0} + C_{B_0})t}}.$$

This is one of the forms of so-called equation of *logistic growth*, and its graph has a very characteristic form with plateau at the beginning and a part with quick

$$\frac{d}{dt}x(t) = k\cdot(C_{A0} - x(t))\cdot(C_{B0} - x(t)) \qquad x(0) = 0$$

$$\int_0^x \frac{1}{[(C_{A0} - xx)\cdot(C_{B0} + xx)]}\,dxx = \int_0^t k\,dyy \text{ solve}, x \rightarrow (-C_{B0})\cdot C_{A0}\cdot\frac{e^{(-t)\cdot k\cdot C_{B0} - t\cdot k\cdot C_{A0}} - 1}{e^{(-t)\cdot k\cdot C_{B0} - t\cdot k\cdot C_{A0}}\cdot C_{A0} + C_{B0}}$$

$$C_{A0} := 0.8 \qquad C_{B0} := 0.001 \qquad k := 0.2$$

$$x(t) := (-C_{B0})\cdot C_{A0}\cdot\frac{e^{(-t)\cdot k\cdot C_{B0} - t\cdot k\cdot C_{A0}} - 1}{e^{(-t)\cdot k\cdot C_{B0} - t\cdot k\cdot C_{A0}}\cdot C_{A0} + C_{B0}} \qquad C_A(t) := C_{A0} - x(t) \quad C_B(t) := C_{B0} + x(t)$$

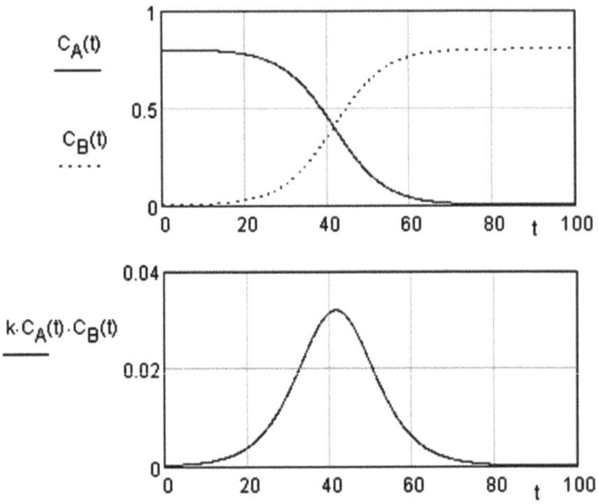

$$\frac{d}{dt}(k\cdot C_A(t)\cdot C_B(t)) = 0 \;\Big|\; \begin{matrix}\text{solve}, t\\ \text{float}, 3\end{matrix} \rightarrow 41.7$$

Fig. 1.18 Solution of direct kinetic problem for self-catalyzed reaction

growth of function, finishing with limiting value for the function. Logistic curve equation could be found in many branches of natural science. Many processes and effects pass due to the logistic law of growth: dissemination of data flows in science, origin and slow force of new ideas' way with their more or less rapid development and so on. By analogy kinetic curve of the reagent could be called a curve of logistic slump.

Graph of time-dependence of self-catalyzed reaction rate, given in Fig. 1.18, shows the presence of maximum. Location of maximum point could be found from the condition equality of second derivative $d^2C_B(t)/dt$ to zero:

$$t_{\max} = \frac{\ln(C_{A_0}/C_{B_0})}{k(C_{A_0} + C_{B_0})}.$$

It is important to emphasize, that substance B is still not a catalyst in the true sense of this term. Here, as we see, this substance accumulates in quiet substantial quantities during reaction, which is not typical for the effects of ordinary "classic" catalysis.

Let us remark, that while deriving an equation we were guided by kinetics of the first stage of total reaction only, completely ignoring next stage of interaction between enol and iodide. Such approach in chemical kinetics is based upon the principle of *limitative stage*: if total process is realized through a number of successive stages, then kinetic patterns of the process in common are defined by rate of the slowest stage in successive chain of transformations.

Chapter 2
Multi-Step Reactions: The Methods for Analytical Solving the Direct Problem

2.1 Developing a Mathematical Model of a Reaction

If a reaction proceeds by a large number of elementary steps and involves many different substances, developing its mathematical model "by hand" turns into a quite exhausting procedure fraught with different possible errors, especially provided complicated reaction stoichiometry. This stage can be considerably simplified by using matrix algebra suits.

Let us consider a reversible reaction consisting of two elementary steps:

$$\text{(I)}\quad A + B \xrightarrow{k_1} AB,$$

$$\text{(II)}\quad AB \xrightarrow{k_2} A + B.$$

A rate of each of the steps is written as

$$\vec{r} = k_1 C_A(t) C_B(t),$$

$$\overleftarrow{r} = k_2 C_{AB}(t).$$

Obviously, a reaction mathematical model is an equation system

$$\frac{dC_A(t)}{dt} = -\vec{r} + \overleftarrow{r} = -k_1 C_A(t) C_B(t) + k_2 C_{AB}(t),$$

$$\frac{dC_B(t)}{dt} = -\vec{r} + \overleftarrow{r} = -k_1 C_A(t) C_B(t) + k_2 C_{AB}(t),$$

$$\frac{dC_{AB}(t)}{dt} = \vec{r} - \overleftarrow{r} = k_1 C_A(t) C_B(t) - k_2 C_{AB}(t).$$

V.I. Korobov and V.F. Ochkov, *Chemical Kinetics with Mathcad and Maple*,
DOI 10.1007/978-3-7091-0531-3_2, © Springer-Verlag/Wien 2011

Let us define a *stoichiometric matrix* $\underline{\underline{\alpha}}$ for the given kinetic scheme as

$$\underline{\underline{\alpha}} = \begin{bmatrix} -1 & -1 & 1 \\ 1 & 1 & -1 \end{bmatrix}.$$

The number of rows in a stoichiometric matrix corresponds to the number of elementary steps and the number of columns is equal to the total number of substances taking part in a reaction. Each matrix element is a stoichiometric factor of a definite substance at a given step. At the same time, a minus sign is ascribed to stoichiometric factors of reactants and ones of products have a positive sign. If a substance is not involved in a given stage, a corresponding stoichiometric factor is equal to zero. Thus, the middle column of the stoichiometric matrix reflects participation of substance B in the overall process: it is consumed at the step (I) and accumulated at the step (II). Let us place expressions for the rates of each of the steps in a *rate vector* $\underline{r}$

$$\underline{r} = \begin{bmatrix} \vec{r} \\ \overleftarrow{r} \end{bmatrix} = \begin{bmatrix} k_1 C_A C_B \\ k_2 C_{AB} \end{bmatrix}.$$

Let us now find a product of the transposed matrix $\underline{\underline{\alpha}}$ and the vector $\underline{r}$

$$\underline{\underline{\alpha}}^T \cdot \underline{r} = \begin{bmatrix} -1 & 1 \\ -1 & 1 \\ 1 & -1 \end{bmatrix} \begin{bmatrix} k_1 C_A(t) C_B(t) \\ k_2 C_{AB}(t) \end{bmatrix} = \begin{bmatrix} -k_1 C_A(t) C_B(t) + k_2 C_{AB}(t) \\ k_1 C_A(t) C_B(t) - k_2 C_{AB}(t) \\ k_1 C_A(t) C_B(t) - k_2 C_{AB}(t) \end{bmatrix}.$$

It is easy to see that this yields a vector formed by right parts of a differential equation system for a kinetic scheme under consideration.

Thereby, a kinetic equation for a multi-step reaction in a matrix form is written as:

$$\underline{\underline{\frac{d}{dt} C(t)}} = \underline{\underline{\alpha}}^T \cdot \underline{r(t)}. \tag{2.1}$$

Naturally, the above multiplication of a stoichiometric matrix and a rate vector can be performed by suits of mathematical packages. Figure 2.1 shows the operation sequence of such automatic developing the mathematical model of the considered reaction in the Mathcad environment.

Another way of developing a mathematical model can also be indicated. If an overall chain of conversions contains reversible steps, they can be considered as formally elementary. As a result, dimensions of a stoichiometric matrix and a rate vector can be reduced (Fig. 2.2).

So as to perform the above operations with a stoichiometric matrix and a rate vector in Maple, it is necessary to preliminarily link the library of built-in commands and functions of linear algebra `LinearAlgebra` (Fig. 2.3).

ORIGIN := 1

$$2\,A \underset{k_2}{\overset{k_1}{<==>}} 2\,B + C; \quad A + B \underset{k_4}{\overset{k_3}{<==>}} 2\,D; \quad B + D \underset{k_6}{\overset{k_5}{<==>}} E$$

$$A := \begin{pmatrix} \text{"Steps/Reactants"} & \text{"A"} & \text{"B"} & \text{"C"} & \text{"D"} & \text{"E"} \\ \text{"-"} & \text{"1"} & \text{"2"} & \text{"3"} & \text{"4"} & \text{"5"} \\ \text{"1"} & -2 & 2 & 1 & 0 & 0 \\ \text{"2"} & 2 & -2 & -1 & 0 & 0 \\ \text{"3"} & -1 & -1 & 0 & 2 & 0 \\ \text{"4"} & 1 & 1 & 0 & -2 & 0 \\ \text{"5"} & 0 & -1 & 0 & -1 & 1 \\ \text{"6"} & 0 & 1 & 0 & 1 & -1 \end{pmatrix} \qquad r(k,C) := \begin{bmatrix} k_1\cdot(C_1)^2 \\ k_2\cdot(C_2)^2\cdot C_3 \\ k_3\cdot C_1\cdot C_2 \\ k_4\cdot(C_4)^2 \\ k_5\cdot C_2\cdot C_4 \\ k_6\cdot C_5 \end{bmatrix}$$

$$\alpha := \text{submatrix}(A, 3, \text{rows}(A), 2, \text{cols}(A))$$

$$RHS(k,C) := \alpha^T\cdot r(k,C) \rightarrow \begin{bmatrix} -2\cdot k_1\cdot(C_1)^2 + 2\cdot k_2\cdot(C_2)^2\cdot C_3 - k_3\cdot C_1\cdot C_2 + k_4\cdot(C_4)^2 \\ 2\cdot k_1\cdot(C_1)^2 - 2\cdot k_2\cdot(C_2)^2\cdot C_3 - k_3\cdot C_1\cdot C_2 + k_4\cdot(C_4)^2 - k_5\cdot C_2\cdot C_4 + k_6\cdot C_5 \\ k_1\cdot(C_1)^2 - k_2\cdot(C_2)^2\cdot C_3 \\ 2\cdot k_3\cdot C_1\cdot C_2 - 2\cdot k_4\cdot(C_4)^2 - k_5\cdot C_2\cdot C_4 + k_6\cdot C_5 \\ k_5\cdot C_2\cdot C_4 - k_6\cdot C_5 \end{bmatrix}$$

Fig. 2.1 Developing the mathematical model of the reaction in Mathcad

ORIGIN := 1

$$2\,A \underset{k_2}{\overset{k_1}{<==>}} 2\,B + C; \quad A + B \underset{k_4}{\overset{k_3}{<==>}} 2\,D; \quad B + D \underset{k_6}{\overset{k_5}{<==>}} E$$

$$A := \begin{pmatrix} \text{"Steps/Reactants"} & \text{"A"} & \text{"B"} & \text{"C"} & \text{"D"} & \text{"E"} \\ \text{"-"} & \text{"1"} & \text{"2"} & \text{"3"} & \text{"4"} & \text{"5"} \\ \text{"1-2"} & -2 & 2 & 1 & 0 & 0 \\ \text{"3-4"} & -1 & -1 & 0 & 2 & 0 \\ \text{"5-6"} & 0 & -1 & 0 & -1 & 1 \end{pmatrix} \qquad r(k,C) := \begin{bmatrix} k_1\cdot(C_1)^2 - k_2\cdot(C_2)^2\cdot C_3 \\ k_3\cdot C_1\cdot C_2 - k_4\cdot(C_4)^2 \\ k_5\cdot C_2\cdot C_4 - k_6\cdot C_5 \end{bmatrix}$$

$$\alpha := \text{submatrix}(A, 3, \text{rows}(A), 2, \text{cols}(A))$$

$$RHS(k,C) := \alpha^T\cdot r(k,C) \rightarrow \begin{bmatrix} 2\cdot C_3\cdot k_2\cdot(C_2)^2 - k_3\cdot C_1\cdot C_2 - 2\cdot k_1\cdot(C_1)^2 + k_4\cdot(C_4)^2 \\ 2\cdot k_1\cdot(C_1)^2 - k_3\cdot C_1\cdot C_2 - 2\cdot C_3\cdot k_2\cdot(C_2)^2 - k_5\cdot C_2\cdot C_4 + k_4\cdot(C_4)^2 + C_5\cdot k_6 \\ (C_1)^2\cdot k_1 - (C_2)^2\cdot C_3\cdot k_2 \\ C_5\cdot k_6 - C_2\cdot k_5\cdot C_4 - 2\cdot k_4\cdot(C_4)^2 + 2\cdot C_1\cdot C_2\cdot k_3 \\ C_2\cdot C_4\cdot k_5 - C_5\cdot k_6 \end{bmatrix}$$

Fig. 2.2 Using the reduced stoichiometric matrix

```
> restart;
> with(LinearAlgebra):
> alpha:=Matrix([[-2,2,1,0,0],[2,-2,-1,0,0],[-1,-1,0,2
  ,0],[1,1,0,-2,0],[0,-1,0,-1,1],[0,1,0,1,-1]]):
> r:=Vector([k[1]*C[1]^2,k[2]*C[2]^2*C[3],k[3]*C[1]*C[
  2],k[4]*C[4]^2,k[5]*C[2]*C[4],k[5]*C[5]]):
> Rhside:=Transpose(alpha).r:
> alpha,r;
```

$$\begin{bmatrix} -2 & 2 & 1 & 0 & 0 \\ 2 & -2 & -1 & 0 & 0 \\ -1 & -1 & 0 & 2 & 0 \\ 1 & 1 & 0 & -2 & 0 \\ 0 & -1 & 0 & -1 & 1 \\ 0 & 1 & 0 & 1 & -1 \end{bmatrix}, \begin{bmatrix} k_1 C_1^2 \\ k_2 C_2^2 C_3 \\ k_3 C_1 C_2 \\ k_4 C_4^2 \\ k_5 C_2 C_4 \\ k_5 C_5 \end{bmatrix}$$

```
> Rhside;
```

$$\begin{bmatrix} -2k_1 C_1^2 + 2k_2 C_2^2 C_3 - k_3 C_1 C_2 + k_4 C_4^2 \\ 2k_1 C_1^2 - 2k_2 C_2^2 C_3 - k_3 C_1 C_2 + k_4 C_4^2 - k_5 C_2 C_4 + k_5 C_5 \\ k_1 C_1^2 - k_2 C_2^2 C_3 \\ 2k_3 C_1 C_2 - 2k_4 C_4^2 - k_5 C_2 C_4 + k_5 C_5 \\ k_5 C_2 C_4 - k_5 C_5 \end{bmatrix}$$

Fig. 2.3 Developing the mathematical model using the Maple suits

The most significant stoichiometric matrix tool, which plays an important role in solving kinetic problems, is its *rank*. As it is known, a matrix rank defines the number of its linearly independent rows or columns. Using of the notion of a matrix rank allows to reduce the number of differential equations in a reaction mathematical model and, thereby, to make solving the direct and inverse kinetic problems easier. For example, let us consider a reaction scheme:

$$A + B \underset{k_2}{\overset{k_1}{\rightleftarrows}} C;$$

$$C + C \underset{k_4}{\overset{k_3}{\rightleftarrows}} D.$$

Let us develop a mathematical model for this reaction in the form of right parts of the overall differential equation system:

$$\begin{pmatrix} -1 & 1 & 1 & 0 \\ 1 & 1 & -1 & 0 \\ 0 & 0 & -2 & 1 \\ 0 & 0 & 2 & -1 \end{pmatrix}^T \cdot \begin{pmatrix} k_1 C_A C_B \\ k_2 C_C \\ k_3 C_C{}^2 \\ k_4 C_D \end{pmatrix} = \begin{pmatrix} -k_1 C_A C_B + k_2 C_C \\ -k_1 C_A C_B + k_2 C_C \\ -k_1 C_A C_B - k_2 C_C - 2k_3 C_C{}^2 + 2k_4 C_D \\ k_3 C_C{}^2 - k_4 C_D \end{pmatrix}.$$

If to calculate the stoichiometric matrix rank (calculation of a rank in Mathcad and Maple is implemented by means of the `rank` function), it will appear that it is equal to two. This means that it is possible to leave only two differential equations without prejudice to kinetic description of the mathematical model. Let us leave the equations describing a concentration change of substances C and D:

$$\begin{cases} dC_C/dt = -k_1 C_A C_B - k_2 C_C - 2k_3 C_C{}^2 + 2k_4 C_D \\ dC_D/dt = k_3 C_C{}^2 - k_4 C_D \end{cases}.$$

However, as we can see, the current concentrations of reactants A and B appear in one of the equations entering the obtained system. It is necessary to express them in terms of the current concentration of reactants C and D, otherwise it will be impossible to solve the equation. Such replacement can be performed using main balance relationships for complex reactions. Since we consider the process which occurs in a closed system, according to the mass conservation law we can write

$$\sum_i m_{i_0} = \sum_i m_i$$

where m_{i_0} and m_i are initial and current masses of the ith component, respectively. In turn, under isochoric conditions these masses are connected with initial and current concentrations C_{i_0} and C_i by

$$m_{i_0} = C_{i_0} M_i; \;\; m_i = C_i M_i,$$

where M_i is a molar mass of the ith substance taking part in a reaction. Then an equation of system material balance can be written in a matrix form as

$$\underline{\underline{\alpha}} \cdot \underline{M} = \underline{0}$$

where $\underline{M}$ is a vector of molar masses and $\underline{0}$ is a zero vector.

Figure 2.4 shows the Mathcad document intended for simplifying the calculations while developing the reduced system of the differential equations. Performing the elementary symbolic transformations of the vectors and matrices involving the enumerated balance relationships allows to write the equation system in the final form:

$$\begin{cases} dC_C/dt = -k_1 (C_{A_0} - C_C - 2C_D)(C_{B_0} - C_C - 2C_D) - k_2 C_C - 2k_3 C_C{}^2 + 2k_4 C_D \\ dC_D/dt = k_3 C_C{}^2 - k_4 C_D \end{cases}.$$

$$\alpha := \begin{pmatrix} -1 & -1 & 1 & 0 \\ 0 & 0 & -2 & 1 \end{pmatrix} \qquad \mathrm{FullSys}(k_1,k_2,k_3,k_4,A,B,C,D) := \alpha^T \cdot \begin{pmatrix} k_1 \cdot A \cdot B - k_2 \cdot C \\ k_3 \cdot C^2 - k_4 \cdot D \end{pmatrix}$$

$$\boxed{\mathrm{rank}(\alpha) = 2}$$

$$\alpha \cdot \begin{pmatrix} M_A \\ M_B \\ M_C \\ M_D \end{pmatrix} \rightarrow \begin{pmatrix} -M_A - M_B + M_C \\ -2 \cdot M_C + M_D \end{pmatrix}$$

$$-M_A - M_B + M_C = 0 \ \mathrm{solve}, M_C \ \rightarrow M_A + M_B$$

$$-2 \cdot M_C + M_D = 0 \ \left| \begin{array}{l} \mathrm{substitute}, M_C = M_A + M_B \\ \mathrm{solve}, M_D \end{array} \right. \rightarrow 2 \cdot M_A + 2 \cdot M_B$$

$$\sum \begin{pmatrix} m_{A0} \\ m_{B0} \\ 0 \\ 0 \end{pmatrix} \left| \begin{array}{l} \mathrm{substitute}, m_{A0} = A_0 \cdot M_A \\ \mathrm{substitute}, m_{B0} = B_0 \cdot M_B \end{array} \right. \rightarrow A_0 \cdot M_A + B_0 \cdot M_B$$

$$\sum \begin{pmatrix} m_A \\ m_B \\ m_C \\ m_D \end{pmatrix} \left| \begin{array}{l} \mathrm{substitute}, m_A = A \cdot M_A \\ \mathrm{substitute}, m_B = B \cdot M_B \\ \mathrm{substitute}, m_C = C \cdot (M_A + M_B) \\ \mathrm{substitute}, m_D = 2 \cdot D \cdot (M_A + M_B) \\ \mathrm{collect}, M_A, M_B \end{array} \right. \rightarrow (C + A + 2 \cdot D) \cdot M_A + (2 \cdot D + B + C) \cdot M_B$$

$$\mathrm{Given} \qquad A_0 = C + A + 2 \cdot D \, B_0 = 2 \cdot D + B + C \qquad \mathrm{Find}(A,B) \rightarrow \begin{pmatrix} A_0 - C - 2 \cdot D \\ B_0 - 2 \cdot D - C \end{pmatrix}$$

$$\mathrm{submatrix}(\mathrm{FullSys}(k_1,k_2,k_3,k_4,A,B,C,D),2,3,0,0) \rightarrow \begin{pmatrix} k_1 \cdot A \cdot B - k_2 \cdot C - 2 \cdot k_3 \cdot C^2 + 2 \cdot k_4 \cdot D \\ k_3 \cdot C^2 - k_4 \cdot D \end{pmatrix}$$

$$\begin{pmatrix} k_1 \cdot A \cdot B - k_2 \cdot C - 2 \cdot k_3 \cdot C^2 + 2 \cdot k_4 \cdot D \\ k_3 \cdot C^2 - k_4 \cdot D \end{pmatrix} \left| \begin{array}{l} \mathrm{substitute}, A = A_0 - C - 2 \cdot D \\ \mathrm{substitute}, B = B_0 - 2 \cdot D - C \end{array} \right. \rightarrow$$

$$\rightarrow \begin{bmatrix} k_1 \cdot (A_0 - C - 2 \cdot D) \cdot (B_0 - 2 \cdot D - C) - k_2 \cdot C - 2 \cdot k_3 \cdot C^2 + 2 \cdot k_4 \cdot D \\ k_3 \cdot C^2 - k_4 \cdot D \end{bmatrix}$$

Fig. 2.4 Developing the reduced mathematical model

Thus, for multi-step reactions which involve a large number of reactants and products, corresponding mathematical models can be quite awkward. In most cases, it is not necessary to have a system of N equations for modeling N kinetic curves. Fulfillment of a condition `rank(M) < N` means that some substances entering a general kinetic scheme can be not excluded from an initial model at all, since their kinetic curves can be calculated on the basis of information about a time-dependent concentration behaviour of other key components which have not been excluded from an overall equation system.

2.2 The Classical Matrix Method for Solving the Direct Kinetic Problem

The direct problem of chemical kinetics always has an analytical solution if a reaction mathematical model is a linear system of ordinary first-order differential equations. Sequences of elementary first-order kinetic steps, including ones complicated with reversible and competitive steps, correspond to such mathematical models. Let us mark off the *classical matrix method* from analytical methods of solving such ODE systems.

For example, suppose a kinetic scheme of a consecutive-competitive reaction with a reversible step

$$B \xleftarrow{k_1} A \xrightarrow{k_2} C \underset{k_4}{\overset{k_3}{\rightleftarrows}} D.$$

The mathematical model corresponding to the above scheme is

$$\frac{d}{dt}\begin{pmatrix} C_A(t) \\ C_B(t) \\ C_C(t) \\ C_D(t) \end{pmatrix} = \begin{pmatrix} -1 & 1 & 0 & 0 \\ -1 & 0 & 1 & 0 \\ 0 & 0 & -1 & 1 \\ 0 & 0 & 1 & -1 \end{pmatrix}^T \begin{pmatrix} k_1 C_A(t) \\ k_2 C_A(t) \\ k_3 C_C(t) \\ k_4 C_D(t) \end{pmatrix}$$

$$= \begin{pmatrix} -(k_1 + k_2) C_A(t) \\ k_1 C_A(t) \\ k_2 C_A(t) - k_3 C_C(t) + k_4 C_D(t) \\ k_3 C_C(t) - k_4 C_D(t) \end{pmatrix}.$$

Let us rewrite the vector of the right parts of the system as

$$\begin{pmatrix} -(k_1 + k_2) C_A(t) + 0 \cdot C_B(t) + 0 \cdot C_C(t) + 0 \cdot C_D(t) \\ k_1 C_A(t) + 0 \cdot C_B(t) + 0 \cdot C_C(t) + 0 \cdot C_D(t) \\ k_2 C_A(t) + 0 \cdot C_B(t) - k_3 C_C(t) + k_4 C_D(t) \\ 0 \cdot C_A(t) + 0 \cdot C_B(t) + k_3 \cdot C_C(t) - k_4 C_D(t) \end{pmatrix}.$$

Let us form a *rate constant matrix* $\underline{\underline{k}}$ of the constant rates and their combinations put before the concentrations

$$\begin{pmatrix} -(k_1 + k_2) & 0 & 0 & 0 \\ k_1 & 0 & 0 & 0 \\ k_2 & 0 & -k_3 & k_4 \\ 0 & 0 & k_3 & -k_4 \end{pmatrix}.$$

It is easy to make sure that multiplication of the matrix $\underline{\underline{k}}$ and the vector of current concentrations also yields the vector of the right parts of the system

$$\begin{pmatrix} -(k_1+k_2) & 0 & 0 & 0 \\ k_1 & 0 & 0 & 0 \\ k_2 & 0 & -k_3 & k_4 \\ 0 & 0 & k_3 & -k_4 \end{pmatrix} \cdot \begin{pmatrix} C_A(t) \\ C_B(t) \\ C_C(t) \\ C_D(t) \end{pmatrix} = \begin{pmatrix} -(k_1+k_2)C_A(t) \\ k_1C_A(t) \\ k_2C_A(t) - k_3C_C(t) + k_4C_D(t) \\ k_3C_C(t) - k_4C_D(t) \end{pmatrix}.$$

Therefore,

$$\frac{d}{dt}\underline{C(t)} = \underline{\underline{k}}\underline{C(t)}. \tag{2.2}$$

Equation (2.2), as (2.1), is a matrix form of a kinetic equation of a multi-step reaction. One should pay attention that a rate constant matrix always is a square matrix. A solution of (2.2) is written as

$$\underline{C(t)} = \exp\left(\underline{\underline{k}}t\right)\underline{C_0},$$

where $\underline{C_0}$ is a vector of substance initial concentrations. However, a problem of calculating a matrix $\exp(\underline{\underline{k}}t)$ arises here. For the purpose of solving it, matrix algebra introduces the notions of an *eigenvalue vector* of a matrix and an *eigenvector* of a matrix. It is considered that a matrix $\underline{\underline{k}}$ has an eigenvector $\underline{\underline{x}}$ and a corresponding eigenvalue λ, if the condition

$$\underline{\underline{k}} \cdot \underline{\underline{x}}^{\langle i \rangle} = \lambda_i \cdot \underline{\underline{x}}^{\langle i \rangle}, \tag{2.3}$$

is met. Here: λ_i is the ith element of an eigenvalues vector λ, $\underline{\underline{x}}^{\langle i \rangle}$ is the ith column of a matrix of eigenvectors.

Thereby, each column of a matrix $\underline{\underline{x}}$ corresponds to a quite definite eigenvalue of a matrix $\underline{\underline{k}}$. Matrix algebra involves the notions of eigenvalues and eigenvectors to prove that

$$\exp\left(\underline{\underline{k}}t\right) = \underline{\underline{x}} \cdot \exp\left(\underline{\underline{\Lambda}}t\right) \cdot \underline{\underline{x}}^{-1}.$$

In the last relationship, $\underline{\underline{\Lambda}}$ is a *diagonal* matrix, whose main diagonal elements are ones of an eigenvalue vector $\underline{\lambda}$. Respectively, $\exp(\underline{\underline{\Lambda}}t)$ also is a diagonal matrix

$$\exp\left(\underline{\underline{\Lambda}}t\right) = \begin{bmatrix} \exp(\lambda_0 t) & 0 & 0 & \ldots \\ 0 & \exp(\lambda_1 t) & 0 & \ldots \\ 0 & 0 & \exp(\lambda_2 t) & \ldots \\ \ldots & \ldots & \ldots & \ldots \end{bmatrix}.$$

Taking this into account, the end form of the differential equation system solution is

$$\underline{C(t)} = \underline{\underline{x}}\exp\left(\underline{\underline{\Lambda}}t\right)\underline{\underline{x}}^{-1}\underline{C_0}.$$

Consequently, seeking analytical expressions for current concentrations of all reactants involved in a multi-step reaction is reduced to formation of matrices $\underline{\underline{x}}$, $\underline{\underline{\Lambda}}$, and a vector of initial concentrations $\underline{C_0}$ and their multiplication in the above-stated order. Note that the multiplication order is very important since, in the general case, a product of matrices is not commutative.

Figure 2.5 demonstrates the Mathcad document designed to form all the vectors and matrices necessary for solving the direct kinetic problem. The built-in function `eigenvals` should be used to find eigenvalues of a constant rate matrix. A matrix of eigenvectors is calculated with the help of `eigenvecs`.

Matrix of rate constants

$$\begin{bmatrix} -(k_1+k_2) & 0 & 0 & 0 \\ k_1 & 0 & 0 & 0 \\ k_2 & 0 & -k_3 & k_4 \\ 0 & 0 & k_3 & -k_4 \end{bmatrix}$$

Eigenvalue vector of matrix of rate constants

$$\mathrm{eigenvals}\left(\begin{bmatrix} -(k_1+k_2) & 0 & 0 & 0 \\ k_1 & 0 & 0 & 0 \\ k_2 & 0 & -k_3 & k_4 \\ 0 & 0 & k_3 & -k_4 \end{bmatrix}\right) \rightarrow \begin{bmatrix} (-k_1)-k_2 \\ 0 \\ 0 \\ (-k_3)-k_4 \end{bmatrix}$$

Diagonal matrix of elements of eigenvalue vector

$$\begin{bmatrix} \exp[(-k_1-k_2)\cdot t] & 0 & 0 & 0 \\ 0 & \exp(0\cdot t) & 0 & 0 \\ 0 & 0 & \exp(0\cdot t) & 0 \\ 0 & 0 & 0 & \exp[(-k_3-k_4)\cdot t] \end{bmatrix}$$

Eigenvector matrix

$$\mathrm{eigenvecs}\left(\begin{bmatrix} -(k_1+k_2) & 0 & 0 & 0 \\ k_1 & 0 & 0 & 0 \\ k_2 & 0 & -k_3 & k_4 \\ 0 & 0 & k_3 & -k_4 \end{bmatrix}\right) \rightarrow \begin{bmatrix} (k_1+k_2)\cdot\dfrac{k_1+k_2-k_3-k_4}{k_3\cdot k_2} & 0 & 0 & 0 \\ (-k_1)\cdot\dfrac{k_1+k_2-k_3-k_4}{k_3\cdot k_2} & 0 & 0 & 1 \\ \dfrac{-(k_1+k_2-k_4)}{k_3} & -1 & \dfrac{1}{k_3}\cdot k_4 & 0 \\ 1 & 1 & 1 & 0 \end{bmatrix}$$

Attention! The right order of the columns

$$\begin{bmatrix} (k_1+k_2)\cdot\dfrac{k_1+k_2-k_3-k_4}{k_3\cdot k_2} & 0 & 0 & 0 \\ (-k_1)\cdot\dfrac{k_1+k_2-k_3-k_4}{k_3\cdot k_2} & 0 & 1 & 0 \\ \dfrac{-(k_1+k_2-k_4)}{k_3} & \dfrac{k_4}{k_3} & 0 & -1 \\ 1 & 1 & 0 & 1 \end{bmatrix}$$

Fig. 2.5 Solving the direct kinetic problem by the matrix method: forming the necessary matrices

Note an important peculiarity of `eigenvecs`. Strictly speaking, its main destination in the Mathcad environment is symbolic but not numerical calculations. In the symbolic calculation mode, an eigenvector matrix is calculated but columns in this matrix are displayed in an arbitrary order. With each new reference to this function we get matrices with different sequence of columns in them. For this reason, it is necessary to check each time whether the *i*th column of an eigenvector matrix corresponds to the *i*th position of an eigenvalues vector. Such check should be performed by multiplication of a rate constant matrix and a chosen column of a matrix $\underline{\underline{x}}$ and comparison of an obtained result with a result of multiplication of a chosen element of eigenvalue vector and the same column.

A chosen column vector of an eigenvector matrix will correspond to a chosen element of an eigenvalue vector only if equality (2.3) is fulfilled. After a performed check, a user will probably have to sort an eigenvector matrix by himself by rearranging columns in order of correspondence.

The final stage of the calculations resulting in the vector of the differential equation system solution is illustrated in Fig. 2.6.

With the help of obtained analytical expressions it is possible to draw curves for different substances provided concrete numerical values of their initial concentration as well as values of rate constants of separate steps are given.

Let us make another one remark concerning working of `eigenvecs` in Mathcad. A user can note that this function works as if in a different way in different versions of the package yielding, at first sight, different eigenvector matrices. But in reality, there is no contradiction: it is considered in linear algebra that a vector proportional to an eigenvector is also an eigenvector. This becomes obvious when carefully analyzing relationship (2.3). Eigenvector matrices obtained by means of symbolic calculations in the Mathcad 11 environment look, as a rule,

$$
\begin{bmatrix}
(k_1+k_2)\cdot\dfrac{k_1+k_2-k_3-k_4}{k_3\cdot k_2} & 0 & 0 & 0 \\
(-k_1)\cdot\dfrac{k_1+k_2-k_3-k_4}{k_3\cdot k_2} & 0 & 1 & 0 \\
\dfrac{-(k_1+k_2-k_4)}{k_3} & \dfrac{k_4}{k_3} & 0 & -1 \\
1 & 1 & 0 & 1
\end{bmatrix}
\cdot
\begin{bmatrix}
e^{[(-k_1)-k_2]\cdot t} & 0 & 0 & 0 \\
0 & 1 & 0 & 0 \\
0 & 0 & 1 & 0 \\
0 & 0 & 0 & e^{[(-k_3)-k_4]\cdot t}
\end{bmatrix}
\cdot
\begin{bmatrix}
(k_1+k_2)\cdot\dfrac{k_1+k_2-k_3-k_4}{k_3\cdot k_2} & 0 & 0 & 0 \\
k_1\cdot\dfrac{(-k_1)-k_2+k_3+k_4}{k_3\cdot k_2} & 0 & 1 & 0 \\
\dfrac{-(k_1+k_2-k_4)}{k_3} & \dfrac{k_4}{k_3} & 0 & -1 \\
1 & 1 & 0 & 1
\end{bmatrix}^{-1}
\cdot
\begin{pmatrix} C_{A0} \\ 0 \\ 0 \\ 0 \end{pmatrix}
\rightarrow
$$

$$
\begin{bmatrix}
e^{[(-k_1)-k_2]\cdot t}\cdot C_{A0} \\
\left[(-k_1)\cdot\dfrac{e^{[(-k_1)-k_2]\cdot t}}{k_1+k_2}+\dfrac{k_1}{k_1+k_2}\right]\cdot C_{A0} \\
\left[[(-k_1)-k_2+k_4]\cdot\dfrac{e^{[(-k_1)-k_2]\cdot t}}{(k_1+k_2)\cdot(k_1+k_2-k_3-k_4)}\cdot k_2+\dfrac{k_4}{k_1+k_2}\cdot\dfrac{k_2}{k_4+k_3}+e^{[(-k_3)-k_4]\cdot t}\cdot\dfrac{k_3}{k_1+k_2-k_3-k_4}\cdot\dfrac{k_2}{k_4+k_3}\right]\cdot C_{A0} \\
\left[e^{[(-k_1)-k_2]\cdot t}\cdot\dfrac{k_3}{(k_1+k_2)\cdot(k_1+k_2-k_3-k_4)}\cdot k_2+\dfrac{k_3}{k_1+k_2}\cdot\dfrac{k_2}{k_4+k_3}-e^{[(-k_3)-k_4]\cdot t}\cdot\dfrac{k_3}{k_1+k_2-k_3-k_4}\cdot\dfrac{k_2}{k_4+k_3}\right]\cdot C_{A0}
\end{bmatrix}
$$

Fig. 2.6 Calculating the vector of the substance current concentrations very awkward and can be simplified by multiplying their separate columns by a definite constant. But a user will have to perform such simplification by himself. The versions 13 and 14, in which columns of an eigenvector matrix are displayed in a maximally simplified form, are free from this disadvantage.

All the enumerated stages of the matrix solution of an ODE system can be, of course, also realized by the Maple suits. However, provided there is the function `dsolve` which copes with linear systems without difficulty, using of the classical matrix method is unnecessary.

2.3 The Laplace Transform in Kinetic Calculations

2.3.1 Brief Notes from Operational Calculus

The *Laplace transformation* is one of central notions of *operational calculus*. The most important application of the latter is analytical seeking of general and particular solutions of some types of differential equations and systems, including ones containing partial derivatives.

In operatonal calculus, a function $F(s)$ determined by the equality:

$$F(s) = \int_0^\infty e^{-st} f(t)dt, \tag{2.4}$$

where s is a positive real number or a complex number with a positive real part, is assigned to a real function $f(t)$ determined at $t \geq 0$.[1] A function $f(t)$ is named an *original function* or *inverse transform* and $F(s)$ is a *Laplace transform*. Transfer from an original function $f(x)$ to a function $F(x)$ according to (2.4) is named the *direct* Laplace transform. Respectively, transformation of $F(x)$ in an original function is named the *inverse* Laplace transform and performed according to the formula

$$f(t) = \frac{1}{2\pi i} \int_{\gamma - i\infty}^{\gamma + i\infty} e^{st} F(s)ds. \tag{2.5}$$

The method of solving linear differential equations and systems using the Laplace transform is named *operational*. The point of the method is that functions and their derivatives entering differential equations (original functions) are transformed into

[1]The restriction $t \geq 0$ does not imply by any means automatic restriction of a class of problems which can be solved using the Laplace transform. In the vast majority of dynamic models, an independent variable t is time. System behaviour at "negative" time values makes no difference to us.

Laplace transforms according to formula (2.4). As a result of such transform, source differential equations of original functions are represented in an operational form. It is quite important that an algebraic equation of Laplace transforms corresponds to a separate linear differential equation of original functions. In fact, the Laplace transform allows to replace the operation of differentiating with multiplication. Next, an analytical solution of a modified equation (an operator solution) is sought and subjected to the inverse Laplace transform, with definite rules being followed (formula (2.5)), thus yielding a solution of a source differential equation or an ODE system.

In the "precomputer" age the above-stated sequence of operations was performed, as a rule, using special tables of original functions and Laplace transforms. Modern mathematical packages, including Mathcad and Maple, are equipped with corresponding tools to perform the direct and inverse Laplace transforms.

In the Mathcad environment, the direct Laplace transform of an original function is taken using either the command menu `Symbolics|Transform|Laplace` or the symbolic editor panel. An argument name is obligatorily indicated after calling the keyword `laplace`. When the inverse transform is being performed, the menu entry `Symbolics|Transform|InverseLaplace` or the operator `invlaplace` in the symbolic commands pallet can be used. In the last case a name of a Laplace transform is indicated. An example of performing the stated operations is given in Fig. 2.7.

$$e^{t}\ \text{laplace},t \rightarrow \frac{1}{s-1} \qquad \frac{1}{s-1}\ \text{invlaplace},s \rightarrow e^{t}$$

$$e^{a\cdot t}\ \text{laplace},t \rightarrow \frac{1}{s-a} \qquad \frac{1}{s-a}\ \text{invlaplace},s \rightarrow e^{a\cdot t}$$

Homogeneity property

$$k\cdot f_1(t)\ \text{laplace},t \rightarrow k\cdot \text{laplace}(f_1(t),t,s)$$

$$\sin(t)\ \text{laplace},t \rightarrow \frac{1}{s^2+1} \qquad 2\cdot\sin(t)\ \text{laplace},t \rightarrow \frac{2}{s^2+1}$$

$$f_1(t)\ \text{laplace},t \rightarrow \text{laplace}(f_1(t),t,s) \qquad f_2(t)\ \text{laplace},t \rightarrow \text{laplace}(f_2(t),t,s)$$

Linear property

$$f_1(t)+f_2(t)\ \text{laplace},t \rightarrow \text{laplace}(f_1(t),t,s)+\text{laplace}(f_2(t),t,s)$$

$$e^{t}+e^{b\cdot t}\ \text{laplace},t \rightarrow \frac{1}{s-1}+\frac{1}{s-b}$$

Original differentiation

$$\frac{d}{dt}f(t)\ \text{laplace},t \rightarrow s\cdot \text{laplace}(f(t),t,s)-f(0)$$

$$\frac{d^2}{dt^2}f(t)\ \text{laplace},t \rightarrow s\cdot(s\cdot \text{laplace}(f(t),t,s)-f(0)) - \left| \begin{array}{l} t \leftarrow 0 \\ \frac{d}{dt}f(t) \end{array} \right.$$

Fig. 2.7 The Laplace transformation in Mathcad

$$\frac{d}{dt}C_B(t) = k_1 \cdot C_{A0} \cdot \exp(-k_1 \cdot t) - k_2 \cdot C_B(t) \qquad B(0) = 0$$

$$\frac{d}{dt}C_B(t) \text{ laplace}, t \rightarrow s \cdot \text{laplace}(C_B(t), t, s) - C_B(0)$$

$$k_1 \cdot C_{A0} \cdot \exp(-k_1 \cdot t) - k_2 \cdot C_B(t) \text{ laplace}, t \rightarrow k_1 \cdot \frac{C_{A0}}{s + k_1} - k_2 \cdot \text{laplace}(C_B(t), t, s)$$

$$s \cdot LB = k_1 \cdot \frac{C_{A0}}{s + k_1} - k_2 \cdot LB \text{ solve}, LB \rightarrow k_1 \cdot \frac{C_{A0}}{(s + k_1) \cdot (s + k_2)}$$

$$k_1 \cdot \frac{C_{A0}}{(s + k_1) \cdot (s + k_2)} \text{ invlaplace}, s \rightarrow k_1 \cdot C_{A0} \cdot \left[\frac{-1}{k_1 - k_2} \cdot e^{(-k_1) \cdot t} + \frac{1}{k_1 - k_2} \cdot e^{(-k_2) \cdot t}\right]$$

Fig. 2.8 Applying the operator method to derivation of the kinetic curve equation for the intermediate of the consecutive reaction

To learn only the most important rules of operational calculus related to properties of original functions and Laplace tramsforms is enough to sucsessfully solve many kinetic problems. These properties are following:

- *Homogeneity property*. If a Laplace tramsform $F(s)$ corresponds to a function $f(t)$, then at a given real or complex constant k a Laplace tramsform $kF(s)$ corresponds to a function $kf(t)$;
- *Linear property*. A linear combination of Laplace tramsforms corresponds to a linear combination of original functions.

These simple rules are illustrated in Fig. 2.7.[2] The examples of applying the Laplace transform to the derivatives of the original functions are also represented here. As one can see, the transform is reduced to replacement of the differentiation operator with the factor containing the argument `s`. In addition, the expression of the Laplas tramsform contains the values of the function and its derivatives at a zero point `x(0)`, `x'(0)`, ... $x^{(n)}(0)$, i.e. the initial conditions for the source differential equation.[3]

Let us consider an example of using the operator method for solving a differential equation describing a concentration change of an intermediate during a first-order consecutive reaction (Fig. 2.8). The original function here is the intermediate current concentration $C_B(t)$. A use of the Mathcad operator `laplace` leads to the Laplace transform of this function in the form `laplace(`C_B`(t),t,s)`. Pay attention to the result of the transform of the derivative $C_B'(t)$. Besides the

[2]Many of the documents represented in this chapter do not work in Mathcad 14. While embedding the symbolic kernel MuPAD, developers did not notice that the command `laplace` do not carry out its tasks to the full extent. By the time this book is published, this mistake might be already corrected.

[3]Instead of initial conditions, arbitrary constants C_1, C_2, ... C_n can be specified. Then it is obvious that the operator method is also applicable to seeking a general solution of an equation.

Laplace transform, it also contains the value of `C_B(0)`, *i.e.* the initial condition for solving the problem. After that, the form `laplace(C_B(t),t,s)` is replaced with the variable `LB` and the resulting algebraic equation is solved for this variable. Next, the operator solution is subjected to the inverse Laplace transform (the operator `invlaplace,s`). The final expression defines the analytical form of the function `C_B(t)`.

The same sequence of the operations can be performed by the Maple suites. But in contrast to Mathcad, where a user has to find a Laplace transform and recover an original function himself, the Maple's operator method for solving an ODE is almost completely automated. If it is necessary to find a solution by means of mathematical apparatus of operational calculus, it is enough to specify an additional option in the body of `dsolve` in the form of the expression `method = laplace`. Let us illustrate this for seeking the general solution of the linear second-order differential equation

$$y(x) = a + \frac{d^2 y(x)}{dx^2},$$

in Maple.

At first sight, this equation seems to be too abstract and not related to chemical kinetics. However, we will need it in the below problems. Thus, we will show below (Chap. 5) that more complicated relationships describing, in particular, diffusion and heat transfer processes can be reduced to an equation of this type.

In the Maple document given in Fig. 2.9 the considered equation is solved twice. First, `dsolve` uses "classical" built-in suits for seeking an analytical solution by default, and then the same equation is solved by means of the Laplace transform (the option `method = laplace`). In addition, the document uses some other functions. Thus, the command `odeadvisor` diagnoses the differential equation and ascertains which type of equations it belongs to. In the both cases, the function `infolevel[dsolve]` helps to observe the process of solving. The both functions enter the command library `DEtools`. To link the library of functions and commands of integral transforms (`inttrans`) is necessary to implement the operator method. Though the form of visualization of the final result is different in the each case, checking the solution `Sol` of the equation `eq` with the help of the command `odetest(Sol,eq)` shows that the both forms are equivalent.

2.3.2 *Derivation of Kinetic Equations for Linear Sequences of First-Order Reactions*

For this type of reactions, ideology of the operator method in respect to solving systems of linear differential equations remains completely the same but in this case, when applying the Laplace transform, we obtain not a separate algebraic

```
> restart;with(DEtools):
> eq:=y(x)=a+diff(y(x),x$2);
```

$$eq := y(x) = a + \left(\frac{d^2}{dx^2}y(x)\right)$$

```
> odeadvisor(eq);
```

$$[[\_2nd\_order, \_missing\_x]]$$

```
> infolevel[dsolve]:=2:s1:=dsolve(eq,y(x));
Methods for second order ODEs:
--- Trying classification methods ---
trying a quadrature
trying high order exact linear fully integrable
trying differential order: 2; linear nonhomogeneous with symmet
ry [0,1]
trying a double symmetry of the form [xi=0, eta=F(x)]
Try solving first the homogeneous part of the ODE
      checking if the LODE has constant coefficients
      <- constant coefficients successful
   <- solving first the homogeneous part of the ODE successful
```

$$s1 := y(x) = e^{(-x)}\_C2 + e^{x}\_C1 + a$$

```
> with(inttrans):s2:=dsolve(eq,y(x),method=laplace
  );
dsolve/inttrans/solveit:   Transform of eqns is   (`laplace/int
ernal`(y(x),x,_s1)-a/_s1-_s1^2*`laplace/internal`(y(x),x,_s1)+D
(y)(0)+_s1*y(0))
dsolve/inttrans/solveit:   Algebraic Solution is   (`laplace/in
ternal`(y(x),x,_s1) = (-a+D(y)(0)*_s1+_s1^2*y(0))/_s1/(-1+_s1^2
))
```

$$s2 := y(x) = a + (-a + y(0))\cosh(x) + D(y)(0)\sinh(x)$$

```
> odetest(s2,eq);
```

$$0$$

```
> odetest(s1,eq);
```

$$0$$

Fig. 2.9 The operator method in Maple

equation but a system of linear equations which can be solved either by `lsolve` or using the solving block `GIVEN/FIND`.

Thus, Fig. 2.10 represents the Mathcad document meant for solving the ODE system using by the operator method. The system describes kinetics of the first-order consecutive reaction with the reversible second step

$$A \xrightarrow{k_1} B \underset{k_3}{\overset{k_2}{\rightleftarrows}} C,$$

at the initial concentration of the source reactant C_{A_0}. The Laplace transform is applied here to the each equation of the source system and then the system of the algebraic equations is formed of the obtained Laplace transforms. Next, its operator solution is subjected to the inverse Laplace transform.

Note that an operator solution of an equation system using the symbolic Mathcad suits is always represented as a vector. The symbolic command `invlaplace`

$$A \xrightarrow{k_1} B \underset{k_3}{\overset{k_2}{\Longleftrightarrow}} C \qquad \frac{d}{dt}C_A(t) = -k_1 \cdot C_A(t) \qquad \frac{d}{dt}C_B(t) = k_1 \cdot C_A(t) - k_2 \cdot C_B(t) + k_3 \cdot C_C(t)$$

$$C_A(0) = C_{A0} \qquad C_B(0) = 0 \qquad \frac{d}{dt}C_C(t) = k_2 \cdot C_B(t) - k_3 \cdot C_C(t) \qquad C_C(0) = 0$$

$$\frac{d}{dt}C_A(t) + k_1 \cdot C_A(t) \left| \begin{array}{l} \text{laplace}, t \\ \text{substitute}, C_A(0) = C_{A0} \end{array} \right. \rightarrow s \cdot \text{laplace}(C_A(t), t, s) - C_{A0} + k_1 \cdot \text{laplace}(C_A(t), t, s)$$

$$\frac{d}{dt}C_B(t) - (k_1 \cdot C_A(t) - k_2 \cdot C_B(t) + k_3 \cdot C_C(t)) \text{ laplace}, t \rightarrow$$

$$\rightarrow s \cdot \text{laplace}(C_B(t), t, s) - C_B(0) - k_1 \cdot \text{laplace}(C_A(t), t, s) + k_2 \cdot \text{laplace}(C_B(t), t, s) - k_3 \cdot \text{laplace}(C_C(t), t, s)$$

$$\frac{d}{dt}C(t) - (k_2 \cdot C_B(t) - k_3 \cdot C_C(t)) \text{ laplace}, t \rightarrow s \cdot \text{laplace}(C(t), t, s) - C(0) - k_2 \cdot \text{laplace}(C_B(t), t, s) + k_3 \cdot \text{laplace}(C_C(t), t, s)$$

$$\text{Given} \qquad s \cdot LA - C_{A0} + k_1 \cdot LA = 0 \qquad s \cdot LB - k_1 \cdot LA + k_2 \cdot LB - k_3 \cdot LC = 0 \qquad s \cdot LC - k_2 \cdot LB + k_3 \cdot LC = 0$$

$$\text{Find}(LA, LB, LC) \text{ collect}, s \rightarrow \begin{bmatrix} \dfrac{C_{A0}}{s + k_1} \\ k_1 \cdot \dfrac{C_{A0}}{s \cdot \left[s^2 + (k_2 + k_1 + k_3) \cdot s + k_2 \cdot k_1 + k_3 \cdot k_1\right]} \cdot (s + k_3) \\ k_1 \cdot C_{A0} \cdot \dfrac{k_2}{s \cdot \left[s^2 + (k_2 + k_1 + k_3) \cdot s + k_2 \cdot k_1 + k_3 \cdot k_1\right]} \end{bmatrix}$$

$$\frac{C_{A0}}{s + k_1} \text{ invlaplace}, s \rightarrow C_{A0} \cdot e^{(-k_1) \cdot t} \qquad \boxed{\gamma_1 + \gamma_2 = k_2 + k_1 + k_3} \qquad \boxed{\gamma_1 \cdot \gamma_2 = k_2 \cdot k_1 + k_3 \cdot k_1}$$

$$k_1 \cdot \frac{C_{A0}}{s \cdot \left[s^2 + (\gamma_1 + \gamma_2) \cdot s + \gamma_1 \cdot \gamma_2\right]} \cdot (s + k_3) \left| \begin{array}{l} \text{invlaplace}, s \\ \text{collect}, e^{(-\gamma_2) \cdot t} \\ \text{collect}, e^{(-\gamma_1) \cdot t} \\ \text{collect}, k_1, C_{A0} \end{array} \right. \rightarrow \left[\left[\frac{1}{\gamma_1 \cdot (\gamma_1 - \gamma_2)} \cdot k_3 - \frac{1}{\gamma_1 - \gamma_2}\right] \cdot e^{(-\gamma_1) \cdot t} + \left[\frac{-1}{\gamma_2 \cdot (\gamma_1 - \gamma_2)} \cdot k_3 + \frac{1}{\gamma_1 - \gamma_2}\right] \cdot e^{(-\gamma_2) \cdot t} + \frac{k_3}{\gamma_1 \cdot \gamma_2}\right] \cdot C_{A0} \cdot k_1$$

$$k_1 \cdot C_{A0} \cdot \frac{k_2}{s \cdot \left[s^2 + (\gamma_1 + \gamma_2) \cdot s + \gamma_1 \cdot \gamma_2\right]} \text{ invlaplace}, s \rightarrow k_1 \cdot C_{A0} \cdot k_2 \cdot \left[\frac{1}{\gamma_1 \cdot \gamma_2} - \frac{1}{\gamma_2 \cdot (\gamma_1 - \gamma_2)} \cdot e^{(-\gamma_2) \cdot t} + \frac{1}{\gamma_1 \cdot (\gamma_1 - \gamma_2)} \cdot e^{(-\gamma_1) \cdot t}\right]$$

Fig. 2.10 Solving the system of the linear differential equations by means of the operator method

cannot be applied to a whole vector on the stage of subsequent transfer to original functions. Each vector element should be treated separately.

Certain inconveniencies of realization of the operator method in Mathcad are connected with an order of display of symbolic results adopted in this package. When recovering original functions corresponding to found Laplace transforms, a user often has to deal with very awkward expressions displayed in one line and using symbolic commands like `simplify` or `collect` does not enable to represent a result in a more compact form. Here, introducing new parameters equal to some combinations of rate constants can help significantly. Thus, for the purpose of obtaining more compact expressions, the operator solution in Fig. 2.10 is somehow modified by introducing the new constants γ_1 and γ_2 satisfying the equalities

$$\gamma_1 + \gamma_2 = k_1 + k_2 + k_3; \;\; \gamma_1 \gamma_2 = k_1(k_2 + k_3).$$

The kinetic curves of the substances taking part in the reaction plotted in Fig. 2.11 reveal a significant difference from the analogous dependencies for the

$C_{A0} := 1 \qquad k_1 := 0.5 \qquad k_2 := 0.25 \qquad k_3 := 0.1$

$$\begin{pmatrix}\gamma_1\\ \gamma_2\end{pmatrix} := \begin{pmatrix}\gamma_1 + \gamma_2 = k_2 + k_1 + k_3\\ \gamma_1 \cdot \gamma_2 = k_2 \cdot k_1 + k_3 \cdot k_1\end{pmatrix} \left|\begin{array}{l}\text{solve}, \begin{pmatrix}\gamma_1\\ \gamma_2\end{pmatrix}\\ \text{float}, 3\end{array}\right. \rightarrow \begin{pmatrix}.500\\ .350\end{pmatrix}$$

$C_A(t) := C_{A0} \cdot \exp(-k_1 \cdot t)$

$$C_B(t) := \left[\left[\frac{1}{\gamma_1 \cdot (\gamma_1 - \gamma_2)} \cdot k_3 - \frac{1}{\gamma_1 - \gamma_2}\right] \cdot e^{(-\gamma_1) \cdot t} + \left[\frac{-1}{\gamma_2 \cdot (\gamma_1 - \gamma_2)} \cdot k_3 + \frac{1}{\gamma_1 - \gamma_2}\right] \cdot e^{(-\gamma_2) \cdot t} + \frac{k_3}{\gamma_1 \cdot \gamma_2}\right] \cdot C_{A0} \cdot k_1$$

$$C_C(t) := k_1 \cdot C_{A0} \cdot k_2 \cdot \left[\frac{1}{\gamma_1 \cdot \gamma_2} - \frac{1}{\gamma_2 \cdot (\gamma_1 - \gamma_2)} \cdot e^{(-\gamma_2) \cdot t} + \frac{1}{\gamma_1 \cdot (\gamma_1 - \gamma_2)} \cdot e^{(-\gamma_1) \cdot t}\right]$$

$$t_{max} := \frac{-\ln\left(\frac{k_3 - \gamma_2}{k_3 - \gamma_1}\right)}{\gamma_1 - \gamma_2}$$

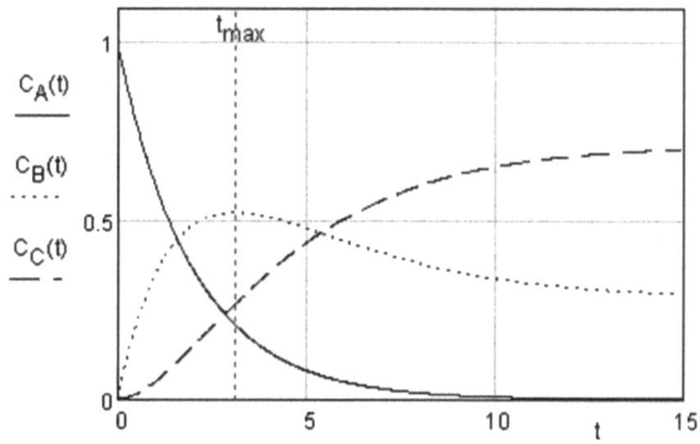

Fig. 2.11 Kinetics of the two-step first-order consecutive reaction with the reversible second step

irreversible consecutive reaction. This difference consists in stabilization of concentrations of the intermediate and the end product with time owing to establishing the equilibrium of the second and third steps.

The intermediate concentration passes through a maximum, as in the case of the irreversible consecutive reaction. Respectively, the end product curve has an inflection. However, the abscissa of the maximum point t_{max} (in this case it coincides with the abscissa of the reflection point) is now determined by a relationship of the rate constants of all the elementary stages. It is easy to make sure that:

$$t_{max} = \frac{\ln\left(\frac{k_1 - k_3}{k_2}\right)}{k_1 - k_2 - k_3}.$$

Thereby, in the symbolic commands `laplace`, `invlaplace` of the Mathcad package we have one more quite reliable instrument for solving the direct problem of chemical kinetics for multi-step reactions described by systems of linear differential equations. The examples of the kinetic equations for some reactions are represented in Table 2.1. Naturally, the classical matrix method can be used to derive these equations. In Maple, capabilities of `dsolve` are enough for this.

Table 2.1 Analytical solutions of the direct problem for some reaction systems with elementary first-order steps

Reaction schemes and initial concentrations	Kinetic curve equations
1	2
1. The irreversible consecutive reaction	
$A \xrightarrow{k_1} B \xrightarrow{k_2} C$ $A(0) = A_0\ B(0) = 0;$ $C(0) = 0$	$A(t) = A_0 e^{-k_1 t};\ B(t) = \dfrac{A_0 k_1}{k_2 - k_1}\left(e^{-k_1 t} - e^{-k_2 t}\right);$ $C(t) = A_0\left(1 - \dfrac{k_2}{k_2 - k_1} e^{-k_1 t} - \dfrac{k_1}{k_1 - k_2} e^{-k_2 t}\right)$
2. The consecutive reaction with the second step being reversible	
$A \xrightarrow{k_1} B \underset{k_3}{\overset{k_2}{\rightleftarrows}} C$ $A(0) = A_0;\ B(0) = 0;$ $C(0) = 0$	$A(t) = A_0 e^{-k_1 t}$ $B(t) = k_1 A_0\left[\dfrac{k_3}{\gamma_1\gamma_2} + \dfrac{\gamma_2 - k_3}{\gamma_1(\gamma_1 - \gamma_2)} e^{-\gamma_2 t} + \dfrac{k_3 - \gamma_1}{\gamma_1(\gamma_1 - \gamma_2)} e^{-\gamma_1 t}\right];$ $C(t) = k_1 k_2 A_0\left[\dfrac{1}{\gamma_1\gamma_2} + \dfrac{1}{\gamma_1(\gamma_1 - \gamma_2)} e^{-\gamma_1 t} - \dfrac{1}{\gamma_2(\gamma_1 - \gamma_2)} e^{-\gamma_2 t}\right];$ where $\gamma_1\gamma_2 = k_1(k_2 + k_3),\ \gamma_1 + \gamma_2 = k_1 + k_2 + k_3$
3. The consecutive reaction with the first step being reversible	
$A \underset{k_2}{\overset{k_1}{\rightleftarrows}} B \xrightarrow{k_3} C$ $A(0) = A_0;\ B(0) = 0;$ $C(0) = 0$	$A(t) = \dfrac{A_0}{\gamma_2 - \gamma_1}\left[(k_2 + k_3 - \gamma_1)e^{-\gamma_1 t} - (k_2 + k_3 - \gamma_2)e^{-\gamma_2 t}\right]$ $B(t) = \dfrac{A_0 k_1}{\gamma_2 - \gamma_1}\left(e^{-\gamma_1 t} - e^{-\gamma_2 t}\right);$ $C(t) = A_0\left[1 + \dfrac{k_1 k_3}{\gamma_1(\gamma_1 - \gamma_2)} e^{-\gamma_1 t} + \dfrac{k_1 k_3}{\gamma_2(\gamma_2 - \gamma_1)} e^{-\gamma_2 t}\right]$ where $\gamma_1\gamma_2 = k_1 k_3;\ \gamma_1 + \gamma_2 = k_1 + k_2 + k_3$
4. The competitive reaction with a reversible step	
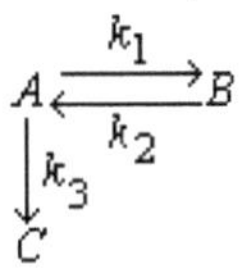 $A(0) = A_0;\ B(0) = 0;$ $C(0) = 0$	$A(t) = A_0\left(\dfrac{k_2 - \gamma_1}{\gamma_2 - \gamma_1} e^{-\gamma_1 t} - \dfrac{k_2 - \gamma_2}{\gamma_2 - \gamma_1} e^{-\gamma_2 t}\right)$ $B(t) = \dfrac{k_1 A_0}{\gamma_2 - \gamma_1}\left(e^{-\gamma_1 t} - e^{-\gamma_2 t}\right);$ $C(t) = A_0\left[1 + \dfrac{k_3(k_2 - \gamma_1)}{\gamma_1(\gamma_1 - \gamma_2)} e^{-\gamma_1 t} + \dfrac{k_3(k_2 - \gamma_2)}{\gamma_2(\gamma_2 - \gamma_1)} e^{-\gamma_2 t}\right]$ $\gamma_1\gamma_2 = k_2 k_3\ \gamma_1 + \gamma_2 = k_1 + k_2 + k_3$
5. The consecutive two-step reaction with both steps being reversible	
$A \underset{k_2}{\overset{k_1}{\rightleftarrows}} B \underset{k_4}{\overset{k_3}{\rightleftarrows}} C$ $A(0) = A_0;\ B(0) = 0;$ $C(0) = 0$	$A(t) = A_0\left[\dfrac{k_2 k_4}{\gamma_1\gamma_2} - \dfrac{k_1(\gamma_1 - k_3 - k_4)}{\gamma_1(\gamma_2 - \gamma_1)} e^{-\gamma_1 t} - \dfrac{k_1(k_3 + k_4 - \gamma_2)}{\gamma_2(\gamma_2 - \gamma_1)} e^{-\gamma_2 t}\right]$ $B(t) = A_0 k_1\left[\dfrac{k_4}{\gamma_1\gamma_2} + \dfrac{k_4 - \gamma_1}{\gamma_1(\gamma_2 - \gamma_1)} e^{-\gamma_1 t} + \dfrac{k_4 - \gamma_2}{\gamma_2(\gamma_2 - \gamma_1)} e^{-\gamma_2 t}\right]$ $C(t) = A_0 k_1 k_3\left[\dfrac{1}{\gamma_1\gamma_2} + \dfrac{1}{\gamma_1(\gamma_2 - \gamma_1)} e^{-\gamma_1 t} - \dfrac{1}{\gamma_2(\gamma_2 - \gamma_1)} e^{-\gamma_2 t}\right],$ where $\gamma_1\gamma_2 = k_2 k_4 + k_1 k_3 + k_1 k_4;$ $\gamma_1 + \gamma_2 = k_1 + k_2 + k_3 + k_4$

(*continued*)

Table 2.1 (continued)

Reaction schemes and initial concentrations	Kinetic curve equations
6. The consecutive-competitive reaction with a reversible step	
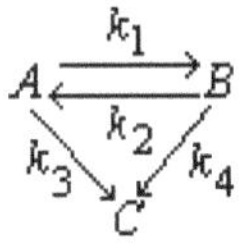 $A(0)=A_0;\ B(0)=0;$ $C(0)=0$	$A(t)=\frac{A_0}{\gamma_2-\gamma_1}[(k_2+k_3-\gamma_1)e^{-\gamma_1 t}-(k_2+k_3-\gamma_2)e^{-\gamma_2 t}];$ $B(t)=\frac{k_1A_0}{\gamma_2-\gamma_1}(e^{-\gamma_1 t}-e^{-\gamma_2 t});$ $C(t)=A_0\left(1-\frac{\gamma_2-k_4}{\gamma_2-\gamma_1}e^{-\gamma_1 t}-\frac{\gamma_1+k_4}{\gamma_1-\gamma_2}e^{-\gamma_2 t}\right),$ where $\gamma_1\gamma_2=k_1k_3+k_2k_4+k_3k_4;$ $\gamma_1+\gamma_2=k_1+k_2+k_3+k_4$
7. The consecutive-competitive reaction with two steps being reversible	
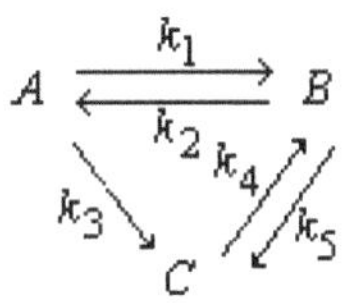 $A(0)=A_0;\ B(0)=0;$ $C(0)=0$	$A(t)=A_0\left(\frac{k_2k_5}{\gamma_1\gamma_2}-\frac{\gamma_2^2-\varepsilon\gamma_2+k_2k_5}{\gamma_2(\gamma_1-\gamma_2)}e^{-\gamma_2 t}-\frac{\gamma_1^2-\varepsilon\gamma_1+k_2k_5}{\gamma_1(\gamma_2-\gamma_1)}e^{-\gamma_1 t}\right)$ $B(t)=A_0\left(\frac{\eta}{\gamma_1\gamma_2}-\frac{k_1\gamma_2-\eta}{\gamma_2(\gamma_2-\gamma_1)}e^{-\gamma_2 t}-\frac{k_1\gamma_1-\eta}{\gamma_1(\gamma_1-\gamma_2)}e^{-\gamma_1 t}\right);$ $C(t)=A_0\left(\frac{\delta}{\gamma_1\gamma_2}-\frac{k_3\gamma_2-\delta}{\gamma_2(\gamma_2-\gamma_1)}e^{-\gamma_2 t}-\frac{k_3\gamma_1-\delta}{\gamma_1(\gamma_1-\gamma_2)}e^{-\gamma_1 t}\right);$ где$\gamma_1+\gamma_2=k_1+k_2+k_3+k_4+k_5;$ $\gamma_1\gamma_2=k_2k_3+k_1k_4+k_2k_5+k_3k_5+k_3k_4+k_1k_5;$ $\delta=\gamma_1\gamma_2-k_2k_5-k_3k_5-k_1k_5;$ $\eta=\gamma_1\gamma_2-k_2k_3-k_1k_4-k_2k_5-k_3k_4;$ $\varepsilon=\gamma_1+\gamma_2-k_1-k_3$
8. The consecutive two-step reaction with three steps being reversible	
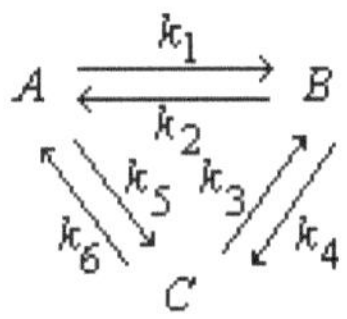 $A(0)=A_0;\ B(0)=0;$ $C(0)=0$	$A(t)=A_0\left(\frac{\beta}{\gamma_1\gamma_2}+\frac{\alpha\gamma_1-\gamma_1{}^2-\beta}{\gamma_2-\gamma_1}e^{-\gamma_1 t}+\frac{\gamma_2^2-\alpha\gamma_2+\beta}{\gamma_2-\gamma_1}e^{-\gamma_1 t}\right)$ $B(t)=A_0\left(\frac{\varepsilon}{\gamma_1\gamma_2}+\frac{k_1\gamma_1-\varepsilon}{\gamma_2-\gamma_1}e^{-\gamma_1 t}+\frac{\varepsilon-k_1\gamma_2}{\gamma_2-\gamma_1}e^{-\gamma_2 t}\right)$ $C(t)=A_0\left(\frac{\delta}{\gamma_1\gamma_2}+\frac{k_5\gamma_1-\delta}{\gamma_2-\gamma_1}e^{-\gamma_1 t}+\frac{\delta-k_5\gamma_2}{\gamma_2-\gamma_1}e^{-\gamma_2 t}\right),$ where $\alpha=k_2+k_3+k_4+k_6;\ \beta=k_2k_4+k_2k_6+k_3k_6;$ $\varepsilon=k_1k_4+k_1k_6+k_4k_5;\ \delta=k_1k_3+k_2k_5+k_3k_5;$ $\gamma_1\gamma_2=\beta+\varepsilon+\delta;\ \gamma_1+\gamma_2=\alpha+k_1+k_5$

2.3.3 *Transient Regime in a System of Flow Reactors*

Heretofore, the kinetic models of the reactions proceeding in closed systems have been considered. In such systems material exchange with surroundings is excluded (batch reactors). In practice, many reactions are carried out in open systems under a regime of continuous feed of reactants to a reactor and withdrawal of formed products from it. There are two continuous reactor models: a *batch* reactor and a *plug flow* one.

Let us consider a stirred tank reactor (STR). In such reactor, a reaction mixture is stirred in the way that current concentrations of reactants taking part in a reaction are the same at any time moment in any point of reaction space. Let the first-order reaction $A \xrightarrow{k} B$ proceed in the STR. The solution of reactant A with a concentration of C_0 is continuously fed to the reactor inlet with a rate of $v\ l\ s^{-1}$. The reaction mixture is withdrawn from the reactor with the same rate. Under these conditions, the reaction space volume remains constant and is Vl. The change of the reactant quantity per unit of time is:

$$\frac{dn_A(t)}{dt} = vC_0 - vC_A(t) - kC_A(t)V. \tag{2.6}$$

Dividing the both parts of the equation by V yields the mathematical model of the STR in the form:

$$\frac{dC_A(t)}{dt} = -kC_A(t) + \frac{v}{V}(C_0 - C_A(t)). \tag{2.7}$$

Thus, in contrast to the kinetic equation describing the first-order reaction in a batch reactor, the last equation contains the summand which characterizes geometrical dimensions of the reactor and the rate of substance feed to the reactor and its withdrawal from it.

Solving (2.8) at the initial reactant concentration (Fig. 2.12) shows that the current reactant concentration depends on the kinetic and macroscopic parameters of the system in a complicated way,

$$C_A(t) = \frac{C_{A0}}{kV + v}\left[v + kVe^{-\left(k+\frac{v}{V}\right)t}\right].$$

The graph zone represented in Fig. 2.12 shows that the kinetic curve of the source reactant has a characteristic feature: in the course of time the concentration becomes time-constant. We met the similar curve behaviour when analyzing reversible reactions; however, in this case, stabilization of the concentration with time is not related to establishing the chemical equilibrium state. In the case under consideration, we have a *steady-state regime* of the process, a state when a reactant loss because of proceeding a reaction is precisely compensated by its gain at the expense of feed of new reactant portions to a reactor. The expression for the steady-state concentration of substance A is easily reduced by equating the derivative in (2.7) to zero

$$C_{A_{st}} = C_{A0}\frac{v}{kV + v}.$$

$$\int_{C_{A0}}^{C_A} \frac{1}{-k\cdot x + \frac{v}{V}\cdot(C_{A0}-x)}\,dx = \int_0^t 1\,dy \text{ solve}, C_A \rightarrow \frac{\left[-e^{\frac{-[t\cdot k\cdot V+t\cdot v-V\cdot \ln[(-C_{A0})\cdot k\cdot V]]}{V}}\right] + C_{A0}\cdot v}{k\cdot V + v}$$

$$-e^{\frac{-[t\cdot k\cdot V+t\cdot v-V\cdot \ln[(-C_{A0})\cdot k\cdot V]]}{V}} \text{ expand}, C_{A0} \rightarrow \frac{1}{e^{t\cdot k}\cdot e^{\frac{1}{V}\cdot t\cdot v}}\cdot C_{A0}\cdot k\cdot V$$

$$\boxed{C_A(t) = \frac{C_{A0}}{k\cdot V + v}\cdot\left[v + k\cdot V\cdot e^{-\left(k+\frac{v}{V}\right)\cdot t}\right]} \qquad -k\cdot C_A + \frac{v}{V}\cdot(C_{A0}-C_A) = 0 \text{ solve}, C_A \rightarrow C_{A0}\cdot\frac{v}{k\cdot V + v}$$

$$\boxed{C_{Ast} = C_{A0}\cdot\frac{v}{k\cdot V + v}} \qquad V := 1000 \quad v := 400 \quad k := 0.1 \quad C_{A0} := 1$$

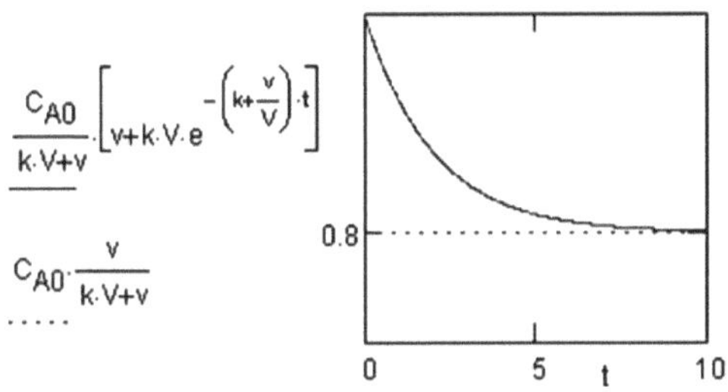

Fig. 2.12 The time dependence of the reactant concentration in the stirred tank reactor

Let us now consider a cascade of three stirred tank reactors of volumes of V_1, V_2, V_3 and with current concentrations of C_1, C_2, C_3, respectively. The scheme of the reaction mixture flows for the each reactor is represented in Fig. 2.13.

The first reactor has two inlets, vC_0 mol s^{-1} of reactant A are fed to the first one from the outside and v_1C_2 mol s^{-1} enter the second one from reactor 2. For substance A, the outlet matter flow of reactor 1 and the inlet matter flow of reactor 2 are equal to $(v+v_1)C_1$ mol s^{-1}. The outlet matter flow of reactor 2 is $(v+v_1)C_2$ mol s^{-1} and next is divided between two paths. Its part v_1C_2 mol s^{-1} returns into reactor 1 and vC_2 mol s^{-1} is fed to the inlet of reactor 3. vC_3 mol L^{-1} of substance A is eventually withdrawn from the system. In connection with the above and taking into account relationships (2.6) and (2.7), we can write the equations describing the time-dependent loss of the reactant concentration in the each reactor

$$dn_1(t)/dt = vC_0 + v_1C_2(t) - (v+v_1)C_1(t) - kC_1(t)V_1,$$

$$dn_2(t)/dt = (v+v_1)C_1(t) - (v+v_1)C_2(t) - kC_2(t)V_2,$$

$$dn_3(t)/dt = vC_2(t) - vC_3(t) - kC_3(t)V_3,$$

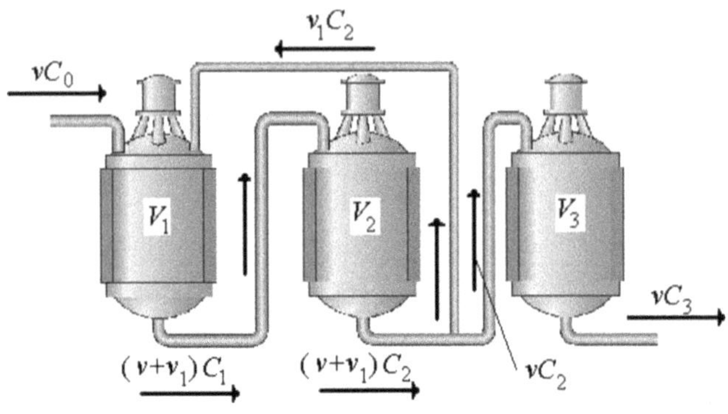

Fig. 2.13 The cascade of the stirred tank reactors

or

$$\frac{dC_1(t)}{dt} = -kC_1(t) + \frac{vC_0 + v_1C_2(t) - (v + v_1)C_1(t)}{V_1},$$

$$\frac{dC_2(t)}{dt} = -kC_2(t) + \frac{(v + v_1)[C_1(t) - C_2(t)]}{V_2},$$

$$\frac{dC_3(t)}{dt} = -kC_3(t) + \frac{v[C_2(t) - C_3(t)]}{V_3}.$$

The obtained system of the linear differential equations can be solved by applying the operational system apparatus.

Let us consider that before starting the reaction, the concentrations of A in the each reactor are the same and equal to C_0 mol L^{-1}. Additionally, let us intentionally complicate the conditions of the direct kinetic problem. Assume that in the each reactor the reaction mixture has different temperature and this temperature remains constant during the process (here, of course, we have to assume that new portions of the solution immediately takes a given reactor temperature). Thus, this means that the process is controlled by the different rate constants k_1, k_2, k_3 in the each reactor.

Figure 2.14 represents the part of the network variant of the Mathcad document which allows to model different situations of a system behaviour at different values of the input parameters, namely the rate constants k_1, k_2, k_3, the reactor volumes V_1, V_2, V_3, and the feed rates of the reaction mixture v, v_1. So as not to clutter the document, constructions containing the Laplace transform are hidden from a user (this approach completely corresponds to the sequence of operations shown in Fig. 2.10). Only the elements of the vector of the operator solution as well as recovering the source original functions corresponding to the Laplace transforms are displayed in the workspace. The final result of the document is analytical expressions for the concentration–time relationships for the each reactor as well as the time dependencies of the degree of conversion of the source reactant in the

$$V_1 := 1000 \qquad V_2 := 1500 \qquad V_3 := 100 \qquad C_0 := 1$$

$$k_1 := 1 \qquad k_2 := 2 \qquad k_3 := 5 \qquad v := 1000 \qquad v_1 := 100$$

$$F_1(s) \text{ factor } \rightarrow \frac{1}{5} \cdot \frac{30 \cdot s^2 + 115 \cdot s + 82}{(6 \cdot s + 17) \cdot (s + 2) \cdot s}$$

$$F_2(s) \text{ factor } \rightarrow \frac{1}{5} \cdot \frac{30 \cdot s^2 + 85 \cdot s + 22}{(6 \cdot s + 17) \cdot (s + 2) \cdot s}$$

$$F_3(s) \text{ factor } \rightarrow \frac{6 \cdot s^3 + 89 \cdot s^2 + 204 \cdot s + 44}{s \cdot (s + 15) \cdot (s + 2) \cdot (6 \cdot s + 17)}$$

$$C_1(t) := F_1(s) \left| \begin{array}{l} \text{factor} \\ \text{invlaplace}, s \\ \text{convert}, \text{rational} \end{array} \right. \rightarrow \frac{-18}{425} \cdot e^{\frac{-17}{6} \cdot t} + \frac{14}{25} \cdot e^{(-2) \cdot t} + \frac{41}{85}$$

$$C_2(t) := F_2(s) \left| \begin{array}{l} \text{factor} \\ \text{invlaplace}, s \\ \text{convert}, \text{rational} \end{array} \right. \rightarrow \frac{132}{425} \cdot e^{\frac{-17}{6} \cdot t} + \frac{14}{25} \cdot e^{(-2) \cdot t} + \frac{11}{85}$$

$$C_3(t) := F_3(s) \left| \begin{array}{l} \text{factor} \\ \text{invlaplace}, s \\ \text{convert}, \text{rational} \end{array} \right. \rightarrow \frac{22}{255} + \frac{3241}{14235} \cdot e^{(-15) \cdot t} + \frac{28}{65} \cdot e^{(-2) \cdot t} + \frac{1584}{6205} \cdot e^{\frac{-17}{6} \cdot t}$$

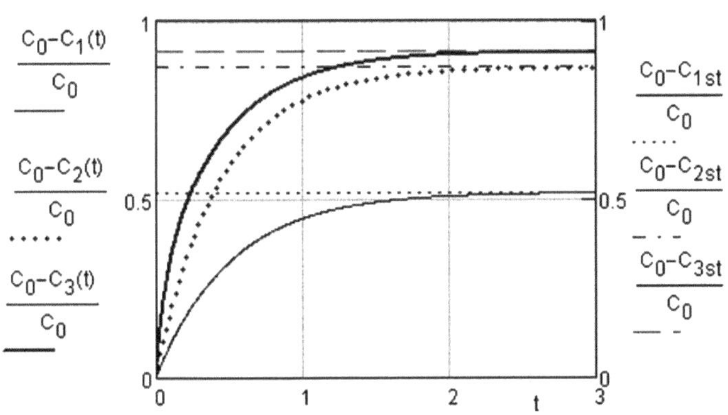

$$C_{1st} = \frac{41}{85} \qquad C_{2st} = \frac{11}{85} \qquad C_{3st} = \frac{22}{255}$$

Fig. 2.14 Modeling the transient regime in the system of three reactors

each reactor. Thereby, a user has an opportunity to observe how a change of one input parameter of another influences the degree of conversion and the time of establishing a steady-state regime of operation of the reactors.

Since symbolic calculations occupy a significant part of the document, we should recommend displaying numerical values in a rational form but not as floating point numbers for obtaining more compact expressions. Otherwise, too long expressions may be yielded what creates definite problems when printing a document. Finally, the symbolic command `convert,rational`, which transforms an expression in a rational form, can be used as shown in Fig. 2.14. Although, the last procedure is not documented and can be used only in the Mathcad versions which use the Maple symbolic kernel.

2.3.4 *Kinetic Models in the Form of Equations Containing Piecewise Continuous Functions*

Unfortunately, in many cases, the Mathcad symbolic processor cannot operate with functions determined using a program block. For example, an attempt to apply `laplace` to a piecewise continuous function which has been defined using the programming operators `if` and `otherwise` fails (Fig. 2.15). Meanwhile, such functions can enter differential equations describing applying different kinds of disturbances to a system under investigation (e.g., a sharp change of reactant feed to a reactor inlet, a random or intentional change of thermal conditions of a reactor, periodical injections of a medicine into an organism etc.). A head-on attempt of applying the Laplace transform to such an expression remains unsuccessful, therefore modification of the expression using the *Heaviside* function is an effective method. If a task consists in analytical solving a problem, a source piecewise continuous function should be replaced with a linear combination of other functions, with the Heaviside function $\Phi(t)$ obviously entering each summand of a new compound function. This function vanishes at negative t values and is equal to unity otherwise. Thus, Fig. 2.15 demonstrates a method of such transfer. The source function $f_1(t)$ is replaced with the equivalent function $f_2(t)$ but a distinctive feature of this new function is that symbolic commands can already be applied to it, including the command of the integral Laplace transform.

Let us consider a real problem of solving an equation with a piecewise continuous function. Suppose that substance A with the initial concentration 1 mol l^{-1} converts into the end product B during a chemical reaction, with reactant B being unstable and decomposing by the action of light to give the source substance A. If the reaction is carried out in the dark, it proceeds without any complication and the rate of decomposition of A is proportional to its current concentration. But if to expose the reaction mixture to light, the system behaviour becomes more complicated: decomposition of A is complicated by its accumulation at the expense of proceeding the reverse process whose kinetic behaviour is determined by a light intensity.

Suppose that the reaction occurs in the dark during 0.5 min and next the system is exposed to a light flux for 1.5 min. Then a light flux intensity is decreased by 75% and maintained constant. The mathematical model of the overall process can be then written in the differential equation form

$$f_1(t) := \begin{cases} 0 & \text{if } 0 < t < \frac{1}{2} \\ 1 & \text{if } \frac{1}{2} \le t \le 2 \\ \frac{1}{5} & \text{if } t > 2 \end{cases}$$

$f_1(t)$ laplace,t →

This value must be a scalar

$$a(t) := 0\cdot\Phi(t) \qquad b(t) := (1-0)\cdot\Phi\left(t-\frac{1}{2}\right)$$

$$c(t) := \left(\frac{1}{5}-1\right)\cdot\Phi(t-2)$$

$$f_2(t) := (a(t)+b(t)+c(t)) \rightarrow \Phi\left(t-\frac{1}{2}\right)-\frac{4}{5}\cdot\Phi(t-2)$$

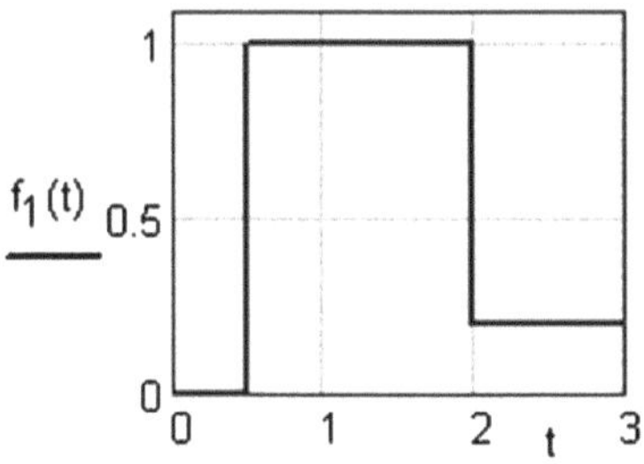

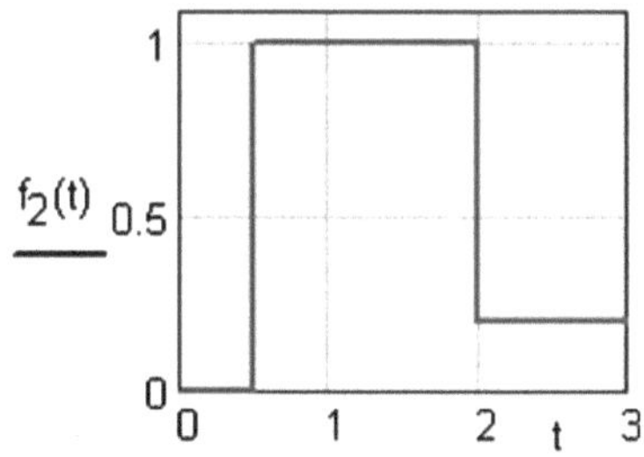

$$f_2(t) \text{ laplace,t} \rightarrow \frac{e^{\frac{-1}{2}\cdot s}}{s}-\frac{4}{5}\cdot\frac{e^{(-2)\cdot s}}{s}$$

Fig. 2.15 Replacing the source function with the linear combination of the functions

$$\frac{dC_A(t)}{dt} = -kC_A(t) + f(t),$$

where the function $f(t)$ is defined by specifying the expression

$$f(t) = \Phi\left(t-\frac{1}{2}\right) + \frac{4}{5}\Phi(t-2).$$

This problem is solved in Fig. 2.16. As we can see, the behaviour of the kinetic curve is quite complicated, however it is possible to obtain its analytical form in this case.

2.4 Approximate Methods of Chemical Kinetics

2.4.1 The Steady-State Concentration Method

Using of approximate methods of chemical kinetics is intended for, first of all, simplifying mathematical models and, respectively, their analysis. The steady-state

$$\frac{d}{dt}A(t) = -k\cdot A(t) + f(t) \qquad A(0) = A_0 \qquad f(t) := \Phi\left(t - \frac{1}{2}\right) - \frac{4}{5}\cdot\Phi(t-2)$$

$$\frac{d}{dt}A(t) \text{ laplace}, t \rightarrow s\cdot \text{laplace}(A(t), t, s) - A(0)$$

$$-k\cdot A(t) + f(t) \text{ laplace}, t \rightarrow (-k)\cdot \text{laplace}(A(t), t, s) + \frac{e^{\frac{-1}{2}\cdot s}}{s} - \frac{4}{5}\cdot\frac{e^{(-2)\cdot s}}{s}$$

$$s\cdot LA - A_0 = (-k)\cdot LA + \frac{e^{\frac{-1}{2}\cdot s}}{s} - \frac{4}{5}\cdot\frac{e^{(-2)\cdot s}}{s} \text{ solve}, LA \rightarrow \frac{1}{5}\cdot\frac{5\cdot A_0\cdot s + 5\cdot e^{\frac{-1}{2}\cdot s} - 4\cdot e^{(-2)\cdot s}}{s\cdot(k+s)}$$

$$A(A_0, k, t) := \frac{1}{5}\cdot\frac{5\cdot A_0\cdot s + 5\cdot e^{\frac{-1}{2}\cdot s} - 4\cdot e^{(-2)\cdot s}}{s\cdot(k+s)} \left|\begin{array}{l}\text{invlaplace}, s\\ \text{collect}, k\\ \text{collect}, \Phi(t-2)\\ \text{collect}, \Phi\left(t-\frac{1}{2}\right)\end{array}\right. \rightarrow$$

$$\rightarrow \frac{1 - e^{(-k)\cdot\left(t-\frac{1}{2}\right)}}{k}\cdot\Phi\left(t - \frac{1}{2}\right) + \frac{\frac{4}{5}\cdot e^{(-k)\cdot(t-2)} - \frac{4}{5}}{k}\cdot\Phi(t-2) + A_0\cdot e^{(-k)\cdot t}$$

$$A_0 := 0.75 \qquad k := 1.4$$

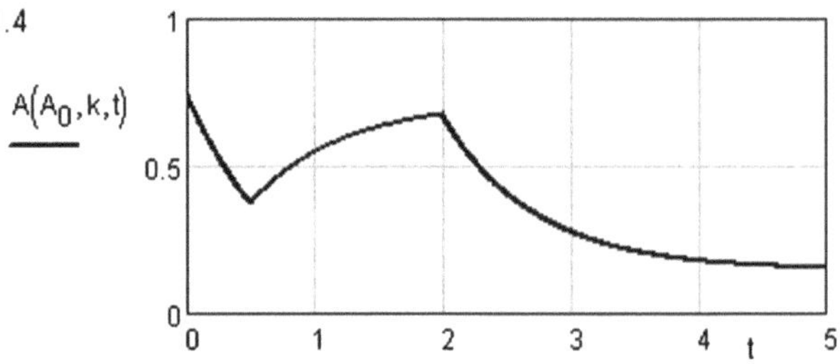

Fig. 2.16 Solving the direct problem for the reaction complicated by the photochemical conversion

concentration method is one of the most widespread approximations. This method is applicable to describing multi-step reactions whose definite steps result in high-reactivity intermediates. The condition of applying the method is a short life time of intermediates compared to a time during which reaction mixture composition can change noticeably (the *Christiansen* condition). It is assumed that time derivatives of concentrations of active intermediates are equal to zero (the *Bodenstein* theorem). Thereby, some differential equations of a system entering a mathematical model can be replaced with algebraic ones. A kinetic analysis of a reaction mechanism is considerably simplified in this case.

Let us consider a simple example to illustrate the point of the method. Let us have the kinetic scheme of the consecutive reaction with the first step being reversible

$$A \underset{k_2}{\overset{k_1}{\rightleftarrows}} B \xrightarrow{k_3} C. \tag{2.8}$$

If to assume that the constant $k_3 \gg k_1$, the rate of intermediate consumption is much higher than one of its formation. Substance B is an unstable intermediate, therefore the principle of steady-state concentrations is applicable to it. Then we obtain

$$dC_B(t)/dt = k_1 C_A(t) - k_2 C_B(t) - k_3 C_B(t) = 0,$$

hence

$$C_B(t) = \frac{k_1}{k_2 + k_3} C_A(t).$$

It is easy to see that the rate of product C formation is then written in the form

$$r_P = \frac{k_1 k_3}{k_2 + k_3} C_A(t). \tag{2.9}$$

Let us carry out a check of the steady-state principle. For this purpose, let us calculate the time dependence of the end product formation rate from the relationships obtained by accurate solving the direct kinetic problem (see Table 2.1). Next, let us compare the result with the calculations from obtained formula (2.9). The corresponding plots represented in Fig. 2.17 show that the behaviour of the both curves coincide after less than 0.5 s at given values of the rate constants satisfying the condition $k_3 \gg k_1$. This indicates applicability of the steady-state concentration method to the considered model of the consecutive reaction.

The steady-state concentration method played an important role in establishing fundamental relationships of chemical kinetics. Earlier, it was noticed that a number of monomolecular reactions involving gases have different orders with respect to a source substance depending on a gas mixture pressure. *Lindemann* assumed that the overall process of conversion of reactant A into end product P consists of a number of steps, the first one being a bimolecular collision of A molecules. Although this step does not result in a chemical conversion, it is accompanied by redistribution of colliding molecule energy and leads to formation of excited particles A^* potentially capable to overcome an energy barrier

$$A + A \xrightarrow{k_1} A^* + A.$$

This step is also called *excitation*. Excited molecules A^* may either loose their energy if colliding with nonexcited molecules (*deexcitation*)

$$A^* + A \xrightarrow{k_2} A + A,$$

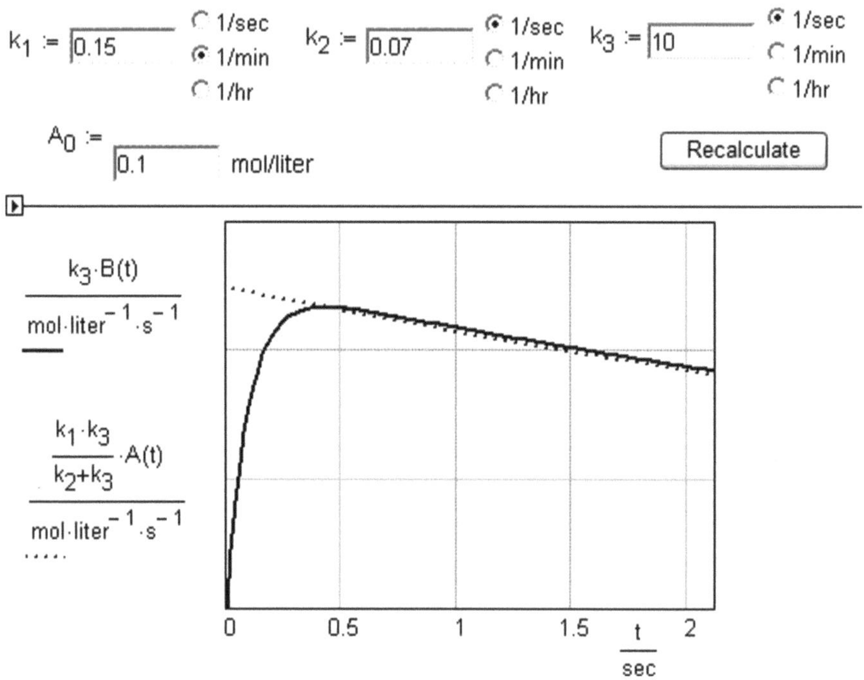

Fig. 2.17 Checking applicability of the steady-state concentration method

or be converted into the end product

$$A^* \xrightarrow{k_3} P.$$

Thereby, if particle A^* is considered as an unstable intermediate, the Lindemann's kinetic scheme can be analyzed in the steady-state approximation assuming $dA^*/dt \approx 0$. The results of this analysis are represened in Fig. 2.18.

In Fig. 2.18, the intermediate concentation is expressed in terms of the concentration of nonexcited molecules A and substituted into the equation for the rate r_P of the third step leading to the end product. The analysis of the obtained expression indicates two limiting cases. In the first one (a low pressure), the current concentrations (or the partial pressures) of A are so small that the term k_2A in the denominator of the expression for r_P can be neglected owing to meeting the condition $k_2A \ll k_3$. This suggests that product formation should obey relationships for a reaction having an order of 2. If transferrring to a high pressure, on the contrary, $k_2A \gg k_3$. If neglecting a k_3 value, it should be concluded that we have a kinetic order of 2 for these conditions. The drawn conclusions also have a physical interpretation. Under condiditons of a low pressure, molecules are too far from each other that makes the deexcitation step unlikely and the limiting step of the process is the excitation step. If transferring to a high pressure, the nature of a limiting step changes. The process rate is now determined by the third step, since waisting energy of excited particles is

$$\frac{d}{dt}A(t) = \begin{pmatrix} 1 \\ -1 \\ -1 \end{pmatrix}^T \cdot \begin{pmatrix} k_1 \cdot A^2 \\ k_2 \cdot \mathbf{A} \cdot A \\ k_3 \cdot \mathbf{A} \end{pmatrix} \left| \begin{array}{l} \text{substitute}, \frac{d}{dt}A(t) = 0 \\ \text{solve}, \mathbf{A} \end{array} \right. \rightarrow k_1 \cdot \frac{A^2}{k_2 \cdot A + k_3}$$

$$\frac{d}{dt}P(t) = k_3 \cdot \mathbf{A} \text{ substitute}, \mathbf{A} = k_1 \cdot \frac{A^2}{k_2 \cdot A + k_3} \rightarrow \frac{d}{dt}P(t) = k_3 \cdot k_1 \cdot \frac{A^2}{k_2 \cdot A + k_3}$$

$$\frac{d}{dt}P(t) = k_3 \cdot k_1 \cdot \frac{A^2}{k_2 \cdot A + k_3} \text{ substitute}, k_2 \cdot A + k_3 = k_2 \cdot A \rightarrow \frac{d}{dt}P(t) = k_3 \cdot k_1 \cdot \frac{A}{k_2}$$

$$\frac{d}{dt}P(t) = k_3 \cdot k_1 \cdot \frac{A^2}{k_2 \cdot A + k_3} \text{ substitute}, k_2 \cdot A + k_3 = k_3 \rightarrow \frac{d}{dt}P(t) = k_1 \cdot A^2$$

Fig. 2.18 The analysis of the monomolecular reaction mechanism

considerably facilitated under the conditions of small distances between molecules, which have no time to react.

It should be noted that, despite the made assumptions, the *Lindemann* theory qualitatively fits the experimental data predicting the change of the reaction order with the pressure change.

The steady-state concentration method is widely used for an analysis of kinetic mechanisms of complex reactions. In particular, this method plays an important role in the chain reaction theory.

For example, a large number of theoretical kinetic models have been suggested for describing ethane thermal cracking. Let us consider one of simplified mechanisms

$$(1)\, C_2H_6 \xrightarrow{k_1} 2\,CH_3^\bullet,$$
$$(2)\, CH_3^\bullet + C_2H_6 \xrightarrow{k_2} CH_4 + C_2H_5^\bullet,$$
$$(3)\, C_2H_5^\bullet \xrightarrow{k_3} C_2H_4 + H^\bullet,$$
$$(4)\, C_2H_6 + H^\bullet \xrightarrow{k_4} C_2H_5^\bullet + H_2,$$
$$(5)\, 2\,C_2H_5^\bullet \xrightarrow{k_5} C_4H_{10},$$
$$(6)\, 2\,C_2H_5^\bullet \xrightarrow{k_6} C_2H_4 + C_2H_6.$$

As one can see, formation of end products – ethylene, butylene, and hydrogen – proceeds involving the intermediates – radicals $CH_3^\bullet$, $H^\bullet$, $C_2H_5^\bullet$. Let us solve the direct problem for this kinetic scheme in the steady-state approximation.

Firstly, it is necessary to derive the expressions for the intermediate steady-state concentrations. For this purpose, let us form the stoichiometric matrix and the rate vector and obtain the vector of right parts of the overall differential equation system for the given kinetic mechanism (Fig. 2.19).

The elements of the right part vector related to the rates of intermediate formation and consumption should be equated with zero and the obtained algebraic equations should be solved for the steady-state concentrations of the radicals. As a result, these operations (Fig. 2.19) yield

$$C_{CH_3^*} = 2\frac{k_1}{k_2},$$

$$C_{C_2H_5^*} = \sqrt{\frac{k_1}{k_5 + k_6}}\sqrt{C_2H_6},$$

$$C_{H\bullet} = \frac{k_3}{k_4}\sqrt{\frac{k_1}{(k_5 + k_6)C_{C_2H_6}}}.$$

Next, the elements related to the stable reactants should be extracted from the vector of the right parts of the differential equations and the intermediate concentrations, which enter the equations, should be replaced with the obtained expressions. This stage of the symbolic evaluation is shown in Fig. 2.20. Thereby, by applying the steady-state concentration method, we obtained the reduced system of the differential equations describing the time-dependent concentration decrease or increase of the stable reactants.

Thereby, each of the equations entering the system has the form

$$\frac{d}{dt}C_{C_2H_6}(t) = -k_1\frac{3k_5 + 2k_6}{k_5 + k_6}C_{C_2H_6}(t) - \frac{\sqrt{k_1}k_3}{\sqrt{k_5 + k_6}}\sqrt{C_{C_2H_6}(t)},$$

$$\frac{d}{dt}C_{CH_4}(t) = 2k_1C_{C_2H_6}(t),$$

$$\frac{d}{dt}C_{C_2H_4}(t) = \frac{\sqrt{k_1}k_3}{\sqrt{k_5 + k_6}}\sqrt{C_{C_2H_6}(t)} + k_1k_6\frac{C_{C_2H_6}(t)}{k_5 + k_6},$$

$$\frac{d}{dt}C_{H_2}(t) = \sqrt{\frac{k_1}{k_5 + k_6}}k_3\sqrt{C_{C_2H_6}(t)},$$

$$\frac{d}{dt}C_{C_4H_{10}}(t) = \frac{k_1k_5}{k_5 + k_6}C_{C_2H_6}(t).$$

It is noticeable that only one function, $C_{C_2H_6}(t)$, enters the right part of the system, i.e. the current concentrations of methane, ethylene, hydrogen, and butylene are determined only by a value of the ethane current concentration.

The analytical solving the system at the initial condition $C_{C_2H_6}(0) = C_{C_2H_60}$ can be easily realized in Maple. Figure 2.21 demonstrates the worksheet part where the

$$\text{ORIGIN} := 1 \quad [C_2 \cdot H_6] = C_1 \quad [CH_3^*] = C_2 \quad [CH_4] = C_3 \quad [C_2 \cdot H_5^*] = C_4 \quad [H^*] = C_5$$

$$[C_2 \cdot H_4] = C_6 \quad [H_2] = C_7 \quad [C_4 \cdot H_{10}] = C_8$$

$$\alpha := \begin{pmatrix} -1 & 2 & 0 & 0 & 0 & 0 & 0 & 0 \\ -1 & -1 & 1 & 1 & 0 & 0 & 0 & 0 \\ 0 & 0 & 0 & -1 & 1 & 1 & 0 & 0 \\ -1 & 0 & 0 & 1 & -1 & 0 & 1 & 0 \\ 0 & 0 & 0 & -2 & 0 & 0 & 0 & 1 \\ 1 & 0 & 0 & -2 & 0 & 1 & 0 & 0 \end{pmatrix} \quad r(k,C) := \begin{bmatrix} k_1 \cdot C_1 \\ k_2 \cdot C_1 \cdot C_2 \\ k_3 \cdot C_4 \\ k_4 \cdot C_1 \cdot C_5 \\ k_5 \cdot (C_4)^2 \\ k_6 \cdot (C_4)^2 \end{bmatrix} \quad rhs(k,C) := \alpha^T \cdot r(k,C)$$

$$rhs(k,C) \rightarrow \begin{bmatrix} (-k_1) \cdot C_1 - k_2 \cdot C_1 \cdot C_2 - k_4 \cdot C_1 \cdot C_5 + k_6 \cdot (C_4)^2 \\ 2 \cdot k_1 \cdot C_1 - k_2 \cdot C_1 \cdot C_2 \\ k_2 \cdot C_1 \cdot C_2 \\ k_2 \cdot C_1 \cdot C_2 - k_3 \cdot C_4 + k_4 \cdot C_1 \cdot C_5 - 2 \cdot k_5 \cdot (C_4)^2 - 2 \cdot k_6 \cdot (C_4)^2 \\ k_3 \cdot C_4 - k_4 \cdot C_1 \cdot C_5 \\ k_3 \cdot C_4 + k_6 \cdot (C_4)^2 \\ k_4 \cdot C_1 \cdot C_5 \\ k_5 \cdot (C_4)^2 \end{bmatrix}$$

$$rhs(k,C)_2 = 0 \text{ solve}, C_2 \rightarrow 2 \cdot \frac{k_1}{k_2} \qquad rhs(k,C)_5 = 0 \text{ solve}, C_5 \rightarrow k_3 \cdot \frac{C_4}{k_4 \cdot C_1}$$

$$rhs(k,C)_4 = 0 \left| \begin{array}{l} \text{substitute}, C_2 = 2 \cdot \dfrac{k_1}{k_2} \\ \text{substitute}, C_5 = k_3 \cdot \dfrac{C_4}{k_4 \cdot C_1} \\ \text{solve}, C_4 \end{array} \right. \rightarrow \begin{bmatrix} \dfrac{1}{k_5 + k_6} \cdot \left[(k_5 + k_6) \cdot k_1 \cdot C_1\right]^{\frac{1}{2}} \\ \dfrac{-1}{k_5 + k_6} \cdot \left[(k_5 + k_6) \cdot k_1 \cdot C_1\right]^{\frac{1}{2}} \end{bmatrix}$$

$$\boxed{[CH_3^*] = 2 \cdot \frac{k_1}{k_2}} \qquad \boxed{[C_2 \cdot H_5^*] = \sqrt{\frac{k_1}{k_5 + k_6}} \cdot (C_1)^{\frac{1}{2}}} \qquad \boxed{[H^*] = \frac{k_3}{k_4} \cdot \sqrt{\frac{k_1}{k_5 + k_6}} \cdot (C_1)^{\frac{-1}{2}}}$$

Fig. 2.19 The symbolic evaluation of the intermediate steady-state concentrations

$$\text{rhs\_new}(k, C) := \text{stack}\left(\text{rhs}(k,C)_1, \text{rhs}(k,C)_3, \text{rhs}(k,C)_6, \text{rhs}(k,C)_7, \text{rhs}(k,C)_8\right)$$

$$\text{rhs\_new}(k, C) \left| \begin{array}{l} \text{substitute}, C_2 = 2 \cdot \dfrac{k_1}{k_2} \\ \text{substitute}, C_4 = \sqrt{\dfrac{k_1}{k_5 + k_6}} \cdot (C_1)^{\frac{1}{2}} \\ \text{substitute}, C_5 = \dfrac{k_3}{k_4} \cdot \sqrt{\dfrac{k_1}{k_5 + k_6}} \cdot (C_1)^{\frac{-1}{2}} \\ \text{collect}, C_1 \end{array} \right. \rightarrow \begin{bmatrix} \left[(-3) \cdot k_1 + k_6 \cdot \dfrac{k_1}{k_5 + k_6}\right] \cdot C_1 - k_4 \cdot (C_1)^{\frac{1}{2}} \cdot \dfrac{k_3}{k_4} \cdot \left(\dfrac{k_1}{k_5 + k_6}\right)^{\frac{1}{2}} \\ 2 \cdot k_1 \cdot C_1 \\ k_3 \cdot \left(\dfrac{k_1}{k_5 + k_6}\right)^{\frac{1}{2}} \cdot (C_1)^{\frac{1}{2}} + k_6 \cdot \dfrac{k_1}{k_5 + k_6} \cdot C_1 \\ k_4 \cdot (C_1)^{\frac{1}{2}} \cdot \dfrac{k_3}{k_4} \cdot \left(\dfrac{k_1}{k_5 + k_6}\right)^{\frac{1}{2}} \\ k_5 \cdot \dfrac{k_1}{k_5 + k_6} \cdot C_1 \end{bmatrix}$$

Fig. 2.20 Deriving the right part of the reduced differential equation system for ethane cracking

```
[> restart;
[> R1:=dsolve({diff(C2H6(t),t)=-alpha*C2H6(t)-beta*C2H6(t)^(1/2),
   C2H6(0)=C2H6[0]}):
[> assign(R1);
[> Ethane(t)=allvalues(C2H6(t));
```

$$\text{Ethane}(t) = \frac{\left(-\beta + e^{\left(-\frac{\alpha t}{2}\right)} \sqrt{C2H6_0}\, \alpha + e^{\left(-\frac{\alpha t}{2}\right)} \beta\right)^2}{\alpha^2}$$

Fig. 2.21 Derivation of the equation for the source reactant kinetic curve

curve equation for ethane is derived by means of `dsolve`. So as to avoid excessively awkward constructions, the following symbols are introduced

$$k_1 \frac{3k_5 + 2k_6}{k_5 + k_6} = \alpha,$$

$$\frac{\sqrt{k_1} \cdot k_3}{\sqrt{k_5 + k_6}} = \beta.$$

The expression for the current ethane concentration is next substituted into the rest of the equations of the reduced ODE system and their solution yields expressions for the last of the stable substances taking part in the overall process

$$C_{CH_4}(t) = 2k_1 \int_0^t C_{C_2H_6}(u) du,$$

$$C_{C_2H_4}(t) = \beta \int_0^t \left(\sqrt{C_{C_2H_6}(u)} + \frac{k_6}{k_3^2} C_{C_2H_6}(u) \right) du,$$

$$C_{H_2}(t) = \beta \int_0^t \sqrt{C_{C_2H_6}(u)} du,$$

$$C_{C_4H_{10}}(t) = \frac{\beta^2 k_5}{k_3^2} \int_0^t C_{C_2H_6}(u) du.$$

Here, u is a auxiliary value. Now, it is possible to evaluate the kinetic curves for all the reactants. Of course, information about the numerical values of the rate constants of all the steps are needed for this purpose. Such information can be found in reference literature but to use kinetic databases available in the global network is the simplest. A quite reliable source is the database of National Institute of Standards and Technology (http://www.kinetics.nist.gov). After a short search, the temperature dependencies of the constants $k_1 - k_4$ are found,

$$k_1 = 4.26 \times 10^{16} e^{-\frac{44579}{T}},$$

$$k_2 = 1.65 \times 10^9 \left(\frac{T}{298} \right)^{4.25} e^{\frac{-3890}{T}},$$

$$k_3 = 8.85 \times 10^{12} e^{-\frac{19469}{T}},$$

$$k_4 = 1.71 \times 10^{12} \left(\frac{T}{298} \right)^{2.32} e^{-\frac{3414}{T}}.$$

The rate constants for ethyl radical recombination in steps (5) and (6) do not depend on temperature,

$$k_5 = 1.15 \times 10^{13}; k_6 = 1.45 \times 10^{12}.$$

Finally, Fig. 2.22 illustrates the kinetic curve behaviour for the source reactant and the end product of ethane thermal decomposition. The calculations assume that the process proceeds at a temperature of 1,100 K and at an initial pressure of the gas mixture 2 atm. One should remember that, in this case, it is considered an approximate solution obtained in the framework of the approximation about a steady-state course of the process.

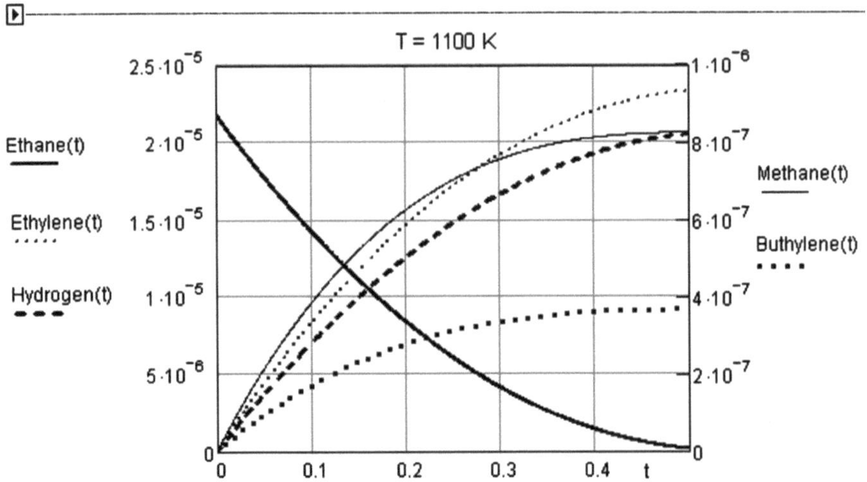

Fig. 2.22 The kinetic curves of the stable participants of the ethane cracking process

2.4.2 *The Quasi-Equilibrium Approximation: Enzymatic Reaction Kinetics*

Let us refer again to kinetic scheme (2.8) describing the consecutive reaction with the first step being reversible. If the equilibrium described by the constants k_1 and k_2 is established quickly, i.e. the step resulting in the end product is slow, the quasi-equilibrium principle is applicable in this case. According to this principle, the concentration of intermediate B practically does not differ from its equilibrium value owing to the hard-to-disturb equilibrium and can be expressed in terms of the ratio of the constants k_1 and k_2

$$K = k_1/k_2 = C_B/C_A.$$

Consequently,

$$C_B = \frac{k_1}{k_2} C_A,$$

and the rate of product C formation is given by

$$r_C = k_2 C_B = k_1 C_A.$$

Thus, the quasi-equilibrium principle should be obeyed provided the condition $k_3 \ll k_2$ is met. The corresponding calculation (Fig. 2.23) confirms this assumption.

The quasi-equilibrium principle gained a widespread practical use in *enzyme kinetics*, a branch of chemical kinetics describing catalytic reactions involving

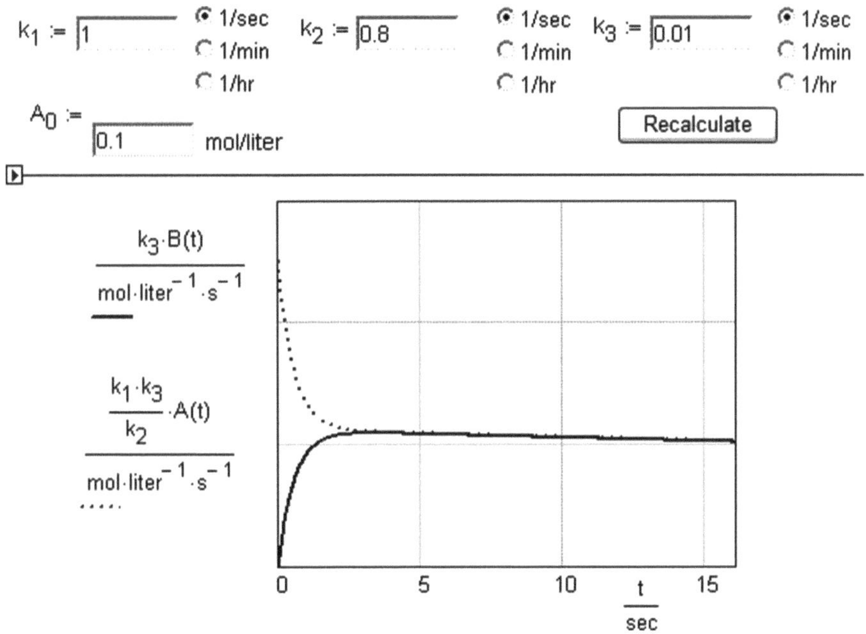

Fig. 2.23 Checking the quasi-equilibrium principle

enzymes. Enzymes are biological catalysts, protein molecules, as a rule. This is a particular type of reactions characterized by high enzyme effectiveness with respect to catalyzed processes and high selectivity of catalysts. Knowledge of enzyme reaction kinetics is necessary for understanding processes occurring into living organisms.

The scheme for the simplest enzyme reaction is

$$E + S \underset{k_{-1}}{\overset{k_1}{\rightleftharpoons}} ES \xrightarrow{k_2} P + E.$$

Here, E is an enzyme, S is a reactant (a substrate), P is a reaction product, ES is an enzyme–substrate complex. The process provides for reversible ES formation followed by its decay into the reaction product with simultaneous enzyme regeneration. Provided the step of product formation is slow ($k_2 \ll k_{-1}$), the reaction kinetics can be described in the framework of the quasi-equilibrium principle.

Thus, the quasi-equilibrium concentration C_{ES} of the enzyme–substrate complex is related to the rate constants k_1 and k_{-1} as well as to the concentrations of the enzyme C_E and the substrate C_S by

$$\frac{k_1}{k_{-1}} = \frac{C_E C_S}{C_{ES}}.$$

The concentration C_{ES} can be deduced from this relationship and substituted into the equation for the formation rate of end product P

$$r_P = k_2 C_{ES}.$$

The enzyme fraction bound in the complex cannot be found experimentally, therefore, the following calculations use the material balance equation

$$C_{ES} + C_E = C_{E_0},$$

where C_{E_0} is an overall enzyme concentration. With the help of the last three relationships, one can obtain the known *Michaelis–Menten* equation

$$r_P = \frac{k_2 C_{E_0} C_S}{K_M + C_S}. \tag{2.10}$$

Derivation of the *Michaelis–Menten* equation using the suits of the Mathcad symbolic redactor is given in Fig. 2.24. Here, K_M is the *Michaelis constant* whose physical meaning corresponds to the dissociation constant for the enzyme–substrate complex, $K_M = k_{-1}/k_1$.

Interpretation of (2.10) leads to quite interesting conclusions. Thus, if the substrate concentration is low $(K_M \gg C_S)$, the equation reduces to the form

$$r_P \approx \frac{k_2 C_{E_0} C_S}{K_M}$$

On the contrary, at high substrate concentrations $K_M \ll C_S$ and

$$r_P \approx k_2 C_{E_0}.$$

Thereby, at low S concentrations the reaction has an order of 1 with respect to the substrate. When transferring to high concentrations, the reaction rate stops depending on a concentration, i.e. the reaction changes its order into zero. The high substrate concentrations favor the maximum reaction rate

$$r_{\max} = k_2 C_{E_0}. \tag{2.11}$$

Figure 2.24 illustrates the general behaviour of the reaction rate *vs*. the current substrate concentration.

Taking into account expression (2.11), (2.10) can be rewritten in the following forms

$$\frac{1}{r_P} = \frac{1}{r_{\max}} + \frac{K_M}{r_{\max}} \cdot \frac{1}{C_S}, \tag{2.12}$$

$$E + S \underset{k_{-1}}{\overset{k_1}{\Longleftrightarrow}} ES \overset{k_3}{\longrightarrow} P + E \qquad r_P = k_2 \cdot C_{ES}$$

$$\frac{k_1}{k_{\_1}} = \frac{C_{ES}}{C_E \cdot C_S} \text{ substitute}, C_E = C_{E0} - C_{ES} \rightarrow \frac{k_1}{k_{\_1}} = \frac{C_{ES}}{C_S \cdot (C_{E0} - C_{ES})}$$

$$\frac{k_1}{k_{\_1}} = \frac{C_{ES}}{C_S \cdot (C_{E0} - C_{ES})} \text{ solve}, C_{ES} \rightarrow \frac{C_S \cdot C_{E0} \cdot k_1}{k_{\_1} + C_S \cdot k_1}$$

$$r_P = k_2 \cdot C_{ES} \text{ substitute}, C_{ES} = \frac{C_S \cdot C_{E0} \cdot k_1}{k_{\_1} + C_S \cdot k_1} \rightarrow r_P = \frac{C_S \cdot C_{E0} \cdot k_1 \cdot k_2}{k_{\_1} + C_S \cdot k_1}$$

$$r_P = \frac{C_S \cdot C_{E0} \cdot k_2}{\frac{k_{\_1}}{k_1} + C_S} \text{ substitute}, \frac{k_{\_1}}{k_1} = K_M \rightarrow r_P = \frac{C_S \cdot C_{E0} \cdot k_2}{C_S + K_M}$$

$$C_{E0} := 1.214 \times 10^{-7} \quad k_2 := 10 \quad K_M := 4.815 \cdot 10^{-4} \quad C_S := 10^{-5}, 0.0001 .. 0.015$$

$$r_P(C_S) := \frac{C_S \cdot C_{E0} \cdot k_2}{C_S + K_M} \qquad r_{max} := k_2 \cdot C_{E0}$$

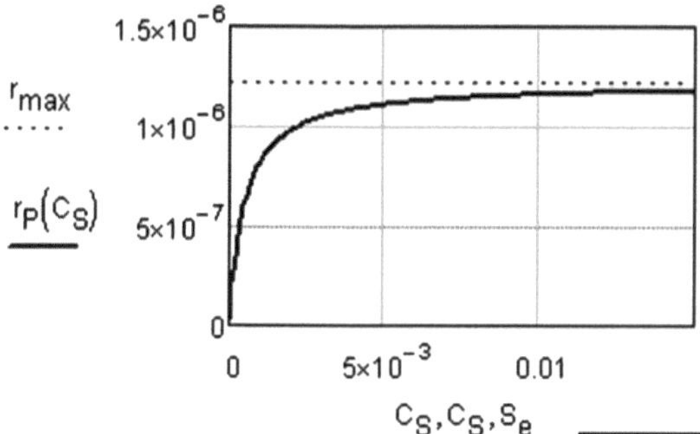

$$\frac{1}{r_P} = \frac{1}{C_{E0} \cdot k_2} + \frac{K_M}{k_2 \cdot C_{E0}} \cdot \frac{1}{C_S}$$

$$r_P = r_{max} - K_M \cdot \frac{r_P}{C_S}$$

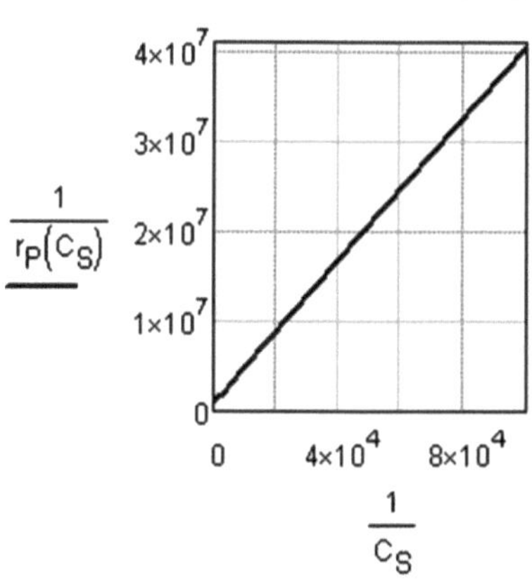

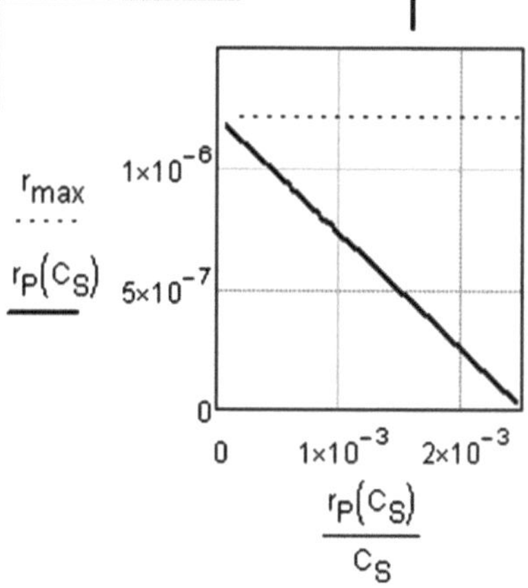

Fig. 2.24 Derivation of the kinetic equations for the enzyme reaction

$$r_P = r_{\max} - K_M \frac{r_P}{C_S}. \tag{2.13}$$

Consequently, the curve of a rate vs. a concentration is linearized on ${r_P}^{-1} - {C_S}^{-1}$ coordinates (*Lineweaver–Burk* coordinates) or on $r_P - r_P/C_S$ coordinates. The both types of coordinates can be used when treating experimental data for solving the inverse kinetic problem. The method of initial rates, when initial reaction rates are measured at different initial substrate concentrations, is used for this purpose.

Chapter 3
Numerical Solution of the Direct Problem in Chemical Kinetics

3.1 Given/Odesolve Solver in Mathcad System

Among the built-in tools of the Mathcad suite an important role belongs to those designed for the numerical solving ordinary differential equations (ODEs) and their systems. Let's look through these tools starting with the solving unit `Given/Odesolve`. The `Odesolve` function first appeared in Mathcad 2000 Pro. On default this function used Runge–Kutta method of the fourth-level of precision.[1]

The organization of the `Given/Odesolve` solver resembles greatly that of the `Given/Find` solving block: it starts with the `Given` keyword. An ODE or a system as well as the initial conditions should be placed in the solver body. The solving is performed with a call up of built-in function `Odesolve` using the following format:

```
Odesolve(x, b, [steps]),
```

where `x` is the unknown, and `b` is the upper integration limit. The last parameter, `steps`, determines the number of integration steps and is optional. If this parameter is absent, the number of steps is set up automatically.

As an example, let's calculate a kinetic curve for an intermediate in a consecutive reaction where the second step is of the second order:

$$A \xrightarrow{k_1} B,$$

$$2B \xrightarrow{k_2} P.$$

Analytical solution of this problem was discussed in the Sect. 1.3.2, so we can compare numerically- and analytically-obtained results.

The way of using `Given/Odesolve` solver and the results of its work are shown in Fig. 3.1. The following values were used: initial concentration $A = 0.5$ M,

[1]In this book we will not discuss the essence of such and such numerical integration methods. This information is available in virtually all handbooks on numerical methods.

V.I. Korobov and V.F. Ochkov, *Chemical Kinetics with Mathcad and Maple*,
DOI 10.1007/978-3-7091-0531-3_3, © Springer-Verlag/Wien 2011

$k_2 := 0.1 \quad k_1 := 0.05 \quad A_0 := 0.5$

$$C_B(t) := \frac{\sqrt{2}}{4} \cdot \frac{I1(2\cdot\sqrt{2})\cdot K1\left(2\cdot\sqrt{2}\cdot e^{\frac{-t}{40}}\right) - I1\left(2\cdot\sqrt{2}\cdot e^{\frac{-t}{40}}\right)\cdot K1(2\cdot\sqrt{2})}{I1(2\cdot\sqrt{2})\cdot K0\left(2\cdot\sqrt{2}\cdot e^{\frac{-t}{40}}\right) + I0\left(2\cdot\sqrt{2}\cdot e^{\frac{-t}{40}}\right)\cdot K1(2\cdot\sqrt{2})} \cdot e^{\frac{-t}{40}}$$

Given $\frac{d}{dt}C_b(t) = k_1 \cdot A_0 \cdot e^{-k_1 \cdot t} - 2\cdot k_2 \cdot C_b(t)^2 \quad C_b(0) = 0 \quad C_b := \mathrm{Odesolve}(t, 200)$

$t_1 := 0, 5 .. 200$

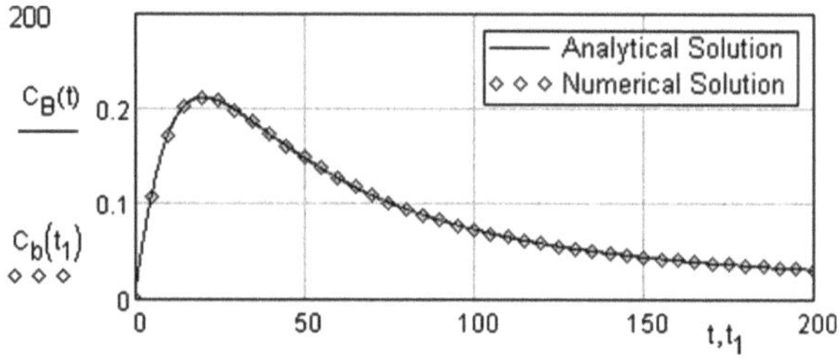

Fig. 3.1 Kinetic curve calculation for an intermediate in a consecutive second-order reaction using Odesolve function (on-line calculation http://twt.mpei.ac.ru/MCS/Worksheets/Chem/Chem-Kin-3-01.xmcd)

rate constants k_1. Referring to Fig. 3.1, the kinetic curve calculated numerically (dots) matches the curve calculated using analytical formula (line). The authors want to point out that during the assigning of the results of numerical computation one should use only the name of the desired function (without argument). For the construction of a graph both function name and its argument are specified (Fig. 3.1).

The results of this example allow one to think that the Odesolve function is a sufficient tool for solving the direct kinetic problem. Now we proceed to a numerical solution of an *ordinary differential equation* (ODE) set using a solver. Solving process for an ODE set does not differ much from that for a single ODE: all equations in the set as well as the starting condition equalities are given in the solver. The only difference is in the number of arguments that Odesolve function should have. An array of desired function names is required for solving sets of equations. The variable, where the calculation result is stored, is an array of function names as well. Below is an example of a numerical calculation of all component concentrations in a hypothetical multi-step reaction.

$$\begin{aligned} 2A &\underset{k_2}{\overset{k_1}{\rightleftarrows}} 2B + C, \\ A + B &\underset{k_4}{\overset{k_3}{\rightleftarrows}} 2D, \\ B + D &\underset{k_6}{\overset{k_5}{\rightleftarrows}} E. \end{aligned} \tag{3.1}$$

We can see that the process goes through six elementary steps with five compounds; and most of the reactions have order higher than first. Finding the analytical solution of the direct problem is barely possible in this case. However, having numerical integration tools we do not need an analytical approach.

The solution of this example is shown in Fig. 3.2. One can see that the use of Odesolve function is rather simple.

Working with Odesolve the user can choose an algorithm for the numerical ODE solution. If the cursor is above Odesolve keyword, the right-click will activate the corresponding context menu (Fig. 3.3). In Mathcad 2001–13 the user could choose between integration with a fixed (Fixed) or adaptive (Adaptive) step as well as a special method for stiff systems integration (Stiff). A new option, Adams/BDF, was added in Mathcad 14–15.

An obvious advantage of the Given/Odesolve solver is that differential equations and ODE sets are written in the usual manner. That is why Mathcad documents designed for kinetic problems are rather clear. Another feature of the Odesolve solver is the possibility to use not only differential equations but also usual algebraic ones.

For example, let's consider solving the direct problem for a parallel reaction with the following mechanism:

$$A + B \xrightarrow{k_1} P_1,$$
$$A + C \xrightarrow{k_2} P_2.$$

$k_1 := 0.5 \quad k_2 := 0.6 \quad k_3 := 1.8 \quad k_4 := 0.6 \quad k_5 := 2.9 \quad k_6 := 0.8 \quad t_{end} := 5$

Given $\quad \frac{d}{dt}A(t) = -2\cdot k_1\cdot A(t)^2 + 2\cdot k_2\cdot B(t)^2\cdot C(t) - k_3\cdot A(t)\cdot B(t) + k_4\cdot D(t)^2$

$\frac{d}{dt}B(t) = 2\cdot k_1\cdot A(t)^2 - 2\cdot k_2\cdot B(t)^2\cdot C(t) - k_3\cdot A(t)\cdot B(t) + k_4\cdot D(t)^2 - k_5\cdot B(t)\cdot D(t) + k_6\cdot E(t)$

$\frac{d}{dt}E(t) = k_5\cdot B(t)\cdot D(t) - k_6\cdot E(t) \quad \frac{d}{dt}D(t) = 2\cdot k_3\cdot A(t)\cdot B(t) - 2\cdot k_4\cdot D(t)^2 - k_5\cdot B(t)\cdot D(t) + k_6\cdot E(t)$

$\frac{d}{dt}C(t) = k_1\cdot A(t)^2 - k_2\cdot B(t)^2\cdot C(t)$

$A(0) = 1 \quad B(0) = 1 \quad C(0) = 0 \quad D(0) = 0 \quad E(0) = 0$

$$\begin{pmatrix} A \\ B \\ C \\ D \\ E \end{pmatrix} := \text{Odesolve}\left[\begin{pmatrix} A \\ B \\ C \\ D \\ E \end{pmatrix}, t, t_{end}, 1000\right]$$

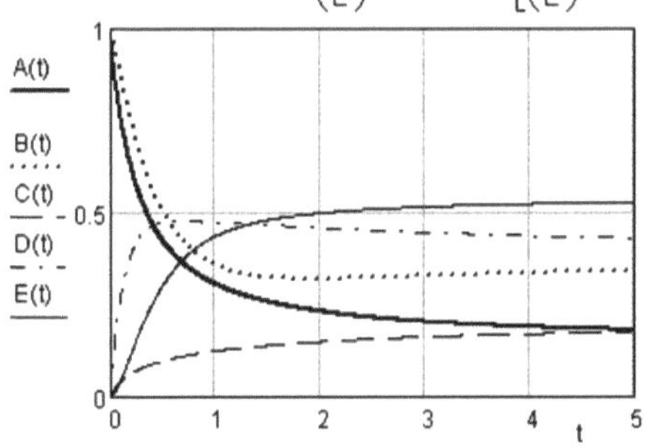

Fig. 3.2 Calculation of the kinetic curves for all components in a multi-step reaction using Odesolve function(on-line calculation http://twt.mpei.ac.ru/MCS/Worksheets/Chem/ChemKin-3-02.xmcd)

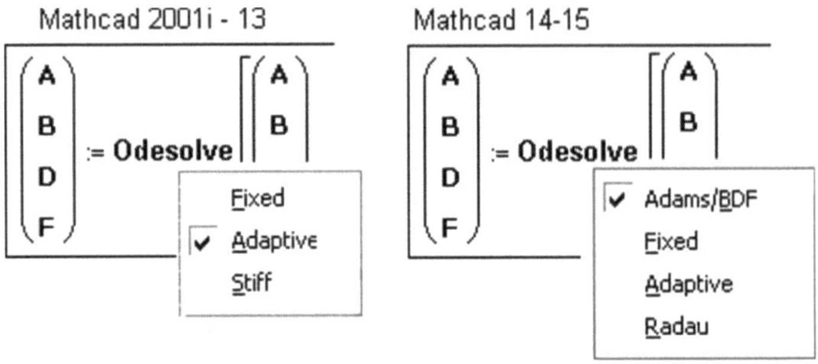

Fig. 3.3 Choosing an Odesolve algorithm in different Mathcad versions

Here reagent A is consumed in two parallel steps through interaction with reagents B and C. Each step has the second order. To find kinetic curves for all the compounds in this reaction it is enough to use a mathematical model consisting of two differential and three algebraic equations (the latter represent the material balance of the reaction system):

$$\frac{dC_B(t)}{dt} = -k_1 C_A(t) C_B(t),$$

$$\frac{dC_C(t)}{dt} = -k_2 C_A(t) C_C(t),$$

$$C_{A_0} - C_A(t) = C_{B_0} - C_B(t) + C_{C_0} - C_C(t),$$

$$C_{P_1}(t) = C_{B_0} - C_B(t),$$

$$C_{P_2}(t) = C_{C_0} - C_C(t).$$

All equations can be put in the solver without changes. The corresponding numerical solution is shown in Fig. 3.4.

So far, we have considered the problem of calculating kinetic curves for compounds. In mathematics this is called a *Cauchy problem*. It is known that for the Cauchy problem entry conditions are given, i.e. the value(s) of the desired function $y(x)$ at point $x = x_0$. In kinetics, as a rule, this point corresponds to the starting point of a reaction. Often one needs to find a partial solution of a differential equation using known function values in several points. Such kind of problems is called a *boundary-value problem (boundary problem)*.

The two-point boundary problem may be exemplified by the calculation of the kinetic curve for an intermediate in a consecutive reaction $A \rightarrow B \rightarrow P$ using known values of concentration at two moments of time (Fig. 3.5). Assume the initial A concentration C_{A_0}, rate constants $k_1 = 0.008\ \mathrm{s}^{-1}$, $k_2 = 0.004\ \mathrm{s}^{-1}$, and the concentration of an intermediate B after 7.2 and 445.1 s from the beginning of the

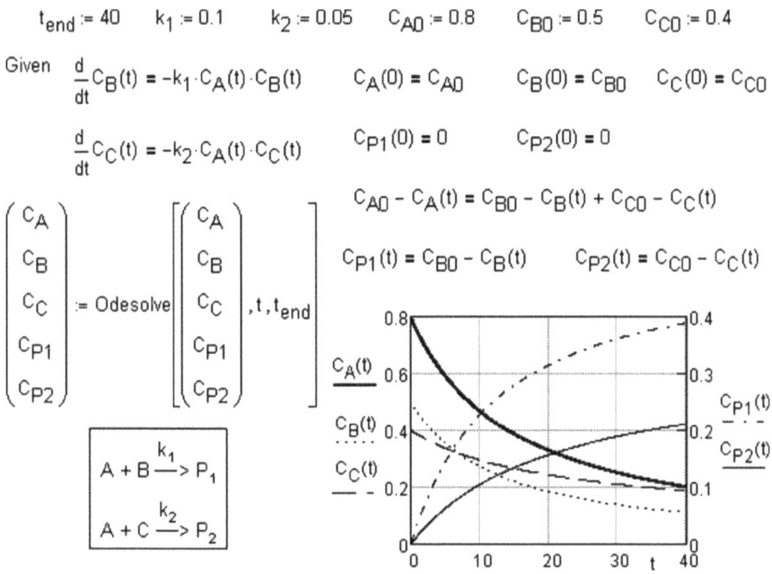

Fig. 3.4 Solving a system of "differential – algebraic" equations (on-line calculation http://twt.mpei.ac.ru/MCS/Worksheets/Chem/ChemKin-3-04.xmcd)

reaction $C_b = 0.05$ M. To calculate the kinetic curve for the compound *B* we will use the equation:

$$dC_B(t)/dt = -k_1 C_{A_0} e^{-k_1 t} - k_2 C_B(t),$$

that is converted into:

$$d^2 C_B(t)/dt = -k_1{}^2 C_{A_0}{}^2 e^{-k_1 t} - k_2 dC_B(t)/dt.$$

The solution of this boundary problem is given in Fig. 3.5 and it does not require additional explanation.

The technology of working with the function Odesolve in Mathcad Prime differs from the same technology with this function in Mathcad 11–15.

Firstly, in Mathcad Prime we must not input the keyword Given, but a *Solve block*.

Secondly, in Mathcad Prime we have an opportunity to solve differential equations with *units*.

All these changes are shown in Fig. 3.6, which shows the solution in Mathcad Prime of a task of acceleration a car with an engine, creating thrust of 300 kgf and acted upon by forces of inertia and air resistance.

The authors do not have time to translate all tasks of the book into Mathcad Prime, but readers can download Mathcad Prime files from the site of the book (http://twt.mpei.ac.ru/TTHB/ChemKin.html).

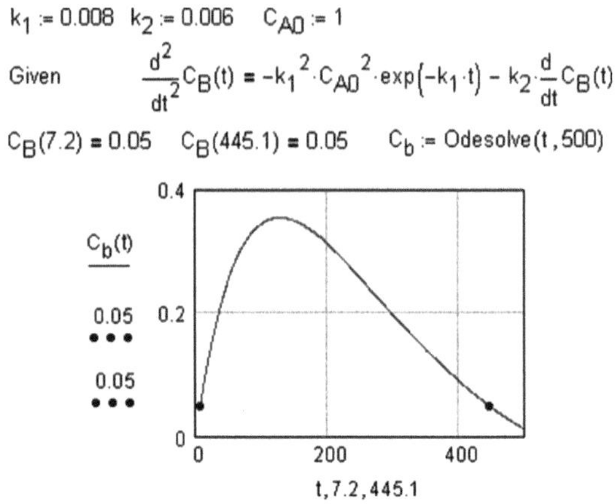

Fig. 3.5 Solving a boundary-value problem with solver

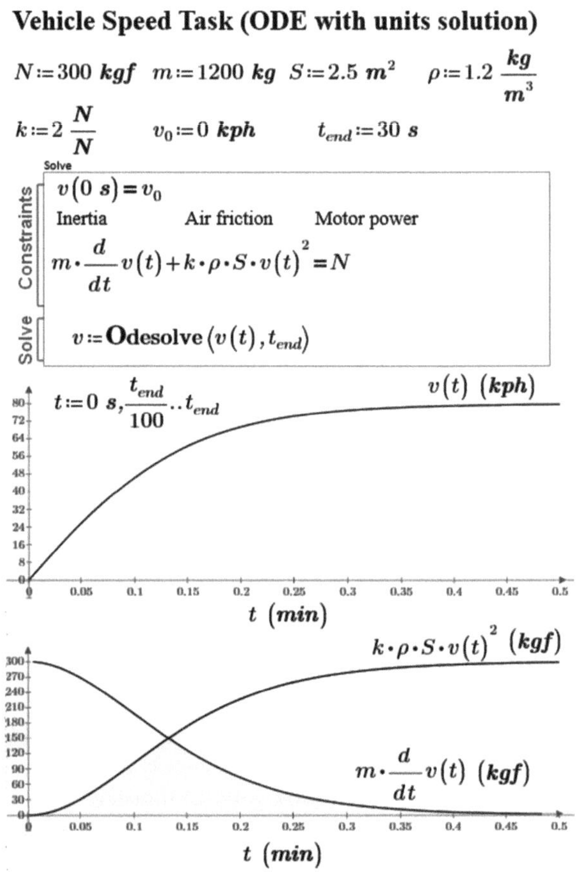

Fig. 3.6 Solving a ODE in Mathcad Prime

3.1.1 Built-In Mathcad Integrators

In addition to Given/Odesolve solver, one can use built-in Mathcad suite integrators for solving the direct kinetic problem. These integrators are designed specifically for the numerical solving of ODEs and ODEs sets. There are several built-in functions of this kind. It is important to mention that each of the integrators requires right-hand member array of the ODEs set. Recall that the mathematical model for a complex reaction is obtained through multiplying the stoichiometric array by the rate array. The result of this operation is the array of the ODE set right-hand values. That is why built-in integrators are very convenient for solving kinetic problems with stoichiometric arrays of high dimensionality.

We will begin the learning of the integrators with the `rkfixed` function. It implements the fourth-order Runge–Kutta method (rk) with fixed step of integration (fixed). According to Mathcad syntax this function has five required arguments:

```
rkfixed(v,x1,x2,npoints,f),
```

where `v` is the initial values vector; `x1` and `x2` – independent variable values that determine the interval of integration; `npoints` – the integration steps quantity; `f` – the array function of the ODE set right-hand values.

Creation of the array function `f` requires following a specific procedure. Usually this function is declared right before `rkfixed` using the following syntax:

```
f(t,x):=[array of the ODE set right-hand values],
```

where `t` is the independent variable and x is the desired functions array. The right-hand part of this construction is an array of the ODE set right-hand values; and each of these functions is represented by an index variable `x`, i.e. x_0, x_1, etc.

The quantity of the array x elements as well as the quantity of the right-hand values is equal to the quantity of equations in the system. Let's clarify how rkfixed works by an example of a numerical solution of the direct problem for a three-step consecutive chemical reaction (Fig. 3.7).

In the beginning, the right-hand value array Model(k,x) is defined. Current concentrations of the compounds *A, B, C, D* are given as x_1, x_2, x_3, x_4 (here `1, 2, 3, 4` are ranging indices, not lettered ones!). After assigning the rate constant values, a function F(t,x) is declared. The first argument must be an independent variable – time. This declaration allows the rkfixed integrator to understand the function F(t,x). The numerical calculation result is stored in a matrix S. Based on the rows and columns quantities in this matrix one may conclude: the solution matrix has (n + 1) columns if the set has n equations. It is easy to see that the first column starts with a value $t_n = 0$ and ends with a value $t_k = 100$. The matrix has (N + 1) rows for N integration steps. The second column contains the value of the variable x_2 at each integration step. In this case these are the values of the reagent A concentration. Correspondingly, the third, fourth and fifth columns represent

$$A \xrightarrow{k_1} B \qquad B \xrightarrow{k_2} C \qquad C \xrightarrow{k_3} D$$

$$\text{ORIGIN} := 1 \qquad \alpha := \begin{pmatrix} -1 & 1 & 0 & 0 \\ 0 & -1 & 1 & 0 \\ 0 & 0 & -1 & 1 \end{pmatrix}$$

$$\text{Model}(k, x) := \alpha^T \cdot \begin{pmatrix} k_1 \cdot x_1 \\ k_2 \cdot x_2 \\ k_3 \cdot x_3 \end{pmatrix} \rightarrow \begin{pmatrix} -k_1 \cdot x_1 \\ k_1 \cdot x_1 - k_2 \cdot x_2 \\ k_2 \cdot x_2 - k_3 \cdot x_3 \\ k_3 \cdot x_3 \end{pmatrix} \qquad \text{inC} := \begin{pmatrix} 4 \\ 0 \\ 0 \\ 0 \end{pmatrix}$$

$$k_1 := 0.15 \qquad k_2 := 0.1 \qquad k_3 := 0.075$$

$$F(t, x) := \text{Model}(k, x) \qquad S := \text{rkfixed}(\text{inC}, 0, 100, 500, F)$$

$$t := S^{\langle 1 \rangle} \quad C_A := S^{\langle 2 \rangle} \quad C_B := S^{\langle 3 \rangle} \quad C_C := S^{\langle 4 \rangle} \quad C_D := S^{\langle 5 \rangle}$$

S =

	1	2	3	4	5
1	0	4	0	0	0
2	0.2	3.882	0.117	$1.174 \cdot 10^{-3}$	$5.902 \cdot 10^{-6}$
3	0.4	3.767	0.228	$4.597 \cdot 10^{-3}$	$4.647 \cdot 10^{-5}$
4	0.6	3.656	0.334	0.01	$1.543 \cdot 10^{-4}$
5	0.8	3.548	0.434	0.018	$3.599 \cdot 10^{-4}$
6	1	3.443	0.53	0.027	$6.917 \cdot 10^{-4}$
7	1.2	3.341	0.62	0.038	$1.176 \cdot 10^{-3}$
8	1.4	3.242	0.705	0.051	$1.838 \cdot 10^{-3}$
9	1.6	3.147	0.786	0.065	$2.7 \cdot 10^{-3}$
10	1.8	3.054	0.863	0.08	$3.782 \cdot 10^{-3}$
11	2	2.963	0.935	0.097	$5.106 \cdot 10^{-3}$
12	2.2	2.876	1.003	0.114	$6.688 \cdot 10^{-3}$

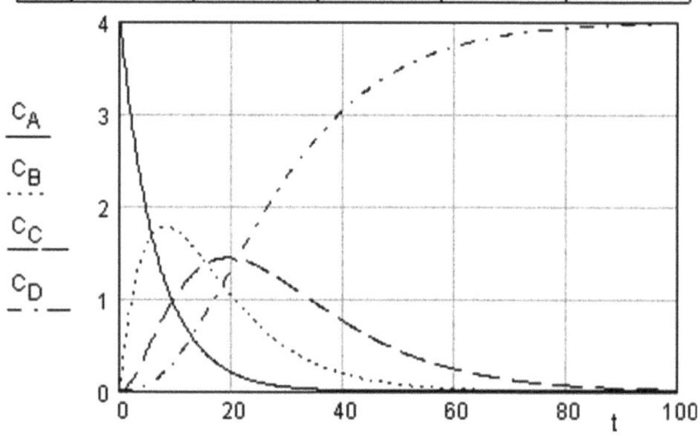

Fig. 3.7 Numerical solution of the direct problem for a consecutive reaction with two intermediates (on-line calculation http://twt.mpei.ac.ru/MCS/Worksheets/Chem/ChemKin-3-06.xmcd)

concentrations of the reagents *B*, *C* and *D*. Thus, the matrix S give a pictorial view of the reaction mixture changes over time. Finally, we can make a graphical representation of the calculation results as kinetic curves for each reagent.

Thereby, Mathcad tools for numerical solution of differential equation sets allow one to calculate quickly kinetic curves of all reactants in a complex chemical process. As the independent variable in chemical problems is time *t*, discussed methods can be used in modeling any time function (having postulated a differential model for the process). One can find many analogies of the kinetic models of chemical reactions in other fields of knowledge (microbiology, sociology, etc.) It is useful to discuss the corresponding examples in order to form some practical skills in creating differential models.

Let's consider one of such problems. Certain microorganisms propagate in proportion to the colony size (with an aspect ratio k) but at the same time produce some excrement, which is a poison for them.

The rate of the colony disappearance is proportional to the amount of poison and current microorganism population with an aspect ratio k_1. The poison formation rate is proportional to the number of microorganisms (with a ratio k_2). Suppose the initial colony size equals N_0, and the amount of poison Z is 0 at the beginning. One is required to make the corresponding set of differential equations and solve it numerically, and present graphically the microorganisms population trends along with the amount of poison in the system. Assume $k = 0.1$; $k_1 = 0.0001$; $k_2 = 0.01, N_0 = 2{,}000$.

First, we should create a differential equation set in accordance with the problem specifications. Changes in the microorganisms population is determined by an increase kN as a result of reproduction and a decrease $-k_1NZ$ due to poisoning. Therefore, the first differential equation of the system will be of this form:

$$dN/dt = kN - k_1NZ.$$

The rate of poison amount change will be described as:

$$dZ/dt = k_2N.$$

A differential model for the process has been defined; now we can create a corresponding Mathcad document to solve the problem (Fig. 3.8).

As the Fig. 3.8 implies, the number of microorganisms first increases with time, achieves the highest value at some point, and after that the colony becomes extinct. The curve $Z(t)$ is a typical saturation one. At the beginning, the poison accumulation rate is small, but it increases with the lapse of time until it reaches the maximum. Certainly, after full disappearance of the microorganisms the amount of poison stabilizes and becomes constant. A chemical analog for this model would be a complex chemical reaction where compounds N and Z participate in an intermediate step. In this case, the compound Z is an autocatalyst for the decomposition of N in accordance with discussed mathematical correlations.

Further in the book we will see that a minor (at first sight) modification of the starting differential equations set and initial conditions may cause a significant change in the dynamic outlook of the solution.

One may often have a case when in order to obtain enough accurate result it is necessary to use a variable integration step: decreasing in the area of large derivative changes and, vice versa, increasing when the derivative changes slowly.

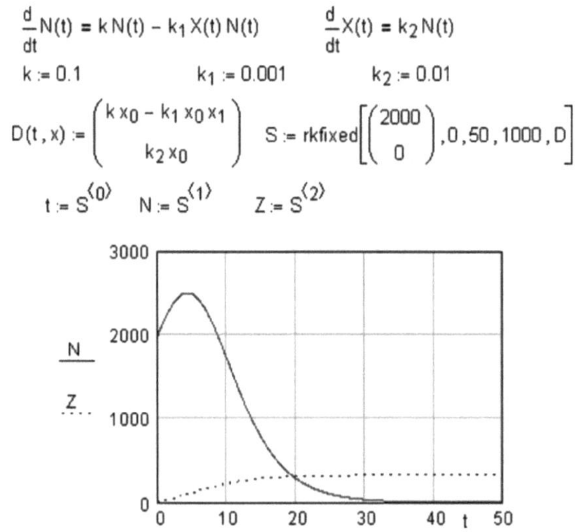

Fig. 3.8 Microorganism population and poison amount trends (on-line calculation http://twt.mpei.ac.ru/MCS/Worksheets/Chem/ChemKin-3-07.xmcd)

This approach is implemented in an algorithm that function `Rkadapt` uses. During the work of this function the step of integration is adapted in accordance with the derivative trend in the selected interval of integration. Let's consider the following ODE system as an example:

$$dX(t)/dt = a - (b+1)X(t) + X(t)^2 Y(t),$$

$$dY(t)/dt = bX(t) - X(t)^2 Y(t).$$

Further, we will discuss minutely this system which is a mathematical model of the widely-known kinetic scheme *bruesselator*. Now we will compare the results of its numerical solution using the functions `rkfixed` and `Rkadapt`.

The corresponding plots are shown in Fig. 3.9. They show that using a fixed step can lead to an instable solution which can be interpreted wrongly from a physical point of view (dashed line). The function `Rkadapt`, as we can see, allows us to eliminate the mistakes of `rkfixed`, and reveals the true behavior of the desired function in the given independent variable range (due to an adaptable step of integration). In practice, `Rkadapt` is preferable in the solving of many direct problems, especially in cases when the starting kinetic model is non-linear.

We also want to mention that the function `Rkadapt` requires the same five arguments specified in an `rkfixed` body. Even though integration utilizes a changeable step, the result will still be represented for evenly distributed points as specified by user.

There is one more circumstance related to a variety of built-in integrators. It is the existence of so-called *stiff* ODE sets. The concept of stiffness may be illustrated by the example of the kinetic equation for a multi-step reaction:

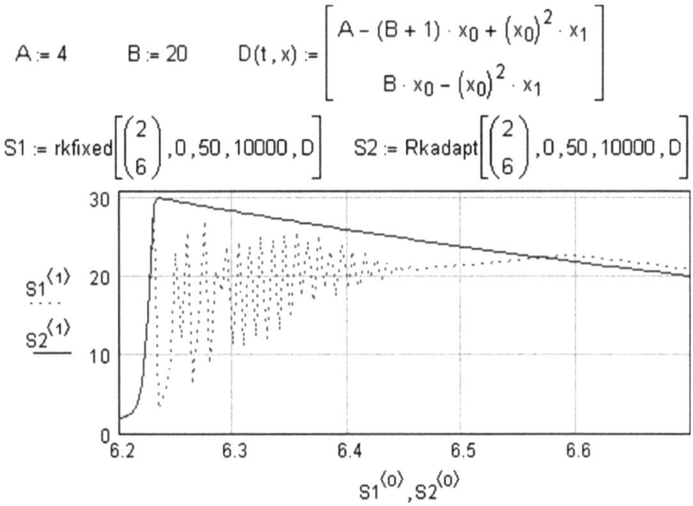

Fig. 3.9 Comparison of the results for calculations with fixed step of integration (*solid line*) to an adaptive one (*dashed line*)

$$\frac{d}{dt}\underline{C(t)} = \underline{\underline{k}}\,\underline{C(t)}.$$

It is considered that the mathematical model is stiff if among the eigenvalues λ_i of the rate constant matrix $\underline{\underline{k}}$ there exist such eigenvalues for which $\mathrm{Re}(\lambda_i) \ll 0$. Usually this condition holds if the rate constants matrix has elements very different in modulo (three and more orders).

Let's consider the following kinetic scheme:

$$A \xrightarrow{k_1=10000} B \xrightarrow{k_2=1} P,$$

that can described with only two differential equations in accordance with the stoichiometric matrix rank. The corresponding ODE set can be written as:

$$\frac{d}{dt}\begin{pmatrix} C_A(t) \\ C_B(t) \end{pmatrix} = \begin{pmatrix} -k_1 & 0 \\ k_1 & -k_2 \end{pmatrix}\begin{pmatrix} C_A(t) \\ C_B(t) \end{pmatrix}.$$

The eigenvalues vector for this matrix is equal to:

$$\underline{\lambda} = \begin{pmatrix} -1000 \\ 1 \end{pmatrix},$$

which allows one to consider the mathematical model as a stiff one. In this case the integration should be performed using special built-in functions for stiff systems. Mathcad suite has integrators `Stiffb`, `Stiffr`, `Radau`, and starting from Mathcad 14 – an integrator `AdamsBDF`. Figure 3.10 shows the Mathcad

$$\begin{pmatrix}-k_1 & 0\\ k_1 & -k_2\end{pmatrix}\cdot\begin{pmatrix}C_A\\ C_B\end{pmatrix}\to\begin{pmatrix}-k_1\cdot C_A\\ k_1\cdot C_A-k_2\cdot C_B\end{pmatrix}\qquad \text{eigenvals}\left(\begin{pmatrix}-k_1 & 0\\ k_1 & -k_2\end{pmatrix}\right)\ \left|\ \begin{matrix}\text{substitute}, k_1=1000\\ \text{substitute}, k_2=1\end{matrix}\right.\to\begin{pmatrix}-1000\\ -1\end{pmatrix}$$

$$k_1 := 1000 \qquad k_2 := 1 \qquad F(t,x_0,x_1) := \begin{pmatrix}-k_1\cdot x_0\\ k_1\cdot x_0-k_2\cdot x_1\end{pmatrix}$$

$$J(t,x_0,x_1) := \begin{pmatrix}\frac{d}{dt}F(t,x_0,x_1)_0 & \frac{d}{dx_0}F(t,x_0,x_1)_0 & \frac{d}{dx_1}F(t,x_0,x_1)_0\\ \frac{d}{dt}F(t,x_0,x_1)_1 & \frac{d}{dx_0}F(t,x_0,x_1)_1 & \frac{d}{dx_1}F(t,x_0,x_1)_1\end{pmatrix}$$

$$F(t,x) := F(t,x_0,x_1)\ \left|\ \begin{matrix}\text{substitute}, x_0=x_0\\ \text{substitute}, x_1=x_1\end{matrix}\right.\to\begin{pmatrix}-1000\cdot x_0\\ 1000\cdot x_0-x_1\end{pmatrix}\qquad S_1 := \text{rkfixed}\left[\begin{pmatrix}1\\0\end{pmatrix},0,100,100,F\right]$$

Found a number with a magnitude greater than 10^307 while trying to evaluate this expression.

$$J(t,x) := J(t,x_0,x_1)\ \left|\ \begin{matrix}\text{substitute}, x_0=x_0\\ \text{substitute}, x_1=x_1\end{matrix}\right.\to\begin{pmatrix}0 & -1000 & 0\\ 0 & 1000 & -1\end{pmatrix}$$

$$S_2 := \text{Rkadapt}\left[\begin{pmatrix}1\\0\end{pmatrix},0,100,100,F\right]\qquad S_3 := \text{Radau}\left[\begin{pmatrix}1\\0\end{pmatrix},0,100,100,F\right]$$

$$S_4 := \text{Stiffb}\left[\begin{pmatrix}1\\0\end{pmatrix},0,100,100,F,J\right]\qquad S_5 := \text{Stiffr}\left[\begin{pmatrix}1\\0\end{pmatrix},0,100,100,F,J\right]$$

$S_2^{\langle 2\rangle}=$	$S_3^{\langle 2\rangle}=$	$S_4^{\langle 2\rangle}=$	$S_5^{\langle 2\rangle}=$
0	0	0	0
0.368	0.368	0.368	0.368
0.135	0.136	0.135	0.135
0.05	0.05	0.05	0.05
0.018	0.018	0.018	0.018
$6.745\cdot10^{-3}$	$6.729\cdot10^{-3}$	$6.728\cdot10^{-3}$	$6.733\cdot10^{-3}$
$2.481\cdot10^{-3}$	$2.498\cdot10^{-3}$	$2.474\cdot10^{-3}$	$2.476\cdot10^{-3}$
$9.128\cdot10^{-4}$	$9.073\cdot10^{-4}$	$9.093\cdot10^{-4}$	$9.104\cdot10^{-4}$
$3.358\cdot10^{-4}$	$3.227\cdot10^{-4}$	$3.343\cdot10^{-4}$	$3.348\cdot10^{-4}$
$1.235\cdot10^{-4}$	$1.404\cdot10^{-4}$	$1.229\cdot10^{-4}$	$1.231\cdot10^{-4}$
$4.545\cdot10^{-5}$	$5.043\cdot10^{-5}$	$4.518\cdot10^{-5}$	$4.527\cdot10^{-5}$
$1.672\cdot10^{-5}$	$1.049\cdot10^{-5}$	$1.661\cdot10^{-5}$	$1.665\cdot10^{-5}$
$6.15\cdot10^{-6}$	$1.588\cdot10^{-6}$	$6.106\cdot10^{-6}$	$6.121\cdot10^{-6}$
$2.263\cdot10^{-6}$	$4.725\cdot10^{-6}$	$2.245\cdot10^{-6}$	$2.251\cdot10^{-6}$
$8.324\cdot10^{-7}$	$3.633\cdot10^{-6}$	$8.252\cdot10^{-7}$	$8.277\cdot10^{-7}$
$3.062\cdot10^{-7}$	$1.86\cdot10^{-6}$	$3.034\cdot10^{-7}$	$3.044\cdot10^{-7}$

Fig. 3.10 An example of a kinetic scheme described with a stiff set of differential equations (Mathcad 11.2)

document illustrating the solution of the direct kinetic problem for a stiff model using several integrators.

As Fig. 3.10 shows, the integrator rkfixed cannot solve the problem at all. It shows a diagnostic message «Found a number with a magnitude greater than 10^307 while trying to evaluate this expression». In this case the integrators `Rkadapt`, `Radau`, `Stiffb`, and `Stiffr` do work but give different computation results for the chosen step of integration. One can find that a fivefold increase in the number of steps virtually levels the results. However, the integration of more complex stiff sets still should be done by using the specially designed built-in functions.

The functions `Stiffb` and `Stiffr` require an additional argument `J(t,x)`, which is a matrix of partial time (`t`, zero column) and the x vector derivatives for a kinetic function. A way to form the vector `J(t,x)` is shown in Fig. 3.10. It is worth mentioning that, in the early versions, this stage was done manually (paying attention to the fact that the symbol editor cannot differentiate expressions with index variables). The user had to transform the index variables into lettered ones and vice versa (see the Fig. 3.10 where the index and lettered variables are formatted with different styles). Because of that the `Mathcad 2001i` version included the `Radau` function that did not require the argument `J(t,x)`. Although it was very convenient, the user had to accept some loss of precision. In the `Mathcad 14` version the functionality of the `Radau` was expanded. In addition, this version had tools for automatization of the matrix `J(t,x)` symbolic calculation.

3.1.2 The Maple System Commands dsolve, odeplot in Numerical Calculations

The command dsolve of the Maple system was previously discussed as a method of analytical solution of the direct problem in chemical kinetics. This command also can be used for a numerical solving of ODEs or their sets. In this case one should use the following syntax:

```
dsolve({ode,ic},numeric,vars,options)
```

Here `ode` is the differential equation (or ODE set) with the initial conditions `ic`. The option `numeric` is a directive for numerical computations (one may use the construction "`type = numeric`" instead of the keyword `numeric`); `vars` is the desired function (or the desired function set in case of ODE set); `options` are additional options given in the `keyword = value` form.

The option `numeric` (or `type = numeric`) indicates that dsolve will return a numerical calculation result. The most important additional option is the `method`. It determines which numerical method will be used in the desired function search. The Maple system gives a choice among a variety of methods (see the list in Table 3.1).

On default, during a solution of the Cauchy problem the `dsolve` function automatically utilizes the Runge–Kutta method of the fourth to fifth order of accuracy. The major options are given in Table 3.2.

An important component of using the dsolve function in numerical calculations is the output order of the results. In particular, when the following construction is being used:

```
>dsol:=dsolve({sys1,invalues},numeric);
```

Table 3.1 Numerical methods for solving ordinary differential equations in Maple® suite

No.	The `method` value	Numeric method used by `dsolve` solver
1.	`rk45`	*Runge–Kutta–Fehlberg* method of the fourth to fifth order of accuracy
2.	`classical` *or* `classical[foreuler]`	*Euler* method
3.	`classical[heunform]`	Modified *Euler* method using *Heun's* formula
4.	`classical[impoly]`	*Euler* method subkind
5.	`classical[rk2]`	*Runge–Kutta* method of the second order of accuracy
6.	`classical[rk3]`	*Runge–Kutta* method of the third order of accuracy
7.	`classical[rk4]`	*Runge–Kutta* method of the fourth order of accuracy
8.	`classical[adambash]`	*Adams–Bashforth* method
9.	`classical[abmoulton]`	*Adams–Bashforth–Moulton* method
10.	`rosenbrock`	*Rosenbrock* method
11.	`bvp`	Numerical method to solve the **b**oundary **v**alue **p**roblem
12.	`dverk78`	*Runge–Kutta* method of the seventh to eighth order of accuracy
13.	`lsode` *or* `lsode[choice]`, *where* `choice` *can be* `adamsfunc`, `adamsfull`, `adamsdiag`, `adamsband`, `backfunc`, `backfull`, `backdiag`, `backband`	Modification of the *Adams* method for the solving of stiff ODE and their sets
14.	`gear`, `gear[bstoer]`, `gear[polyextr]`	*Gear* method and its modifications
15.	`taylorseries`	Method utilizing the *Taylor* series expansion of integrand

Table 3.2 Some major options for the dsolve command used in numerical calculations

Option	Purpose
`'output' = keyword or array`	Controls the output order of the calculation results. Can have the symbol values `procedurelist`, `listprocedure`, as well as `array` or `Array`. By default `'output' = procedurelist`.
`'stop_cond' = list`	Controls the process of computation finishing when the `list` conditions are met.
`'range' = numeric..numeric`	Determines the independent variable range (the length of the integration interval).
`'stiff' = boolean`	This option is given as a Boolean expression. For example, if `'stiff' = true`, the equation is considered stiff and the default `method` value is set to `rosenbrock` instead of `rk45`.

only a message about the successful procedure implementation is displayed:

```
dsol:=proc(rkf45_x)...end proc
```

After that, the user has to decide how the results should be visualized. It is possible to print the answer in the form of individual values of the desired function, an array, etc. However, the most visual output form is the graphical one. The plotting of the results is provided by the command `odeplot` from the graphical library `plots`. Figures 3.11 and 3.12 show a solution of the differential equation set, which describes the kinetics of the first-order reversible reaction $A \Leftrightarrow B$ with arbitrary rate constant values.

3.1.3 Oscillation Processes Modeling

In some reactions, one can see the periodic changes of the reagent concentrations over time. Correspondingly, the rate of the reaction has an oscillating character.

```
> restart;
> sys1:=diff(A(t),t)=-k[1]*A(t)+k[2]*B(t),diff(B(t),t)=-k[2]*B
  (t)+k[1]*A(t);
```

$$sys1 := \frac{d}{dt}A(t) = -k_1 A(t) + k_2 B(t), \frac{d}{dt}B(t) = -k_2 B(t) + k_1 A(t)$$

```
> invalues:=A(0)=1,B(0)=0.1: k[1]:=0.05: k[2]:=0.025:
> dsol1:=dsolve({sys1,invalues},numeric,range=0..4);
```

dsol1 := **proc**(x_rkf45) ... **end proc**

```
> dsol1(0.005);
```

[t = 0.005, A(t) = 0.999762544525681028, B(t) = 0.100237455474319170]

```
> dsol1(0.05);
>
```

[t = 0.05, A(t) = 0.997629447563644246, B(t) = 0.102370552436355650]

```
> dsol2:=dsolve({sys1,invalues},numeric,output=array([1,10,20,
  50,100,200,500,1000]));
```

dsol2 :=

t	A(t)	B(t)
1.	0.954237541297251024	0.145762458702748954
10.	0.665832112872106862	0.434167887127893281
20.	0.507982314965579328	0.592017685034420982
50.	0.381561119707259366	0.718438880292740834
100.	0.367016860823496994	0.732983139176502818
200.	0.366666830014558354	0.733333169985442401
500.	0.366666617033504116	0.733333382966496417
1000.	0.366666860183484712	0.733333139816515600

Fig. 3.11 Numerical solution of the direct kinetic problem using Mathcad tools

```
> with(plots):
Warning, the name changecoords has been redefined
```

```
> fncs:={A(t),B(t)}; p:=dsolve({sys1,invalues},fncs,
  type=numeric,range=0..500):
```

$$fncs := \{A(t), B(t)\}$$

```
> odeplot(p,[[t,A(t)],[t,B(t)]],0..60,numpoints=200,axes=boxed
  ,thickness=2);
```

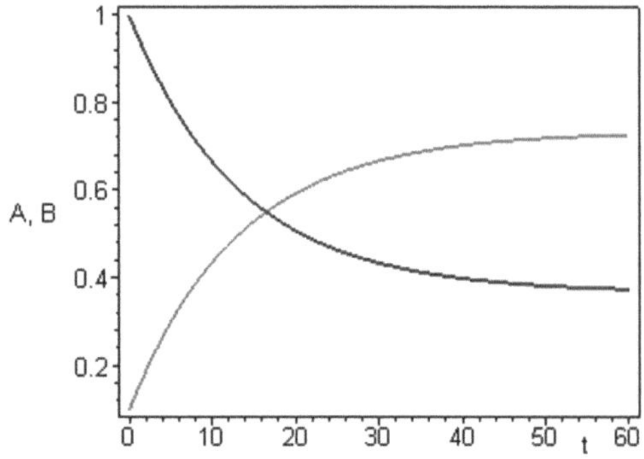

Fig. 3.12 Kinetic curves for reversible reaction participants calculated using numerical calculation results

Such reactions are called *oscillating* or *periodic*. Nowadays several dozens of homogeneous and heterogeneous oscillating reactions have been explored. Investigations of the kinetic models for these complex processes have allowed formulating a series of general conditions, which are required for the stable oscillations of the reaction rates and intermediate concentrations:

- Stable oscillations appear usually in open systems, where reagent concentrations can be maintained constant
- The complex reaction must include autocatalytic steps as well as product-inhibited ones
- The reaction mechanism must include steps with the order higher than first

These conditions are required but not sufficient for the oscillation to occur in the system. An important role is played also by the ratio between the rate constants of certain steps and starting reagent concentrations. An investigation of the oscillating reactions is still an important chemical kinetics problem because it is crucial in understanding catalysis, periodic process laws for living systems, and chemical technology.

Sometimes chemical problems can be answered using the knowledge from other sciences that are not related to chemistry at first sight. For example, some information about a complex reactions flow can be gained from the mathematical models of the interspecific competition. A classical example is the "predator–prey" model, which describes the population trends for predators and prey in living conditions (the *Lotka–Volterra* model). The analogies between this model and many reaction schemes are evident.

The main point of the model is the following. Let's consider some closed ecological system that includes two populations: "predators" and herbivorous "prey". The population sizes are K and B correspondingly. The prey population is reproduced by means of nutrition. The prey consume grass only. The amount of grass is characterized by the T values. Predators eat solely the prey. Their population trend is determined by the amount of herbivores. There are no natural enemies for the predators. Instead, the latter experience the natural loss (diseases, age-specific death). Thereby, the conditions can be expressed with the following scheme:

$$T + K \xrightarrow{k_1} 2K,$$

$$K + B \xrightarrow{k_2} 2B,$$

$$B \xrightarrow{k_3} B_x.$$

Here B_x corresponds to the number of dead predators. Using chemical terminology, one can say that the resulting process is a conversion of the starting reagent T into the product B_x. The process is accompanied by the formation of intermediates K and B, which serve for different kinetic functions.

The stages of population expansion are autocatalytic with the reproduction factors (rate constants) k_1 and k_2, but the presence of predator mortality (k_3) prevents the unlimited growth of both populations. Undoubtedly, the overall kinetics of the process is affected by the amount of grass necessary for the prey population increase. Assume that we have an unlimited amount of grass, i.e. $T(t)$ = const. Then, based on the given conditions, one can write the following differential equation set:

$$\frac{dK(t)}{dt} = k_1TK(t) - k_2K(t)B(t) = kK(t) - k_2K(t)B(t),$$

$$\frac{dB(t)}{dt} = k_2K(t)B(t) - k_3B(t).$$

If the constants for each step and initial population sizes are given, then the numerical solution will allow one to predict prey and predator population trends (Fig. 3.13).

As Fig. 3.13 shows, the population trends for both populations exhibit a continuous wave pattern. For the given initial conditions these oscillations have a constant period and amplitude. There is an interdependence between the two population sizes: increase of one of them impedes the growth of other. In the case of some chemical process described by the Lotka–Volterra model, the concentrations of the intermediates K and B would be oscillating.

Often it is more convenient to present the solutions of such systems not as the "concentration over time" trends but rather as the dependence of one of the concentration on the other. In the second part of Fig. 3.13 (see Fig. 6.52 in Chap. 6 too), we have

$Prey_0 := 10 \quad Predator_0 := 10 \quad k := .1 \quad k_2 := .02 \quad k_3 := .09$

$$D(t,y) := \begin{pmatrix} k \cdot y_0 - k_2 \cdot y_0 \cdot y_1 \\ k_2 \cdot y_0 \cdot y_1 - k_3 \cdot y_1 \end{pmatrix} \quad n := \begin{pmatrix} Prey_0 \\ Predator_0 \end{pmatrix} \quad N := 1000 \quad t_{max} := 250$$

$$s := rkfixed(n, 0, t_{max}, N, D) \quad t := s^{\langle 0 \rangle} \quad Prey := s^{\langle 1 \rangle} \quad Predator := s^{\langle 2 \rangle}$$

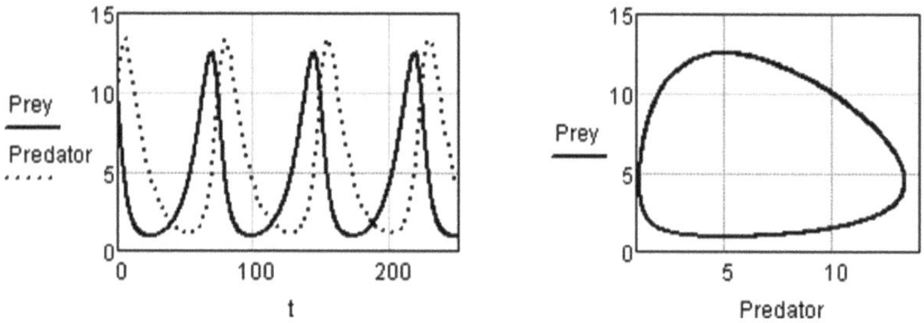

Fig. 3.13 Population trends for predators (dashed line) and prey (solid line) in the Lotka–Volterra model (on-line calculation http://twt.mpei.ac.ru/MCS/Worksheets/Chem/ChemKin-3-12.xmcd)

shown the prey population as a function of the predator population – the so called "phase path". The population dynamics can be represented as a movement along the phase path. The position of the point corresponds to the population sizes at a given moment of time. One can see that the phase path for these initial conditions is a closed graph. In the case of continuous oscillations, the point follows the phase path repeatedly.

Now let's change the form of the Lotka–Volterra model by dividing both parts of the equations by k_2, and assuming that $k_2 dt = d\tau$, $k/k_2 = a$, $k_3/k_2 = b$:

$$\frac{dX(t)}{d\tau} = aX(t) - X(t)Y(t),$$
$$\frac{dY(t)}{d\tau} = X(t)Y(t) - bY(t).$$

The system has only two parameters now, a and b. Both of them are positive according to the physical meaning. If one solves the resulting system for a number of initial conditions, he will end up with a phase path set called the *phase portrait* of the system. For the given case we will have the phase portrait as a set of closed concentric graphs (Fig. 3.14). Due to the $a > 0$, $b > 0$ conditions, all the phase paths are situated in the first quadrant of the coordinate plane. The oscillations occur near some stationary point, which can be determined by putting $dX(t)/dt$ equal to zero. One can easily find that $X_{st} = b$. The corresponding point $(X_{st};Y_{st})$ is called the *critical point* of the system.

In the case when all the phase paths are closed and "incapsulated" one into the other, the critical point is called a *center*.

If the initial grass-eater and predator populations equal b and a correspondingly, the simulation will not reveal oscillations in the system. Any deviation from the stationary values will lead to oscillations.

$$\begin{pmatrix} a\cdot X_{st} - X_{st}\cdot Y_{st} \\ X_{st}\cdot Y_{st} - b\cdot Y_{st}\end{pmatrix} \text{solve}, X_{st}, Y_{st} \rightarrow \begin{pmatrix}0 & 0\\ b & a\end{pmatrix} \qquad a := 2 \qquad b := 3$$

$$f(t,x) := \begin{pmatrix} a\cdot x_0 - x_0\cdot x_1 \\ x_0\cdot x_1 - b\cdot x_1\end{pmatrix} \qquad S(x0,y0) := \text{rkfixed}\left[\begin{pmatrix}x0\\ y0\end{pmatrix},0,10,1000,f\right]$$

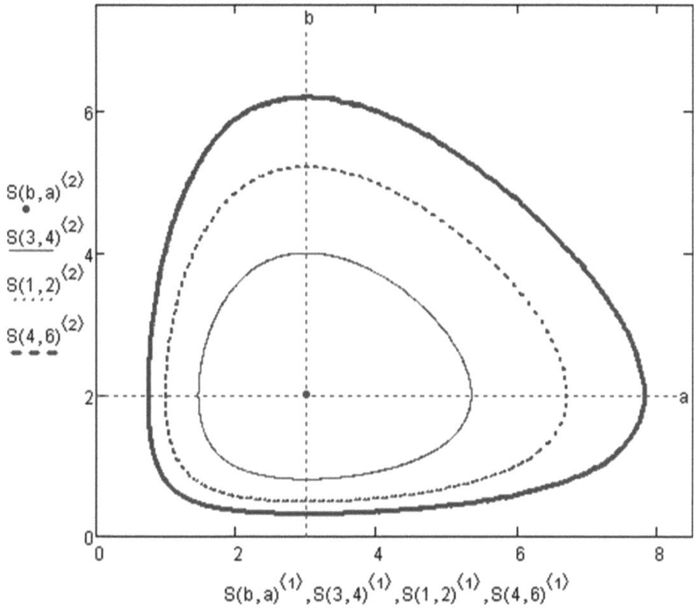

Fig. 3.14 Phase portrait of the Lotka–Volterra system with a critical point (on-line calculation http://twt.mpei.ac.ru/MCS/Worksheets/Chem/ChemKin-3-13.xmcd)

Even though the "predator–prey" model is rather idealized, many kinetic models for real chemical systems are based on it. For example, *D. A. Frank–Kamenetsky* used the Lotka–Volterra model to explain the processes of higher hydrocarbon oxidation.

The hydrocarbon oxidation kinetics is extremely complex because it includes many consequent-parallel steps. Thus, the full mechanism description is problematic. In such cases researchers confine themselves to model descriptions. Each of the model steps may represent a series of elementary stages, and each of the model symbols may correspond to a whole set of compounds playing the same kinetic function.

For example, in the case of a continuous supply of a gasoline–air mixture into the reactor (heated to certain temperature), one can see periodical flashes of the cold flame appearing with a constant frequency. In this case the full combustion does not occur. The oxidation products include aldehydes, organic peroxides and other compounds. Some regularities have been established for this process. In particular, the flash frequency increases with the increase of oxygen concentration and temperature. In order to explain this effect, *D.A. Frank–Kamenetsky* suggested the following kinetic scheme involving two types of intermediates, X and Y:

$$A + X \xrightarrow{k_1} B + 2X,$$
$$X + Y \xrightarrow{k_2} B + 2Y,$$
$$A + Y \xrightarrow{k_3} B.$$

Here A is the starting compound, B is the product, X is the superoxide type molecules or radicals, and Y means the aldehyde type molecules or radicals. One can see that the scheme postulates an autocatalysis with the two intermediates. Let's assume the reagent concentration does not depend on time (A = const), i.e. its consumption rate is compensated by its insertion into the reactor. That will give us the following equation set:

$$\frac{dX(t)}{dt} = k_1 AX(t) - k_2 X(t)Y(t),$$
$$\frac{dY(t)}{dt} = k_2 X(t)Y(t) - k_3 AY(t).$$

After the division of both equation parts by k_2 we end up with the Lotka–Volterra type set:

$$\frac{dX(\tau)}{d\tau} = aX(\tau) - X(\tau)Y(\tau),$$
$$\frac{dY(\tau)}{d\tau} = X(\tau)Y(\tau) - bY(\tau).$$

where $a = k_1 A/k_2$, $b = k_3 A/k_2$, $k_2 dt = d\tau$. It was shown before that the solution for this system has a behavior oscillating in time

Let's show the numerical solution of the Lotka–Volterra model using Maple suite tools (Fig. 3.15). Here the DEplot command from the DEtools library was used in addition to the dsolve. In this case, in addition to the integral curves set for the phase paths, the *directional field* is visualized. The directional field is a series of arrows, each of which represents the motion direction along the phase path at the given point. In many cases the directional field increases the clearness of the phase portrait. The directional field can also be plotted using the Maple commands `phaseportrait` and `dfieldplot`.

To construct the directional field in the Mathcad environment we recommend the user function `field`, which was created by *T. Gutman* (Fig. 3.16).

There are other types of critical points. For example, consider the kinetic scheme consisting of the elementary steps:

$$A \xrightarrow{k_1} X \xrightarrow{k_2} Y \xrightarrow{k_3} P,$$

```
> restart;
> with(DEtools):a:=2:b:=3:LV_model := { diff(x(t), t) =
  a*x(t)-x(t)*y(t),diff(y(t), t) =y(t)*x(t)-b*y(t) };
```

$$LV\_model := \{\frac{d}{dt}x(t) = 2\,x(t) - x(t)\,y(t), \frac{d}{dt}y(t) = x(t)\,y(t) - 3\,y(t)\}$$

```
> DEplot( LV_model, [x(t), y(t)], t=0..10,
  [[x(0)=1,y(0)=2], [x(0)=2,
  y(0)=1], [x(0)=3,y(0)=1]],stepsize=0.05,
  title="Model Lotka-Volterra', color=grey,
  linecolor=black,
  arrows=LARGE );
```

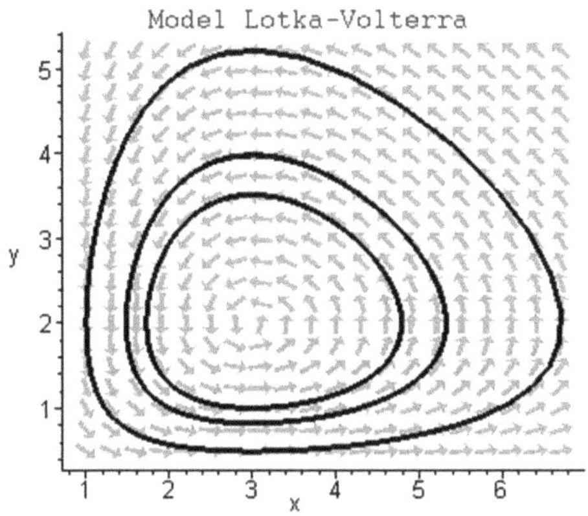

Fig. 3.15 "Predator–prey" model analysis using Maple

in which the first step has the zero order. The concentration changes for the intermediates are described by the equation set:

$$dX(t)/dt = k_1 - k_2X(t),$$

$$dY(t)/dt = k_2X(t) - k_3Y(t).$$

Assume $k_1 = 2.0, k_2 = 1.0, k_3 = 0.5$. We can plot a series of phase portraits for the different initial concentrations of X and Y based on the numerical solution of the direct kinetic problem (Fig. 3.17).

The computation results, presented in Fig. 3.17, show that all phase paths converge at one point. Its coordinates are determined by the values of the intermediates stationary concentrations, X_{st} and Y_{st}:

$$k_1 - k_2X_{st} = 0,$$

$$k_2X_{cm} - k_3Y_{st} = 0,$$

```
arrow := (-.5  .5  0 + .2i  .5  0 - .2i)^T
field(ranges, deriv, scale) := | f ← 0
                               | for  x ∈ ranges_0
                               |   for  y ∈ ranges_1
                               |     | d ← deriv(x, y)
                               |     | d ← stack(1, d)  if  IsScalar(d)
                               |     | d ← (1  i)·d
                               |     | f^<cols(f)> ← (arrow·scale·d/|d|) + x + i·y  if  |d| > 10^-10
                               | f

k := .1      k2 := .02      k3 := .09          D(t, y) := ( k·y0 - k2·y0·y1   )
                                                          ( k2·y0·y1 - k3·y1 )

d(x, v) := D[0, (x v)] → ( 0.1·x + -0.02·v·x   )      vf := field[((0..15) (0..15)), d, .60]
                         ( -0.09·v + 0.02·v·x  )

L := rkfixed[(1 4), 0, 250, 250, D]

( X0  Y0 )    ( k·y0 - k2·y0·y1 = 0   )                      ( 0    0   )
( X1  Y1 ) := ( k2·y0·y1 - k3·y1 = 0  )  solve, y0, y1  →    ( 4.5  5.0 )
```

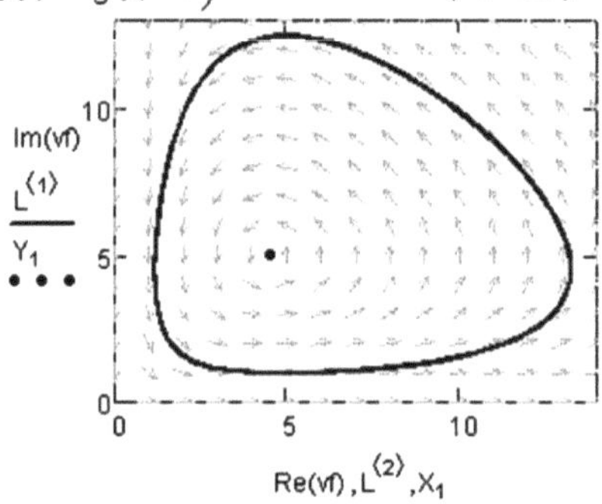

Fig. 3.16 Phase portrait of the Lotka–Volterra system using a directional field (on-line calculation http://twt.mpei.ac.ru/MCS/Worksheets/Chem/ChemKin-3-15.xmcd)

which gives $X_{st} = k_1/k_2, Y_{st} = k_1/k_3$, or, taking into account the given rate constant values, $X_{st} = 2, Y_{st} = 4$. Such type of critical point is called *knot*, and oscillations are impossible in this system.

Now consider the following equation set:

$$dX(t)/dt = aX(t) - bX(t)Y(t),$$
$$dY(t)/dt = aY(t) - bX(t)Y(t).$$

This system is also often used in the differential biological models. If one analyzes the corresponding phase portrait with the directional field (Fig. 3.18),

```
[> restart;
> with(DEtools):k[1]:=2.0:k[2]:=1:k[3]:=0.5:ODE := {
  diff(x(t), t) = k[1]-k[2]*x(t),diff(y(t), t)
  =k[2]*x(t)-k[3]*y(t) };
```

$$ODE := \{\frac{d}{dt}x(t) = 2.0 - x(t), \frac{d}{dt}y(t) = x(t) - 0.5\,y(t)\}$$

```
> DEplot( ODE,[x(t), y(t)], t=0..10,
  [[x(0)=1,y(0)=1],[x(0)=1,
  y(0)=3],[x(0)=3,y(0)=8],[x(0)=1,y(0)=6],[x(0)=3,y(
  0)=6],[x(0)=2.5,y(0)=1],[x(0)=3,y(0)=4],[x(0)=1.25
  ,y(0)=8]],stepsize=0.05,
  title="A-k1->X-k2-Y-k3->P", color=BLUE,
  linecolor=black,
  arrows=MEDIUM );
```

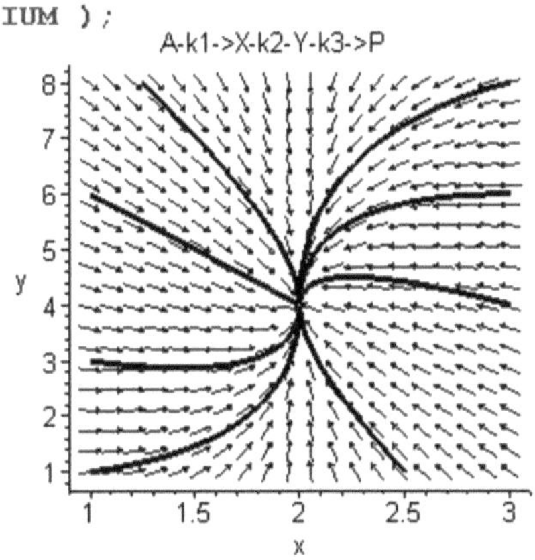

Fig. 3.17 Phase portrait of the system with "knot" – type critical point

one can see that all the phase paths approach a critical point and then move away. In this case, we have the *"saddle"* critical point.[2]

Finally, there is one more remarkable critical point type – *focus*. In order to illustrate it, we will examine one of the kinetic models of photosynthesis.

In the past there was suggested a mechanism for the dark steps cycle of photosynthesis. Sugars with different numbers of carbon atoms, 3–7 (trioses, tetroses, pentoses etc.), participate in this cycle. Having labelled the number with a subscript, one can create the scheme of the process:

[2]In the document shown in Fig. 3.18 (as well as some other documents in this chapter) there were used user functions `IntCurves`, `VField` (*T. Gutman*). The reader can find the corresponding documents on the book site.

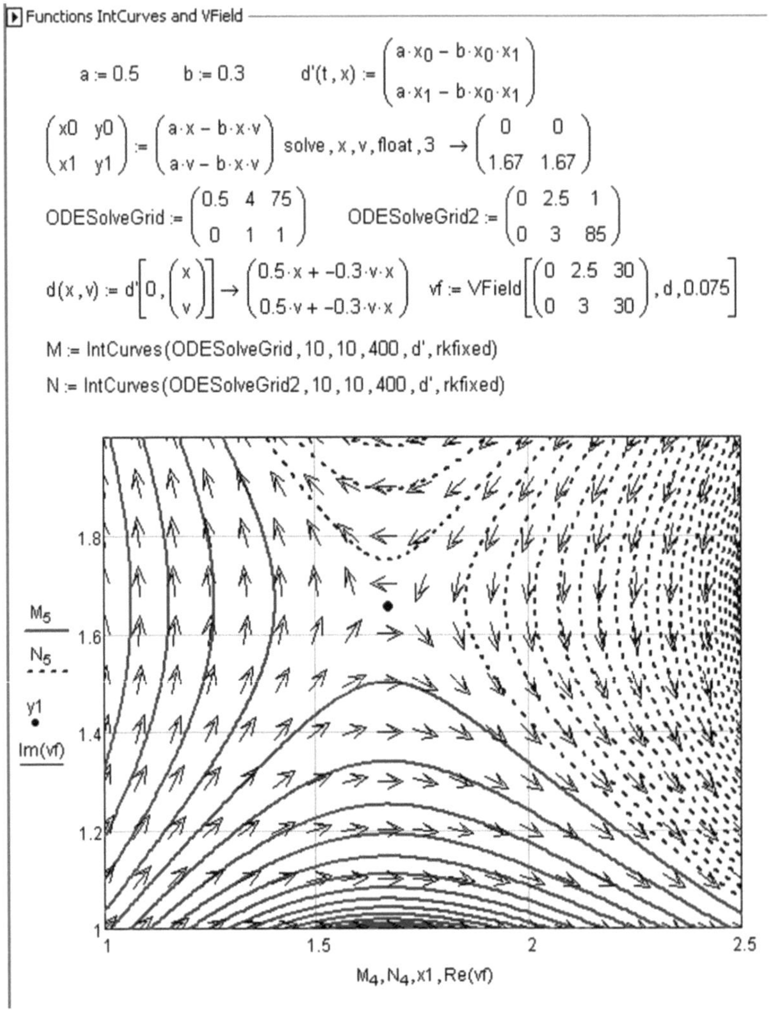

Fig. 3.18 System with a "saddle" critical point

$$
\begin{aligned}
&C_5 + C_1 + X \rightarrow 2C_3,\\
&2C_3 \rightarrow C_6,\\
&C_6 + C_3 \Leftrightarrow C_6,\\
&C_4 + C_3 \Leftrightarrow 2C_7,\\
&C_7 + C_3 \Leftrightarrow 2C_5.
\end{aligned}
$$

Here X means triphosphorpyrydinenucleotide and C_1 is carbon dioxide. This kinetic scheme was analysed by *D.S. Chernavsky*, who assumed some concentrations remaining constant and ended up with the following differential equation set:

$k_1 := 0.1 \quad k_2 := 0.2 \quad k_0 := 0.6 \quad k_3 := 0.025 \quad k_4 := 0.018 \quad k_5 := 0.017$

$$D(t,x) := \begin{bmatrix} k_1\cdot(x_0)^2 - k_2\cdot x_0\cdot x_1 + k_0 \\ k_3\cdot(x_0)^2 - k_4\cdot(x_1)^2 - k_5\cdot x_0\cdot x_1 \end{bmatrix} \qquad S := \mathrm{rkfixed}\left[\begin{pmatrix} 2 \\ 2.8 \end{pmatrix}, 0, 500, 10000, D\right]$$

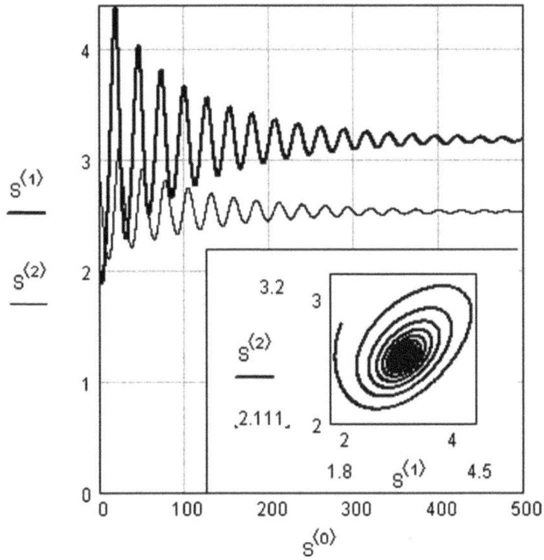

Fig. 3.19 Modelling the photosynthesis kinetics (on-line calculation http://twt.mpei.ac.ru/MCS/Worksheets/Chem/ChemKin-3-18.xmcd)

$$\frac{dC_3}{dt} = a_1 C_3{}^2 - a_2 C_3 C_6 + a_0,$$
$$\frac{dC_6}{dt} = b_1 C_3{}^2 - b_2 C_6{}^2 - b_3 C_3 C_6.$$

Let's solve the set with Mathcad tools (Fig. 3.19), using the numerical integration method with an adaptive step. The constants a_0, a_1, a_2, b_1, b_2, b_3 values have been chosen arbitrarily. The results show that there are periodical concentration oscillations, which decay over time.

The phase path is of the spiral form in this case. The spiral "wraps" around a critical point called focus.

The investigation of the critical point character is closely related to the question of the system stability. Here the chemical kinetics borrows some terms from the dynamic system theory, such as *Lyapunov's stability criteria*.

Without a deep discussion of the mathematical apparatus, we will show how the mathematical suites allow one to determine the critical point type. Assume the mathematical model of a process described by a set of two differential equations. In order to find the critical point type one has to:

- Calculate the critical point coordinates on a phase plane. For this one has to solve the corresponding algebraic equation set, which is obtained through the equating of desired functions derivatives to zero
- Compute the Jacobian matrix for the system using the critical point coordinates
- Find the eigenvalues λ_1 of the latter matrix

Obtained eigenvalues allow one to establish the critical point type and the stability of the stationary state. Six cases are possible here. They are outlined in Fig. 3.20.

As an example we will consider the previously discussed microorganism propagation model with slight modifications. Let's assume that the poison produced during the microorganism life can decompose (for example, by means of the Sun's radiation). The poison decomposition will represent an elementary zero-order reaction with the rate constant k_3. The new mathematical model will look this way:

$$dN/dt = kN - k_1NX,$$
$$dX/dt = k_2N - k_3.$$

The solution of the direct kinetic problem is shown in Fig. 3.21. We can see that in the case of the assumed rate constants the microorganism population is oscillating. The critical point type is the *node*, because all the Jacobian eigenvalues are imaginary.

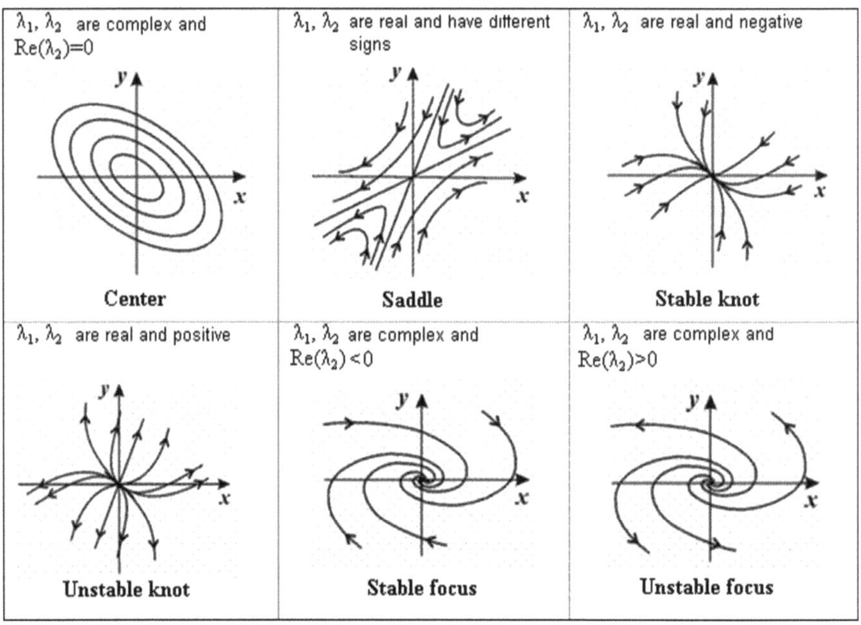

Fig. 3.20 Possible critical point types and phase portraits versus different Jacobian matrix eigenvalues

$k := 0.1 \qquad k_1 := 1 \qquad k_2 := 0.005 \qquad k_3 := 10$

$$\begin{pmatrix} x0_{st} & x1_{st} \end{pmatrix} := \begin{pmatrix} k\,x0 - k_1\,x0\,x1 \\ k_2\,x0 - k_3 \end{pmatrix} \text{solve}, x0, x1 \rightarrow \begin{pmatrix} 2000.0 & 0.1 \end{pmatrix}$$

$$D(t,x) := \begin{pmatrix} k\,x_0 - k_1\,x_0\,x_1 \\ k_2\,x_0 - k_3 \end{pmatrix} \quad J := \text{Jacob}(D(t,x),x) \;\left|\begin{matrix} \text{substitute}, x_0 = x0_{st} \\ \text{substitute}, x_1 = x1_{st} \end{matrix}\right. \rightarrow \begin{pmatrix} 0 & -2000.0 \\ 0.005 & 0 \end{pmatrix}$$

$$\text{eigenvals}(J) \rightarrow \begin{pmatrix} -3.1622776601683793320i \\ 3.1622776601683793320i \end{pmatrix} \qquad S := \text{Adams}\left[\begin{pmatrix} 2000 \\ 0.0 \end{pmatrix}, 0, 40, 1000, D\right]$$

$t := S^{\langle 0 \rangle} \qquad N := S^{\langle 1 \rangle} \qquad Z := S^{\langle 2 \rangle}$

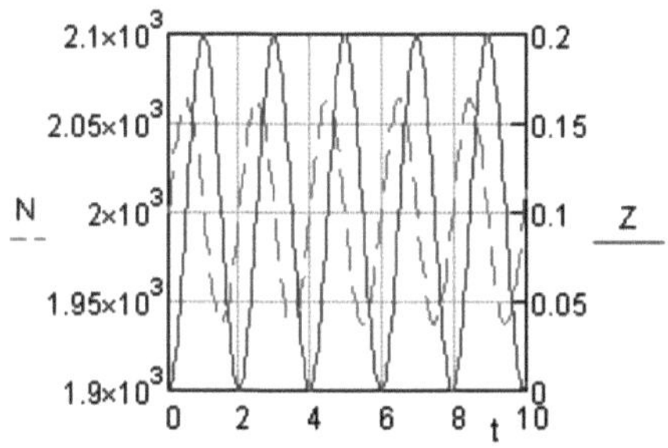

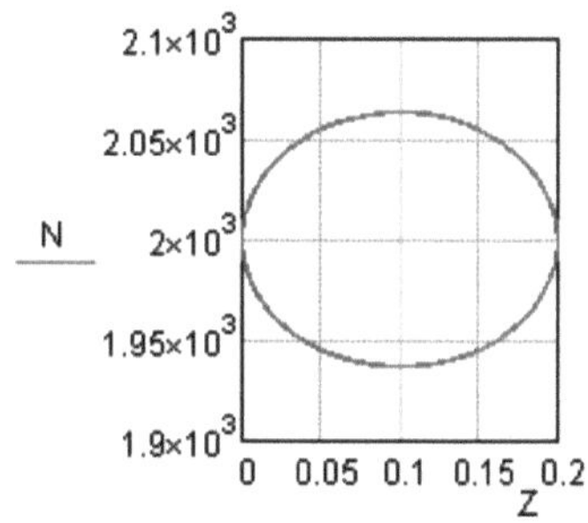

Fig. 3.21 Oscillation mode of the population trend in microorganism colony (on-line calculation http://twt.mpei.ac.ru/MCS/Worksheets/Chem/ChemKin-3-20.xmcd)

The first "chemically grounded" model of a oscillating reaction is a model, which was proposed by *I. Prigozhin* and is called *bruesselator*. The model is based on a hypothetical reaction with the following mechanism:

$$A \xrightarrow{k_1} X,$$

$$B + X \xrightarrow{k_2} Y + D,$$

$$2X + Y \xrightarrow{k_3} 3X,$$

$$X \xrightarrow{k_4} E.$$

It is assumed that the concentrations of the reagents A and B do not change over time. The concentrations of D and E are not included in the mass action law. That is why one need only two equations for the formal kinetic description of the reaction:

$$\begin{aligned} dX(t)/dt &= k_1 A - k_2 B X(t) + k_3 X(t)^2 Y(t) - k_4 X(t) \\ &= k_{1ef} - k_{2ef} X(t) + k_2 X(t)^2 Y(t) - k_4 X(t), \\ dY(t)/dt &= k_2 B X(t) - k_3 X(t)^2 Y(t) = k_{2ef} X(t) - k_3 X(t)^2 Y(t). \end{aligned}$$

It is possible to reduce the number of the controlling parameters in this system by substituting some variables: $\tau = k_4 t$, $x = \sqrt{k_3/k_4}X$, $Y = \sqrt{k_3/k_4}Y$. After these changes the system takes the form:

Table 3.3 Critical point types for the bruesselator

Parameters a and b ratio	Critical point type
$b < (a - 1)^2$	Stable node
$b < a^2 + 1$	Stable focus
$b = a^2 + 1$	Center
$b > a^2 + 1$	Unstable focus (limit cycle)
$b > (a + 1)^2$	Unstable node

$$dx(\tau)\Big/d\tau = a - (b+1)x(\tau) + x(\tau)^2 y(\tau),$$

$$dy(\tau)/d\tau = bx(\tau) - x(\tau)^2 y(\tau),$$

where $a = k_{1ef}\sqrt{k_3/k_4{}^3}$, $b = k_{2ef}/k_4$.

A remarkable feature of the bruesselator is the variety of the critical point types and, consequently, of the phase portraits depending on the a and b parameters ratio (Table 3.3).

The case $b > a^2 + 1$ requires additional examination. The critical point type is the unstable focus. One can see the appearance of the so-called *limit cycle* in the phase portrait. In this case, any point in the phase plane will end up following the same closed phase path regardless of the initial conditions. This means that stable asymptotic concentration oscillations (auto-oscillations) of the same amplitude and frequency will appear with the course of time. Correspondingly, this case is essentially different from the Lotka–Volterra model. In the latter, one can find closed phase paths as well, but there is no the only path that does not depend on the initial conditions. The point set, which "attracts" all phase paths, was called by I. Prigozhin an *attractor*. Thus, the bruesselator has the attractor, while the Lotka–Volterra system does not have. The appearance of the bruesselator limit cycle can be seen in Fig. 3.22. The Mathcad tools were used to plot the phase portrait assuming $a = 1$, $b = 3.25$.

Nowadays many real chemical systems are known, in which processes, accompanied by the concentration oscillations, take place. These can be both heterogeneous and homogeneous reactions. In particular, the hydrogen peroxide reduction on the mercury drop surface can progress periodically in specific conditions. The conjugate process of mercury surface oxidation is accompanied by a change of the surface tension. It leads to the drop shape changes. The oscillating mode of the reaction can be observed through the periodical changes of the mercury drop shape, which resembles a heartbeat ("mercury heart").

The oscillating reactions in homogeneous aqueous media are of a special interest. Probably, the oxidation of organic acids and their esters by the bromate ion is investigated the most. *B.P. Belousov* (1951) observed the periodic color changes during the oxidation of citric acid by bromate ion in sulfuric acid solution in the presence of cerium ions. The detailed investigation of this process was done by *A.M. Zhabotinsky*. The discovery of this reaction stimulated the investigation of periodical processes in chemical systems. It became evident that homogeneous oscillating reactions underlie the most important biochemical processes: generation of biorhythms and nerve impulses, muscles contraction, etc. As of today, the reaction of

Function IntCurves

$a := 1 \qquad b := 3.25 \qquad d'(t,x) := \begin{bmatrix} a - (b+1)\cdot x_0 + (x_0)^2\cdot x_1 \\ b\cdot x_0 - (x_0)^2\cdot x_1 \end{bmatrix}$

$\mathrm{eigenvals}(\mathrm{Jacob}(d'(t,x),x)) \;\Big|\; \mathrm{substitute}, x_0 = a;\ \mathrm{substitute}, x_1 = \frac{b}{a};\ \mathrm{float}, 3 \rightarrow \begin{pmatrix} 0.625 - 0.781i \\ 0.625 + 0.781i \end{pmatrix}$

$\mathrm{ODESolveGrid} := \begin{pmatrix} 0 & 5. & 25 \\ 0 & 5.8 & 1 \end{pmatrix} \qquad \mathrm{ODESolveGrid1} := \begin{pmatrix} 0 & 5. & 1 \\ 0 & 5.8 & 25 \end{pmatrix}$

M := IntCurves(ODESolveGrid, 10, 10, 500, d', rkfixed)

N := IntCurves(ODESolveGrid1, 10, 10, 400, d', rkfixed)

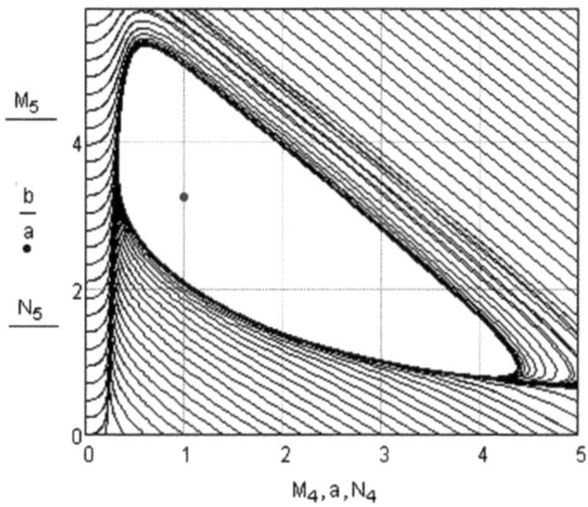

Fig. 3.22 Brusselator phase portrait with a limit cycle (on-line calculation http://twt.mpei.ac.ru/MCS/Worksheets/Chem/ChemKin-3-21.xmcd)

catalytic oxidation of different reducing agents by bromic acid *($HBrO_3$)*, following the auto-oscillating mode, is called the *Belousov–Zhabotinsky reaction*. This reaction goes in the acidic water solution and is accompanied by the concentration oscillations for the oxidized and reduced catalyst forms and intermediates. As a catalyst one can use transition metal ions, such as manganese or cerium. The reducing agents can be different organic compounds (malonic acid, acetylacetone, etc.)

We want to mention that, in spite of many publications dealing with Belousov–Zhabotinsky reaction, the true mechanism of this process in still unknown. Many kinetic schemes were proposed to explain the existence of the concentration oscillations. One of the possible mechanisms is shown in the Table 3.4.

Here one can find several important conjugated processes.

1. During the step (1) $HBrO_2$ is formed. It acts as an autocatalyst in the following reactions.
2. The extensive chain reaction of the oxidant BrO_3^- with the autocatalyst provides the conditions for the Me^+ ions oxidation (steps 4–7).

Table 3.4 Possible mechanism of the Belousov–Zhabotinsky reaction

No	Step number	Reaction	Kinetic parameters values
1	1–2	$BrO_3^- + Br^- + 2H^+ \Leftrightarrow HBrO_2 + HOBr$	$k_1 = 2.1$; $k_2 = 1.0 \times 10^4$
2	3	$HBrO_2 + Br^- + H^+ \rightarrow 2HOBr$	$k_3 = 3.0 \times 10^6$
3	4–5	$BrO_3^- + HBrO_2 + H^+ \Leftrightarrow 2BrO_2^* + H_2O$	$k_4 = 4.2$; $k_5 = 4.2 \times 10^7$
4	6–7	$BrO_2^* + Me^+ + H^+ \Leftrightarrow HBrO_2 + Me^{2+}$	$k_6 = 8.0 \times 10^4$ $k_7 = 8.9 \times 10^3$
5	8	$2HBrO_2 \rightarrow BrO_3{}^- + HOBr + H^+$	$k_8 = 3.0 \times 10^3$
6	9–10	$HOBr + Br^- + H^+ \Leftrightarrow Br_2 + H_2O$	$k_9 = 8.0 \times 10^9$ $k_{10} = 1.1 \times 10^2$
7	11	$RH + Br_2 \rightarrow RBr + Br^- + H^+$	$k_{11} = 4.6 \times 10^3$
8	12	$HOBr + R^* \rightarrow ROH + Br^*$	$k_{12} = 1 \times 10^6 \ldots 1 \times 10^7$
9	13	$RH + Br^* \rightarrow Br^- + H^+ + R^*$	$k_{13} = 1.0 \times 10^6$
10	14	$RH + Me^{2+} \rightarrow Me^+ + H^+ + R^*$	$k_{14} = 2.0 \times 10^{-1}$
11	15	$2R^* + H_2O \rightarrow RH + ROH$	$k_{15} = 3.20 \times 10^9$

3. The oxidation is inhibited due to the chain termination (step 3).
4. The oxidized form of the catalyst is reduced during step 14.

The way other reagents react can be deduced from the given scheme. We have to admit that in spite of the large number of steps, this kinetic model should be considered as simplified. However, the solution of the direct kinetic problem for this scheme at the given conditions (see Table 3.4) shows the presence of stable concentration oscillations. A fragment of the corresponding Mathcad document is shown in Fig. 3.23. Here the mathematical model was developed in compliance with the kinetic scheme given in the Table 3.4. It was assumed that the hydrogen ion concentration is constant during the reaction. One can find the corresponding document on the book's site.

Somewhat different scheme for the Belousov–Zhabotinsky reaction was suggested by *Field*, *Koròs* and *Noyes*. The model is called *oregonator*. It includes following stages:

$$A + Y \xrightarrow{k_1} X + P,$$

$$X + Y \xrightarrow{k_2} 2P,$$

$$A + X \xrightarrow{k_3} 2X + 2Z,$$

$$2X \xrightarrow{k_4} A + P,$$

$$B + Z \xrightarrow{k_5} (1/2)fY.$$

Here A corresponds to the $BrO_3{}^-$ ion; B corresponds to all organic reagents that can be oxidized; P is $HOBr$; X is $HBrO_2$; Y is Br^- ion; Z is the reduced form of the catalyst. The mathematical model can be written as a set of three differential equations:

$$\frac{dX(t)}{dt} = k_1AY(t) - k_2X(t)Y(t) + k_3AX(t) - 2k_4X(t)^2,$$

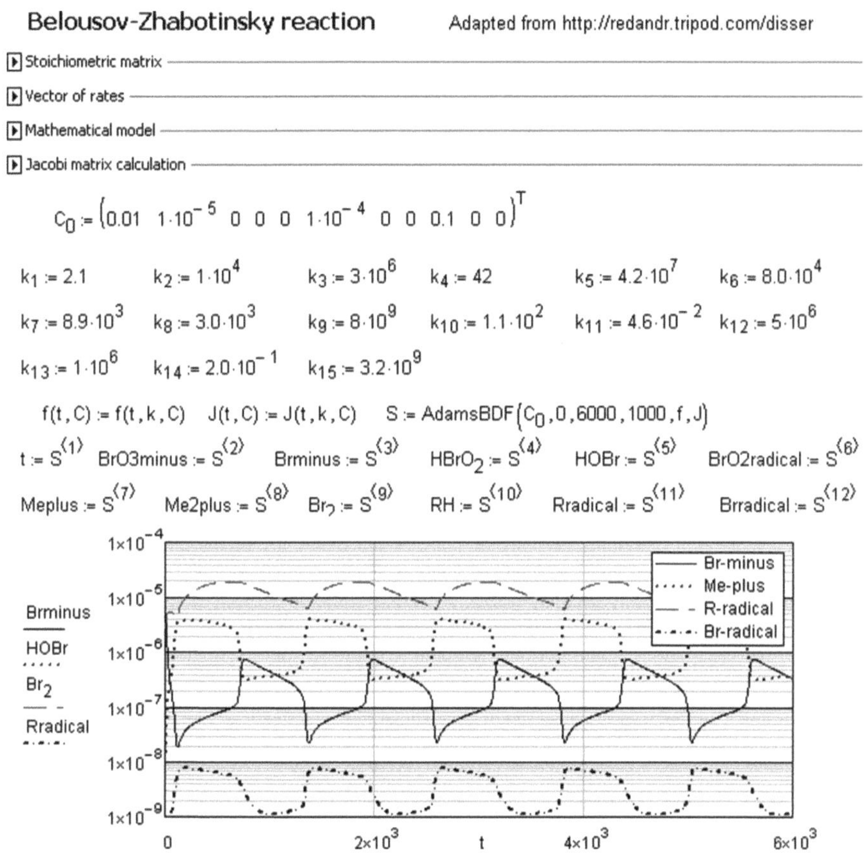

Fig. 3.23 Concentration oscillations in the Belousov–Zhabotinsky reaction (on-line calculation http://twt.mpei.ac.ru/MCS/Worksheets/Chem/ChemKin-3-22.xmcd)

$$\frac{dY(t)}{dt} = -k_1AY(t) - k_2X(t)Y(t) + \frac{1}{2}fk_5BZ(t),$$
$$\frac{dZ(t)}{dt} = 2k_3AX(t) - k_5BZ(t).$$

It is assumed that the concentrations of the compounds A and B remain constant during the reaction. By using dimensionless variables, one can transform the set:

$$\frac{dx(\tau)}{d\tau} = \frac{qy(\tau) - x(\tau)y(\tau) + x[1 - x(\tau)]}{\varepsilon},$$
$$\frac{dy(\tau)}{d\tau} = \frac{-qy(\tau) - x(\tau)y(\tau) + fz(\tau)}{\varepsilon'},$$
$$\frac{dz(\tau)}{d\tau} = x(\tau) - z(\tau),$$

where $x = \frac{2k_4}{k_5 A} X$; $y = \frac{k_2}{k_5 A} Y$; $z = \frac{k_4 k_5 B}{(k_3 A)^2} Z$; $\tau = k_5 B t$. Now the model has three controlling parameters:

$$\varepsilon = \frac{k_5 B}{k_3 A}, \varepsilon' = \frac{2k_4 k_5}{k_2 k_3 A}, q = \frac{2k_1 k_4}{k_2 k_3},$$

values of which influence greatly the system dynamic behavior.

The document shown in Fig. 3.24 can be used as a template in the computer modeling of the oregonator model. By changing the controlling parameters, one can see a variety of the reaction modes with different amplitudes and oscillation frequencies, an appearance of limit cycles, changes in phase paths trajectories, etc. The value of the stoichiometric factor *f* is also of great importance. Compared to bruesselator, this model is more complex in analysis of possible stationary states and plotting of the phase portraits. The reader can find more details in the specialized literature.

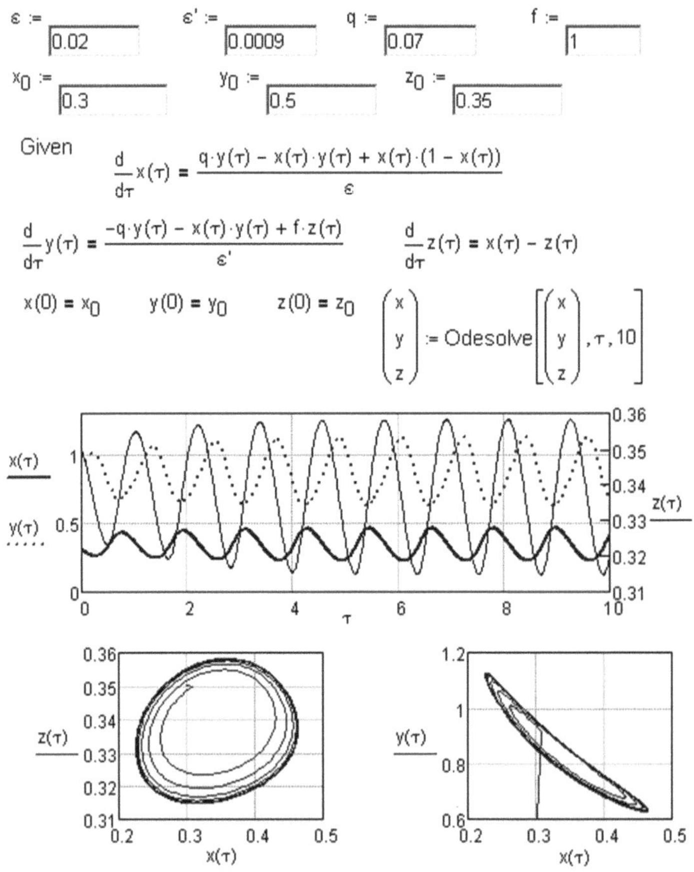

Fig. 3.24 One of the direct problem solutions for the oregonator problem (on-line calculation http://twt.mpei.ac.ru/MCS/Worksheets/Chem/ChemKin-3-23.xmcd)

3.1.4 Some Points on Non-Isothermal Kinetics

By this point, we were considering only the chemical kinetics problems in case of constant temperature of the reaction mixture. But the temperature can change due to ambient conditions (forced heating of cooling) as well as due to internal factors (heat liberation or adsorption during reaction). Previously discussed methods are not sufficient in this case. The mathematical model of the reaction becomes complicated because the temperature is now a function of time. While equations describing a material balance of the system are sufficient for isothermic kinetics, in case of altering temperature one have to consider energy balance as well.

If the reaction has a thermal activation character, changes in temperature lead to changes the rate constant. This relationship is often described by the Arrhenius equation:

$$k = k_0 e^{-\frac{E_a}{RT}},$$

where E_a is the activation energy (J mol^{-1}), k_0 is the pre-exponential factor. The Arrhenius equation is based on the *collision theory*. The theory exploits ideas of an energy barrier and effective collisions of the reacting particles, which happen in a unit of space over a unit of time. The k_0 value is proportional to the total collision number. The activation energy determines the energetic conditions for an active collision – the collision when a transformation of the reagents into the products is possible. It is a certain excess of energy in comparison with the average reactants energy, which have to be applied for the reacting species to react. The Arrhenius equation implies that the reaction rate increases when the temperature rises. The smaller the activation energy is, the greater such increase will be.

So, the rate constant depends on temperature. In the case of the altering temperature the rate constant also becomes a function of time. Consequently, when solving the direct kinetics problem, we have to add the corresponding equations (the temperature over time relationships) to the reaction model.

Let's consider one of the non-isothermic kinetics cases. Some amount of germanium (IV) chloride is being heated. The heating is accompanied by a consequent decomposition:

$$GeCl_4 \xrightarrow{k_1} GeCl_2 + Cl_2,$$

$$GeCl_2 \xrightarrow{k_2} Ge + Cl_2.$$

Assume the heat exchange is organized in a way that the heating appears with a constant rate $\gamma = 10\,\mathrm{K\,min^{-1}}$. One is asked to establish how the gross mass of solids will change if 0.002 mol of $GeCl_4$ is being heated. The initial temperature is 298 K. It is known that the Arrhenius' relationships for the rate constants have the following form:

$$k_1(T) = 3 \times 10^{12} e^{-\frac{29000}{RT}},\ k_2(T) = 6 \times 10^{14} e^{-\frac{48000}{RT}}.$$

In these equations the pre-exponential factors have dimensions of min^{-1}, activation energies are given in cal mol^{-1}.

A mathematical model of the process consists of three differential equations that describe the changes in reactant amounts:

$$\frac{dn_{GeCl_4}(t)}{dt} = -k_1(t)n_{GeCl_4}(t) = -3\times10^{12}\exp\left[-\frac{29000}{R(T_0+\gamma t)}\right]n_{GeCl_4}(t),$$

$$\frac{dn_{GeCl_2}(t)}{dt} = k_1(t)n_{GeCl_4}(t) - k_2(t)n_{GeCl_2}(t) = 3\times10^{12}\exp\left[-\frac{29000}{R(T_0+\gamma t)}\right]n_{GeCl_4}(t) - 6\times10^{14}\exp\left[-\frac{48000}{R(T_0+\gamma t)}\right]n_{GeCl_2}(t),$$

$$\frac{dn_{Ge}(t)}{dt} = k_2(t)n_{GeCl_2}(t) = 6\times10^{14}\exp\left[-\frac{48000}{R(T_0+\gamma t)}\right]n_{GeCl_2}(t),$$

as well as of a heating rate equation:

$$\frac{dT(t)}{dt} = \gamma.$$

Initial conditions for the given ODE set are: $n_{GeCl_4}(0) = 0.002$; $n_{GeCl_2}(0) = 0$; $n_{Ge}(0)$; $T(0) = 298$.

To integrate this ODE set one can use any of the previously discussed mathematical suite built-in tools. In Fig. 3.25 we have shown how to solve the problem using the Mathcad built-in function `AdamsBDF`. The plots allow one to track the trends of compound concentrations over time. As one can see, during the first 15 min of heating the amount of the starting material virtually remains constant. After that, decomposition occurs with a notable rate. When $GeCl_4$ has decomposed, the solid phase of the reaction mixture consists solely of $GeCl_2$. This composition remains unchanged until approximately the 30th minute of heating, when the intermediate begins to decompose into the final product. Finally, some time later the mixture will consist of pure germanium. One can track the change in mass of the initial sample in the same way.

When describing the processes of non-isothermic kinetics, it is convenient to use a unitless variable – conversion of the starting compound X. For example, if we have a single first-order reaction under the programmed temperature changes conditions, we can describe its kinetics with a set of two equations:

$$\frac{dX(t)}{dt} = k_0 e^{-E_a/RT}[1-X(t)], \frac{dT(t)}{dt} = \gamma,$$

with initial conditions $X(t) = 0$, $T(t) = T_0$. The numerical solution of this set for given kinetic parameters values and linear heating mode (Fig. 3.25) shows that a conversion vs. time plot has a distinct S-shape. A slope of such a curve changes depending on the given temperature change rate. It important to note that such curve can be obtained experimentally with a special device called derivatograph. Information about the system behavior within a given temperature range is

$$R := 1.986 \qquad T_0 := 298 \qquad \gamma := 10$$

$$n_0 := \begin{pmatrix} 0.002 \\ 0 \\ 0 \\ T_0 \end{pmatrix} \qquad F(t,x) := \begin{pmatrix} -3\cdot 10^{12}\cdot \exp\left(\dfrac{-29000}{R\cdot x_3}\right)\cdot x_0 \\ 3\cdot 10^{12}\cdot \exp\left(\dfrac{-29000}{R\cdot x_3}\right)\cdot x_0 - 6\cdot 10^{14}\cdot \exp\left(\dfrac{-48000}{R\cdot x_3}\right)\cdot x_1 \\ 6\cdot 10^{14}\cdot \exp\left(\dfrac{-48000}{R\cdot x_3}\right)\cdot x_1 \\ \gamma \end{pmatrix}$$

$$N := \text{AdamsBDF}(n_0, 0, 50, 1000, F) \qquad t := N^{\langle 0 \rangle}$$

$$[GeCl_4] := N^{\langle 1 \rangle} \qquad [GeCl_2] := N^{\langle 2 \rangle} \qquad [Ge] := N^{\langle 3 \rangle} \qquad T := N^{\langle 4 \rangle}$$

$$M_{Ge} := 73 \qquad M_{Cl2} := 35.5 \qquad M_{GeCl4} := M_{Ge} + 4\cdot M_{Cl2} \qquad M_{GeCl2} := M_{Ge} + 2\cdot M_{Cl2}$$

$$m_{GeCl4} := [GeCl_4]\cdot M_{GeCl4} \qquad m_{GeCl2} := [GeCl_2]\cdot M_{GeCl2} \qquad m_{Ge} := [Ge]\cdot M_{Ge}$$

$$m := m_{GeCl4} + m_{GeCl2} + m_{Ge}$$

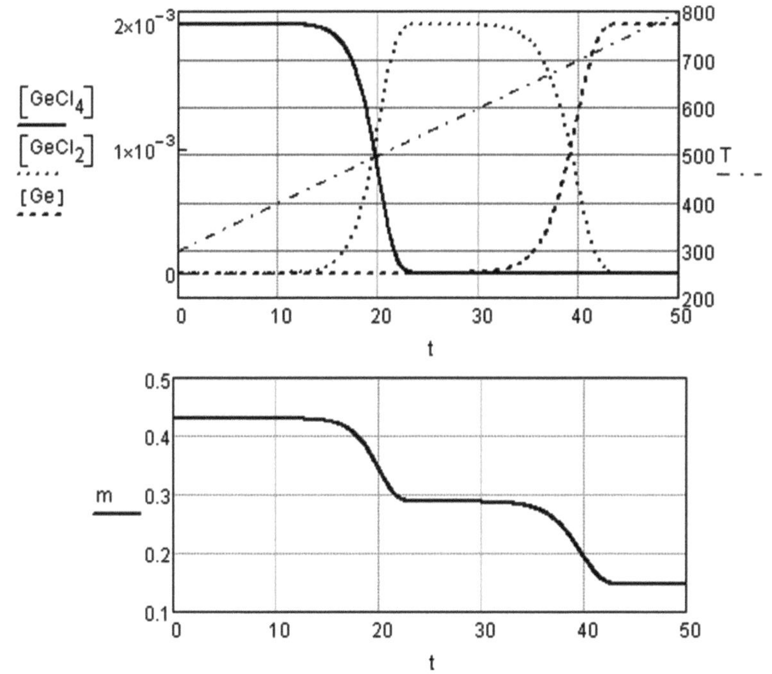

Fig. 3.25 Solution of the $GeCl_4$ decomposition problem (on-line calculation http://twt.mpei.ac.ru/MCS/Worksheets/Chem/ChemKin-3-24.xmcd)

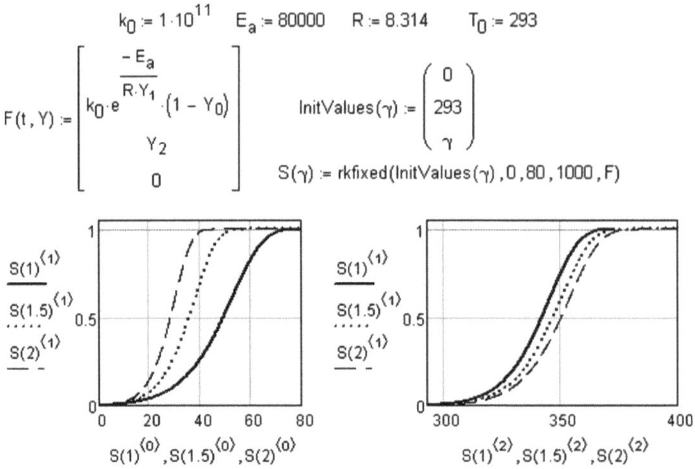

Fig. 3.26 Conversion versus time and temperature for different heating rates (on-line calculation http://twt.mpei.ac.ru/MCS/Worksheets/Chem/ChemKin-3-25.xmcd)

completely enough for solving the inverse problem, i.e. for the determination of the reaction kinetic parameters using experimental data.

We want to point out that so far we considered only the cases when the temperature changes were controlled by external factors (i.e. was determined by the experiment conditions) (Fig. 3.26).

If an exothermic reaction takes place in an isolated system, in other words, when the heat exchange with environment is absent (adiabatic reactor), a temperature will apparently increase over time. The rate of this increase depends both on the kinetic parameters (rate constant) and on the thermodynamic properties of the system (thermal conditions of the reaction, heat capacity). For a well-mixed periodic reactor, where a single first-order reaction $A \to B$ occurs, the mathematical model is described by this set of equations:

$$r_A(t) = k_0 e^{-E_a/RT(t)} C_A(t),$$

$$\frac{dC_A(t)}{dt} = -r_A(t),$$

$$\rho C_p \frac{dT(t)}{dt} = \Delta H r_A(t).$$

Here ρ is density, kg m^{-3}, and C_p is specific heat capacity of the reaction mixture, J (kgK)$^{-1}$. ΔH is the reaction heat effect (taking the sing into account), J mol^{-1}. To be specific, these parameters depend on temperature. In addition, heat capacity and density can change as the reaction goes. One should account for that when performing important calculations. In order not to overcomplicate, we assume that these values are constant. We will define an additional parameter: $J = \Delta H/\rho C_p$. One can see that:

$$\frac{dT(t)}{dt} = -J\frac{dC_A(t)}{dt}.$$

An integration of this equation with the initial conditions $T(0) = T_0$, $C_A(0) = C_{A_0}$ gives:

$$T - T_0 = J(C_{A_0} - C).$$

In the case of adiabatic conditions, the system will warm up and reach the final temperature that corresponds to an exhaustion of the reagent:

$$T_{ad} = T_0 + JC_{A_0}.$$

The temperature T_{ad} is called *adiabatic temperature.*

The modeling results for the behavior of this system are shown in Fig. 3.27. The initial parameters were: $k_0 = 1 \times 10^5\ \text{s}^{-1}$, $E_T = E_a/R = 5{,}000$ K, $C_{A_0} = 1\,\text{kmol}\,\text{m}^{-3}$, $J = 100\ \text{Km}^3\ \text{kmol}^{-1}$. The reader can see that performing the calculations is rather simple.

In the real conditions, some heat from the reaction is liberated into environment through the reactor walls. A differential equation for the temperature changes is given in the following form in this case:

$$\frac{dT(t)}{dt} = Jr_A(t) + \frac{hS(T_s - T(t))}{\rho C_p V}.$$

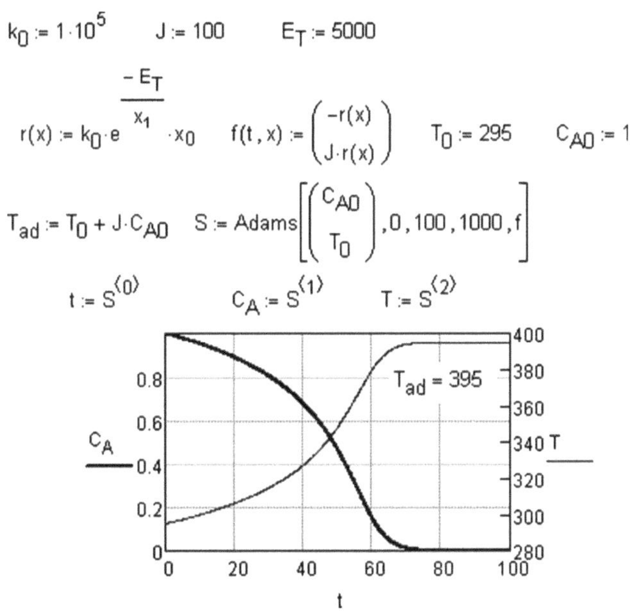

Fig. 3.27 Temperature and reagent concentration changes in a periodic adiabatic reactor (on-line calculation http://twt.mpei.ac.ru/MCS/Worksheets/Chem/ChemKin-3-26.xmcd)

$k_0 := 1 \cdot 10^5 \qquad J := 100 \qquad E_T := 5000$

$\alpha := 0.1 \qquad S := 2 \qquad V := 1 \qquad \rho := 1 \qquad T_S := 295 \qquad C_p := 4$

$$r(x) := k_0 \cdot e^{\frac{-E_T}{x_1}} \cdot x_0 \qquad f(t, x) := \begin{bmatrix} -r(x) \\ J \cdot r(x) + \dfrac{\alpha \cdot S \cdot (T_S - x_1)}{\rho \cdot C_p \cdot V} \end{bmatrix}$$

$$T_0 := 295 \qquad C_{A0} := 1 \qquad S := \mathrm{Adams}\left[\begin{pmatrix} C_{A0} \\ T_0 \end{pmatrix}, 0, 300, 1000, f\right]$$

$$t := S^{\langle 0 \rangle} \qquad C_A := S^{\langle 1 \rangle} \qquad T := S^{\langle 2 \rangle}$$

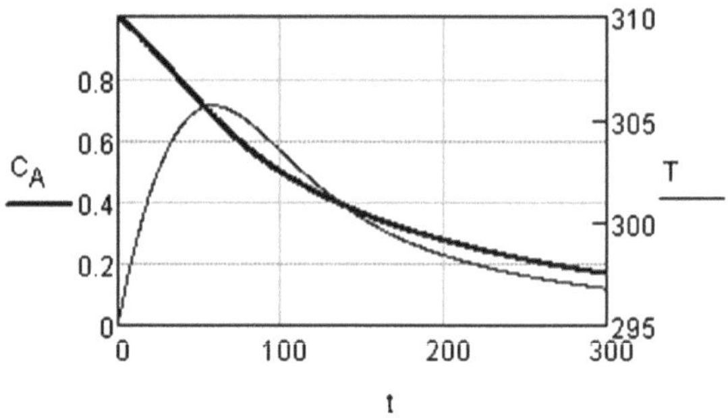

Fig. 3.28 Operation dynamics of a periodic non adiabatic reactor (on-line calculation http://twt.mpei.ac.ru/MCS/Worksheets/Chem/ChemKin-3-27.xmcd)

In order to follow operational trends of such non-adiabatic reactor, one have to introduce additional parameters: heat transfer coefficient *h*, $W/(m^2K)$, reactor volume *V*, m^3, wall surface *S*, m^2. In Fig. 3.28 one possible way to compute the reagent concentration and reaction mixture temperature as a function time is shown.

For a well-mixed flow reactor working in an adiabatic mode:

$$\frac{dC_A(t)}{dt} = \frac{C_{A_0} - C_A(t)}{\tau} - r_A(t),$$

$$\frac{dT(t)}{dt} = \frac{T_0 - T(t)}{\tau} + Jr_A(t).$$

Here $\tau$is, and T_0 is the reagent temperature when entering the reactor. The solution of Cauchy problem for this reactor type allows one to conclude: the dynamic portrait can change strikingly depending on temperature of the initial mixture. Such situation is illustrated in Fig. 3.29. One can see that for the time $\tau = 60$ s many different kinetic curves as well as temperature–time relationships are possible, even though the initial temperatures differ only for 1 K. In both cases a stationary state is reached. However, for $T_0 = 274$ K the stationary conversion is

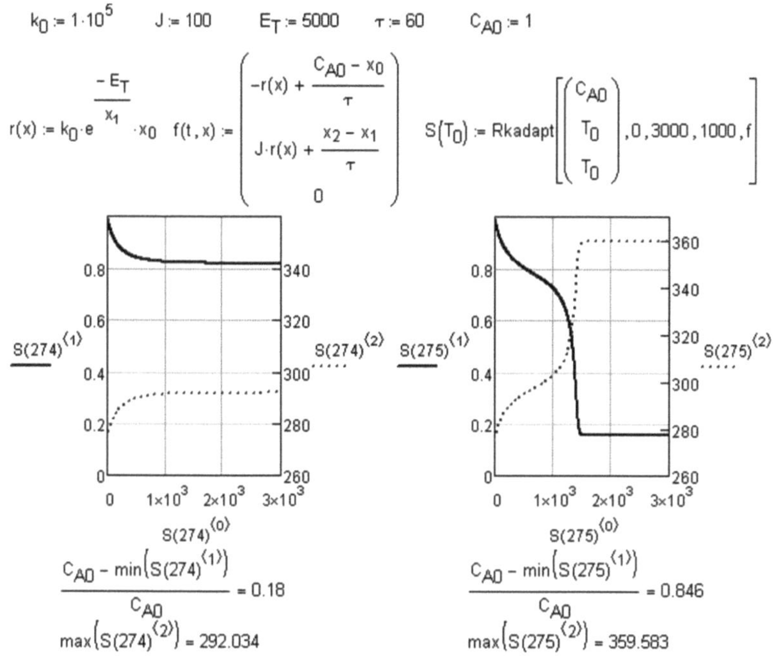

Fig. 3.29 Temperature and concentration trends in a flow adiabatic reactor (on-line calculation http://twt.mpei.ac.ru/MCS/Worksheets/Chem/ChemKin-3-28.xmcd)

low and does not exceed 18%. If the initial temperature equals 275 K, other stationary state is reached. The latter corresponds to rather high conversion (>84%).

Thus, the processes taking place in technological reactors can have a *multi-stationarity* even for relatively simple kinetic schemes (in this case we consider a simple non-reversible first-order reaction). In practice, reactors work usually in the conditions close to stationary. Therefore, a problem of optimal organization of the reaction conditions becomes of great importance. In the discussed example the first stationary state is undesirable from the efficiency point of view.

For example, assuming $A = 1 \times 10^5\ \mathrm{s}^{-1}$, $E_a/R = 5{,}000$ K, $J = 100\ \mathrm{Km^3\,kmol^{-1}}$, $\tau = 60$ s, initial concentration $C_{A_0} = 1\ \mathrm{kmol\,m^{-3}}$, and initial temperature $T_0 = 270$ K, we can find three stationary states. Their quantitative properties are determined by the solutions of an algebraic equation set, to which the differential equation set is transformed when both $dC_A(t)/dt$ and $dT(t)/dt$ equal zero:

$$\frac{C_{A_0} - C_A}{\tau} - Ae^{-\frac{E_T}{T}}C_A = 0,$$

$$\frac{T_0 - T(t)}{\tau} + JAe^{-\frac{E_T}{T}}C_A = 0.$$

$$C_{A0} := 1 \quad \tau := 60 \quad A := 1\cdot 10^{5} \quad E_T := 5000 \quad T_0 := 270 \quad J := 100$$

Given

$$\frac{C_{A0} - C_A}{\tau} - A\cdot e^{-\frac{E_T}{T}}\, C_A = 0 \qquad \frac{T_0 - T}{\tau} + J\cdot A\cdot \exp\left(\frac{-E_T}{T}\right)\cdot C_A = 0$$

$$S(T, C_A) := \mathrm{Find}(T, C_A)$$

$$\begin{pmatrix} T_{1st} \\ C_{1st} \end{pmatrix} := S(279, 0.9) \qquad \begin{pmatrix} T_{2st} \\ C_{2st} \end{pmatrix} := S(322, 0.5) \qquad \begin{pmatrix} T_{3st} \\ C_{3st} \end{pmatrix} := S(346, 0.2)$$

$$\begin{pmatrix} T_{1st} \\ C_{1st} \end{pmatrix} = \begin{pmatrix} 278.996 \\ 0.91 \end{pmatrix} \qquad \begin{pmatrix} T_{2st} \\ C_{2st} \end{pmatrix} = \begin{pmatrix} 322.089 \\ 0.479 \end{pmatrix} \qquad \begin{pmatrix} T_{3st} \\ C_{3st} \end{pmatrix} = \begin{pmatrix} 346.28 \\ 0.237 \end{pmatrix}$$

$$\mathrm{Jacobian}(T, C_A) := \begin{pmatrix} -\frac{1}{\tau} - A\cdot e^{-\frac{E_T}{T}} & -\frac{A\cdot C_A\cdot E_T\cdot e^{-\frac{E_T}{T}}}{T^2} \\ A\cdot J\cdot e^{-\frac{E_T}{T}} & \frac{A\cdot C_A\cdot E_T\cdot J\cdot e^{-\frac{E_T}{T}}}{T^2} - \frac{1}{\tau} \end{pmatrix}$$

$$\mathrm{eigenvals}(\mathrm{Jacobian}(T_{1st}, C_{1st})) = \begin{pmatrix} -0.017 \\ -8.683\times 10^{-3} \end{pmatrix}$$

$$\mathrm{eigenvals}(\mathrm{Jacobian}(T_{2st}, C_{2st})) = \begin{pmatrix} -0.017 \\ 7.055\times 10^{-3} \end{pmatrix}$$

$$\mathrm{eigenvals}(\mathrm{Jacobian}(T_{3st}, C_{3st})) = \begin{pmatrix} -0.017 \\ -0.017 \end{pmatrix}$$

Fig. 3.30 Computations of possible stationary states and analysis of their stability (on-line calculation http://twt.mpei.ac.ru/MCS/Worksheets/Chem/ChemKin-3-29.xmcd)

$$C_{A0} := 1 \quad \tau := 60 \quad A := 1\cdot 10^{5} \quad E_T := 5000 \quad T_0 := 270 \quad J := 100$$

$$T := 260 .. 360$$

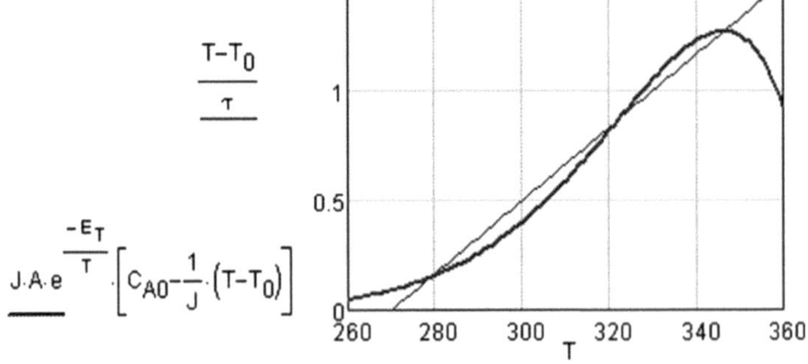

Fig. 3.31 Graphical representation of possible stationary states (on-line calculation http://twt.mpei.ac.ru/MCS/Worksheets/Chem/ChemKin-3-30.xmcd)

▶ Function IntCurves

$C_{A0} := 1 \quad \tau := 60 \quad A := 1 \cdot 10^5 \quad E_T := 5000 \quad T_0 := 270 \quad J := 100$

$\begin{pmatrix} T_{1st} \\ C_{1st} \end{pmatrix} := \begin{pmatrix} 278.996 \\ 0.91 \end{pmatrix} \quad \begin{pmatrix} T_{2st} \\ C_{2st} \end{pmatrix} := \begin{pmatrix} 322.089 \\ 0.479 \end{pmatrix} \quad \begin{pmatrix} T_{3st} \\ C_{3st} \end{pmatrix} := \begin{pmatrix} 346.28 \\ 0.237 \end{pmatrix}$

$$d'(t,x) := \begin{pmatrix} \dfrac{C_{A0} - x_0}{\tau} - A \cdot e^{\frac{-E_T}{x_1}} x_0 \\ \dfrac{T_0 - x_1}{\tau} + J \cdot A \cdot \exp\left(\dfrac{-E_T}{x_1}\right) \cdot x_0 \end{pmatrix} \qquad \text{ODESolveGrid} := \begin{pmatrix} 0.0 & 1. & 1 \\ 250 & 450 & 32 \end{pmatrix} \qquad \text{ODESolveGrid2} := \begin{pmatrix} 1 & 0. & 1 \\ 315 & 300 & 3 \end{pmatrix}$$

M := IntCurves(ODESolveGrid, 10, 350, 1000, d', rkfixed)

N := IntCurves(ODESolveGrid2, 10, 350, 1000, d', rkfixed)

$S_1 := \text{Rkadapt}\left[\begin{pmatrix} 1 \\ 273 \end{pmatrix}, 0, 2000, 1000, d'\right] \qquad S_2 := \text{Rkadapt}\left[\begin{pmatrix} 1 \\ 310 \end{pmatrix}, 0, 3000, 1000, d'\right]$

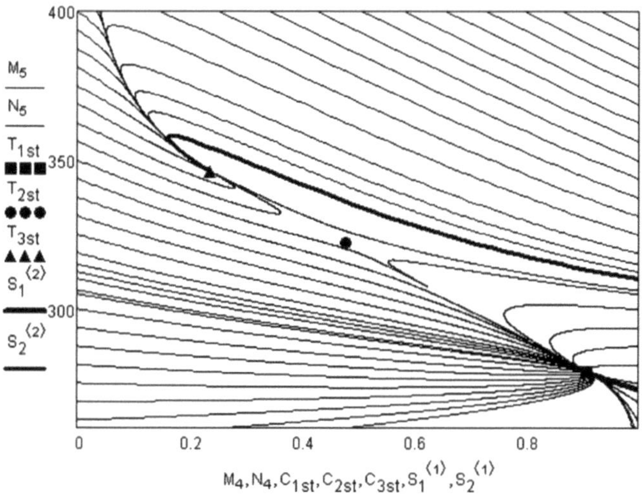

Fig. 3.32 Phase portrait for exothermic reaction in an adiabatic flow reactor

The presence of multistationarity can be illustrated in other way. The last two equations allow one to conclude that:

$$\frac{T - T_0}{\tau} = JAe^{-\frac{E_T}{T}}\left(C_{A_0} - \frac{1}{J}(T - T_0)\right).$$

A left part of the obtained correlation depends linearly on the temperature T. Its value is proportional to the cooling rate caused by a hot ($T > T_0$) airflow out of the reactor. A right part corresponds to the heat generation rate in the reactor due to the reaction exothermicity. It is a non-linear function of temperature. If one plots the temperature dependences of the equation right and left parts, one will see their interception in points corresponding to the calculated stationary temperatures and concentrations (Fig. 3.31).

When considering systems with many stationary states, it is important to investigate the stability of the latter. Stability of a stationary state is directly connected to the thermal stability of the reactor. It may happen that a small perturbation of the system takes it out of the unstable state. The process will convert into the other one, now stable. In this case calculations (Fig. 3.30) show that two out of three possible stationary states are stable: for them the Jacobian matrix eigenvalues are real and of the same sign (stable node). The third stationary state has real, but negative, Jacobian eigenvalues (saddle point). A comparison of these results with the plots shown in Fig. 3.31 allows one to conclude: a stationary state is stable if a slope of the heat elimination curve is smaller than a slope of the heat liberation.

Finally, we can prove the conclusions by plotting a phase portrait of the system (Fig. 3.32). Here dots correspond to the possible stationary states. The phase paths for different initial temperatures are marked with bold lines.

Figure 3.32 allows us to see that one or another stationary state is realized according to the initial conditions.

The discussed examples by no means cover all possible problems of chemical kinetics as well as other differential models for chemical-engineering processes, on which the reactor theory is based.

Chapter 4
Inverse Chemical Kinetics Problem

4.1 Features of the Inverse Problem

As opposed to direct chemical kinetics problem, inverse problem consists of finding the kinetic parameters of the reactive system (reaction orders, reaction rates of separate steps, activation energies) based on experimental data. Generally speaking, the goal of solving the inverse problem is to determine the mechanism of a complex chemical process (again, the term "mechanism" is used here in a formal kinetics meaning as opposed to the elementary step mechanism). The problem can be formulated mathematically as the following. A set of kinetic data points is obtained experimentally. Usually it is a set of the concentration values at specific moments of time. Additionally, a kinetic scheme of the process is postulated in a form of hypothesized elementary steps. This scheme is represented with a specific mathematical model that includes the rate constants of separate steps. Consequently, it is necessary to calculate the constants so that they fit the experiment. One can see here the main feature of inverse problems: the initial data is obtained experimentally and, thus, contains measurement errors.

An attempt to process experimental kinetic data without computer aid can be a very laborious or even impossible task. In this chapter, we will discuss general approaches to the solution of inverse problems using mathematical suites (mostly Mathcad).

Chemical kinetics give the following equation for the concentration C of a reagent i over time t, at constant volume and temperature:

$$C_i(t) = C_{i_0} + \sum_j v_{ij} k_j \varphi_j(t), \tag{4.1}$$

where

$$\varphi_j(t) = \int_0^t \prod_i C_i(t)^{n_{ij}} dt. \tag{4.2}$$

V.I. Korobov and V.F. Ochkov, *Chemical Kinetics with Mathcad and Maple*,
DOI 10.1007/978-3-7091-0531-3_4, © Springer-Verlag/Wien 2011

In these equations, i defines the component; j defines the step; n is the reaction order on the component; v is a stoichiometric coefficient; k is a rate constant; and C_{i0} is initial concentration. If the component concentrations $C_{i\,\exp}$ at the different moments of time t_m are known from experiment, than a general approach to finding the unknown constants is based on minimization of a function S that represents the sum of squared errors of the experimental values $C_{i\,\exp}$ as compared to the calculated ones, $\varphi_j(t_m)$:

$$S(k_0, k_1, \ldots, k_j) = \sum_{i=1}^{K} \sum_{m=1}^{N} [C_{i\exp} - C_i(t_m)]^2, \tag{4.3}$$

where N is the number of data points for the component i; K is the number of the components. Based on whether the functions $\varphi_j(t)$ can be represented analytically, we can define two major approaches to the calculations.

The first one is applicable when the mathematical model can be integrated analytically. In this case, the functions $\varphi_j(t)$ and the corresponding functions $C_i(t)$ are defined with specific analytical expressions that allow approximation of the experimental data set. The unknown parameters from (4.1) are calculated from the results of the approximation. The latter can be obtained by two methods. The first (and simpler) one can be implemented if the equation for $c_i(t)$ is linearizable. In this case, one chooses the corresponding linearizing coordinates for the kinetic curve and plots the corresponding straight line. The valid plot must satisfy the criterion given by (4.3). The slope is usually determined by the values of the specific kinetic parameters of the reaction. Linear representations of many kinetic equations have been discussed earlier during the discussion of direct kinetic problems. An advantage of the approach lies in its simple realization: almost all of the mathematical suites have built-in linear regression functions. Additionally, initial guesses for the unknown parameters are not required in this method.

Other computational approach, in case of analytically integrable kinetic equations, is based on non-linear approximation (`genfit`) or optimization tools (`minimize`). One should remember that the overall success of the calculation can depend greatly on reasonable initial guesses.[1]

It is more difficult to solve the inverse problem if the mathematical model is not analytically integrable. In this case, concentrations of the compounds (or any property of the system proportional to the concentration) are not represented analytically. However, even in these cases one can use several approaches based solely on the built-in functions of Mathcad. In the following chapters of the book, we will try to show that these functions are applicable to indirect problems of rather high complexity. So, let us start with some specific examples of inverse kinetic problem.

---

[1] We will not discuss a technique, still considered in education literature – the so called substitution method. In this approach, each of the kinetic data pairs is put into the corresponding kinetic equation and the constant values are then calculated and averaged. This method is obsolete after development of mathematical suites.

4.2 Determination of Kinetic Parameters Using Data Linearization

4.2.1 *Hydrolysis of Methyl Acetate in Acidic Media*

Hydrolysis of methyl acetate in a dilute aqueous solution at pH < 7 follows the first reaction order. Table 4.1 contains experimental reagent concentrations at different moments of time for seven experiments. Each series was measured under constant temperature. Determine the rate constant for each experiment and, based on the temperature–rate constant relationship, find the activation energy and pre-exponential factor.

This reaction is of the first order; consequently, we can use the following kinetic equation:

$$\ln C = \ln C_0 - kt.$$

The expression is linearizable in semi-logarithmic coordinates, $\ln C(t)$. The rate constant is determined from the slope of the regression line. The given operation is

Table 4.1 Experimental data for methyl acetate hydrolysis

t (min)	Ester concentration, M at different temperatures						
	296.15 K	299.15 K	303.15 K	306.15 K	309.15 K	313.15 K	316.15 K
0	0.2010	0.1831	0.1762	0.1821	0.1777	0.1706	0.1793
20	–	–	–	–	–	0.1475	0.1562
30	0.1983	0.1765	0.1572	0.1633	0.1528	–	–
40	–	–	–	–	–	0.1306	0.1327
60	–	0.1622	0.1430	–	0.1251	0.1185	0.1122
65	0.1890	–	–	–	–	–	–
70	–	–	–	0.1458	–	–	–
80	–	–	–	–	–	0.0987	0.0927
90	0.1832	–	0.1294	0.1326	0.1108	–	–
100	–	0.1505	–	–	–	0.0912	0.0832
120	0.1760	–	0.1207	0.1201	–	0.0802	0.0689
130	–	0.1392	–	–	–	–	–
140	–	–	–	–	–	0.0702	0.0625
150	0.1681	–	–	0.1081	0.0843	–	–
160	–	0.1307	–	–	–	0.0601	0.0564
180	0.1590	–	0.1060	0.0996	0.0700	0.0511	–
190	–	0.1251	–	–	–	–	–
210	0.1540	–	–	–	–	–	–
220	–	0.1171	–	–	–	–	–
240	–	–	0.0877	0.0789	0.0578	–	–
280	–	0.1109	–	–	–	–	–
285	0.1375	–	–	–	–	–	–
300	–	–	0.0734	0.0694	0.0451	–	–
330	0.1342	–	–	–	–	–	–
340	–	0.1062	0.0622	–	–	–	–
360	–	–	–	0.0562	0.0291	–	–

performed in Mathcad with the built-in functions `intercept(x,y)` and `slope(x,y)`. In this case, the arguments `x` and `y` of these functions are the vectors of time values and current concentration logarithms. A part of the solution of this problem is shown on Fig. 4.1.

We have to mention that the starting data set (Table 4.1) is rather large; therefore, it was prepared first as an `Excel` table and loaded into `Mathcad` environment using dynamic data exchange tools (Fig. 4.1 shows only a part of the

ORIGIN := 1

$(T\ t_1\ C_1\ t_2\ C_2\ t_3\ C_3\ t_4\ C_4\ t_5\ C_5\ t_6\ C_6\ t_7\ C_7)^T :=$

296,15	299,15	303,15	306,15	309,15	313,15	316,15
0	0,2010	0	0,1831	0	0,1762	0
30	0,1983	30	0,1765	30	0,1572	30
65	0,1890	60	0,1622	60	0,1430	70
90	0,1832	100	0,1505	90	0,1294	90
120	0,1760	130	0,1392	120	0,1207	120
150	0,1681	160	0,1307	180	0,1060	150
180	0,1590	190	0,1251	240	0,0877	180
210	0,1540	220	0,1171	300	0,0734	240

$T := T^T \qquad i := 1 .. \mathrm{rows}(C)$

$cc1_i := \mathrm{corr}(t_i, \ln(C_i)) \quad a_i := \mathrm{intercept}(t_i, \ln(C_i)) \quad b_i := \mathrm{slope}(t_i, \ln(C_i)) \quad k_i := -b_i$

$y_1(x) := a_1 + b_1 \cdot x \quad y_3(x) := a_3 + b_3 \cdot x \quad y_5(x) := a_5 + b_5 \cdot x \quad y_7(x) := a_7 + b_7 \cdot x$

$y_2(x) := a_2 + b_2 \cdot x \quad y_4(x) := a_4 + b_4 \cdot x \quad y_6(x) := a_6 + b_6 \cdot x$

$cc1^T = (-0.996\ \ -0.983\ \ -0.998\ \ -0.999\ \ -0.997\ \ -0.998\ \ -0.997)$

$k^T = (0.00133\ \ 0.00172\ \ 0.00291\ \ 0.00326\ \ 0.00477\ \ 0.00651\ \ 0.00749)$

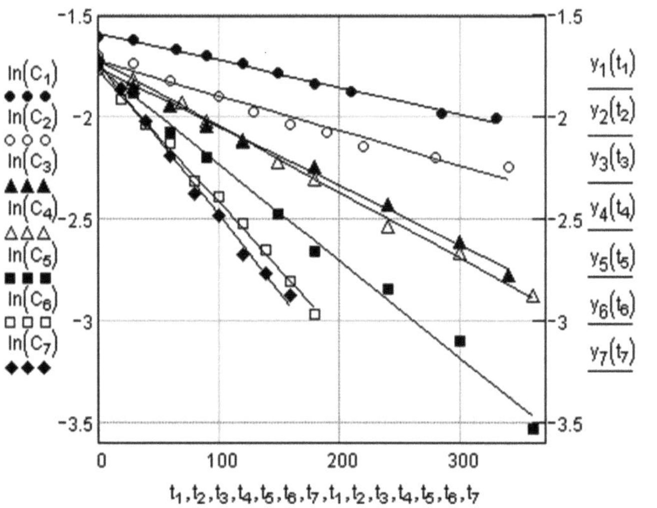

Fig. 4.1 To the rate constant calculation of methyl acetate hydrolysis in acidic media

table). Then the rate constants ($\min^{-1}$) were calculated by the above-mentioned method for each temperature in the given range. Linearity of the relationship between the vectors `t` and `ln(C)` can be checked with a function `corr(x,y)`. The reader can see that the values of this coefficient for each temperature are close to 1; it confirms a linear relationship in this case.

The obtained vector is used in the subsequent calculations, that are based on a linear form of the *Arrhenius* equation:

$$\ln k = \ln k_0 - \frac{E_a}{R}\frac{1}{T}.$$

The expression shows that the relationship between the logarithm of rate constant and T^{-1} is also linear with a slope $-E_a/RT$ and an intercept $\ln k_0$ (Fig. 4.2).

Strictly saying, the given method of pre-exponential factor determination leads to a certain contradiction between the mathematical model and physical meaning of the phenomenon. Defining k_0 as the intercept, we extrapolate the data to the temperatures where $T^{-1} = 0$, i.e. to $T \to \infty$ (more precisely, the "infinite temperature" corresponds to 10^{307} in Mathcad). One cannot imagine that the investigated system would even exist at this temperature. Therefore, it is more correct to calculate the pre-exponential term after the activation energy has been found – using an ordinate of a chosen point on the regression line inside the temperature range under investigation. In the end, one will obtain mathematically identical results but the second one will have physical meaning.

Consequently, the investigated system is characterized by the activation energy of 69.3 kJ mol^{-1}. The pre-exponential term equals $3.8 \times 10^7\ \text{s}^{-1}$.

$$T := (296.15\ \ 299.15\ \ 303.15\ \ 306.15\ \ 309.15\ \ 313.15\ \ 316.15)^T \cdot K$$

$$k := (0.00133\ \ 0.00172\ \ 0.00291\ \ 0.00326\ \ 0.00477\ \ 0.00651\ \ 0.00749)^T \cdot \min^{-1}$$

$$a := \text{intercept}\left(\frac{1}{\frac{T}{K}}, \ln\left(\frac{k}{\min^{-1}}\right)\right) \qquad k_0 := e^a \cdot \min^{-1} \qquad E_a := -R \cdot \text{slope}\left(\frac{1}{T}, \ln\left(\frac{k}{\min^{-1}}\right)\right)$$

$$y(x) := a - \frac{E_a}{R} \cdot x \qquad x := \frac{1}{\max(T)}, \frac{1}{\max(T)} + 0.00001 \cdot K^{-1} .. \frac{1}{\min(T)}$$

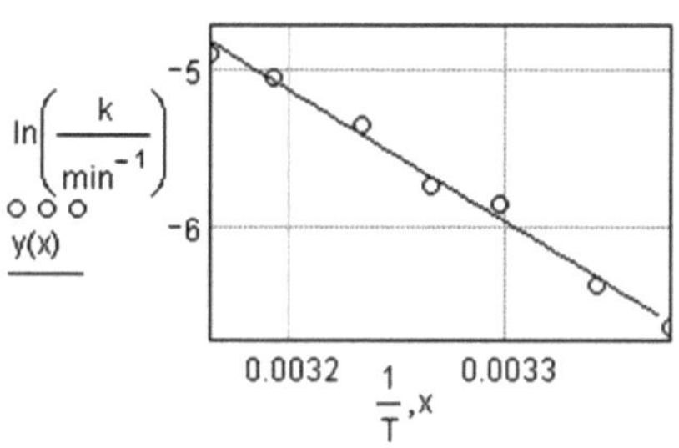

$$E_a = 69.278 \frac{\text{kJ}}{\text{mol}}$$

$$k_0 = 3.771 \times 10^7\ \text{s}^{-1}$$

$$R \equiv 8.3144 \cdot J \cdot \text{mol}^{-1} \cdot K^{-1}$$

Fig. 4.2 Determination of the activation energy and pre-exponential factor

4.2.2 *Butadiene Dimerization: Finding the Reaction Order and the Rate Constant*

Butadiene dimerization occurs through *Diels–Alder* reaction according the equation $2A \xrightarrow{k} B$ (A is a monomer, B is a dimer). The reaction was followed by the registration of total pressure P as certain moments of time in a closed thermostat (Table 4.2). Determine the reaction order on butadiene and the rate constant.

According to the reaction stoichiometry, the concentrations of the reactants are $A = A_0 - 2x$ and $B = x$ (A_0 is the initial concentration of monomer). If the reagent and the product posses the ideal gas properties than the following expressions for the initial (P_0) and current (P) pressure in the system are true:

$$P_0 = A_0RT;$$
$$P = [(A_0 - 2x)RT] + xRT.$$

This gives $x = (P_0 - P)RT$ and $A = A_0 - 2(P_0 - P)/RT$. If the reaction order is not known, we obtain the kinetic equation for the dimerization reaction in a differential form that shows the total pressure change in a reactor over time:

$$\frac{dP(t)}{dt} = -\frac{k}{2}(RT)^{1-n}(2P(t) - P_0)^n \tag{4.4}$$

Suppose the reaction has the first, second or third order on the monomer. Then solution of (4.4) gives different analytical expressions for the kinetic curve plot $P(t)$. This step of the problem solving is shown in part in Fig. 4.3. As the figure suggests, the different orders are represented with the following equations:

$$kt = \ln\left(\frac{P_0}{2P - P_0}\right)(\text{first order}), \tag{4.5}$$

$$kt = -\frac{RT}{P_0} + \frac{RT}{2P - P_0}(\text{second order}), \tag{4.6}$$

$$kt = \frac{R^2T^2}{2(2P - P_0)^2} - \frac{R^2T^2}{2{P_0}^2}(\text{thrid order}). \tag{4.7}$$

Equations (4.5)–(4.7) provide information about kinetic data linearization method. If the reaction is of the first order then the kinetic curve is linearized in

Table 4.2 Kinetic data for butadiene dimerization ($T = 400$ K)

t, s	*0*	*750*	*1,500*	*2,640*	*3,425*	*4,280*	*5,140*	*6,000*	*7,500*	*9,000*	10,500
P, mmHg	626	579	545	510	485	465	450	440	425	410	405

$$\frac{d}{dt}P(t) = \frac{-k}{2}\cdot(R\cdot T)^{1-n}\cdot\left(2\cdot P(t) - P_0\right)^n$$

First order (n=1)

$$\frac{d}{dt}P(t) = \frac{-k}{2}\cdot(R\cdot T)^{1-n}\cdot\left(2\cdot P(t) - P_0\right)^n \text{ substitute}, n = 1 \rightarrow \frac{d}{dt}P(t) = \frac{-1}{2}\cdot k\cdot\left(2\cdot P(t) - P_0\right)$$

$$\int_{P_0}^{P}\frac{1}{2\cdot x - P_0}\,dx = \int_0^t \frac{-k}{2}\,dy \text{ solve}, k \rightarrow \frac{\left(-\ln\left(2\cdot P - P_0\right)\right) + \ln\left(P_0\right)}{t} \qquad \boxed{kt = \ln\left(\frac{P_0}{2\cdot P - P_0}\right)}$$

$$\int_{P_0}^{P}\frac{1}{2\cdot x - P_0}\,dx = \int_0^t \frac{-k}{2}\,dy \text{ solve}, P \rightarrow \frac{1}{2}\cdot P_0\cdot\frac{e^{t\cdot k}+1}{e^{t\cdot k}} \qquad \boxed{P(t) = \frac{1}{2}\cdot P_0\cdot\frac{e^{t\cdot k}+1}{e^{t\cdot k}}}$$

Second order (n=2)

$$\frac{d}{dt}P(t) = \frac{-k}{2}\cdot(R\cdot T)^{1-n}\cdot\left(2\cdot P(t) - P_0\right)^n \text{ substitute}, n = 2 \rightarrow \frac{d}{dt}P(t) = \frac{-1}{2}\cdot\frac{k}{R\cdot T}\cdot\left(2\cdot P(t) - P_0\right)^2$$

$$\int_{P_0}^{P}\frac{1}{\left(2\cdot x - P_0\right)^2}\,dx\cdot 2 = \int_0^t \frac{-k}{2\cdot R\cdot T}\,dy\cdot 2 \rightarrow \frac{-1}{2\cdot P - P_0} + \frac{1}{P_0} = (-t)\cdot\frac{k}{R\cdot T} \qquad \boxed{k\cdot t = \frac{-R\cdot T}{P_0} + \frac{R\cdot T}{2\cdot P - P_0}}$$

$$\int_{P_0}^{P}\frac{1}{\left(2\cdot x - P_0\right)^2}\,dx = \int_0^t \frac{-k}{2\cdot R\cdot T}\,dy \text{ solve}, P \rightarrow \frac{1}{2}\cdot P_0\cdot\frac{2\cdot R\cdot T + t\cdot k\cdot P_0}{R\cdot T + t\cdot k\cdot P_0} \qquad \boxed{P(t) = \frac{1}{2}\cdot P_0\cdot\frac{2\cdot R\cdot T + t\cdot k\cdot P_0}{R\cdot T + t\cdot k\cdot P_0}}$$

Third order (n=3)

$$\frac{d}{dt}P(t) = \frac{-k}{2}\cdot(R\cdot T)^{1-n}\cdot\left(2\cdot P(t) - P_0\right)^n \text{ substitute}, n = 3 \rightarrow \frac{d}{dt}P(t) = \frac{-1}{2}\cdot\frac{k}{R^2\cdot T^2}\cdot\left(2\cdot P(t) - P_0\right)^3$$

$$\int_{P_0}^{P}\frac{1}{\left(2\cdot x - P_0\right)^3}\,dx\cdot 2 = \int_0^t \frac{-k}{2\cdot R^2\cdot T^2}\,dy\cdot 2 \rightarrow \frac{-1}{2\cdot\left(2\cdot P - P_0\right)^2} + \frac{1}{2\cdot P_0^2} = (-t)\cdot\frac{k}{R^2\cdot T^2} \qquad \boxed{k\cdot t = \frac{R^2\cdot T^2}{2\cdot\left(2\cdot P - P_0\right)^2} - \frac{R^2\cdot T^2}{2\cdot P_0^2}}$$

$$\frac{R^2\cdot T^2}{2\cdot\left(2\cdot P - P_0\right)^2} - \frac{R^2\cdot T^2}{2\cdot P_0^2} \text{ simplify} \rightarrow (-2)\cdot R^2\cdot T^2\cdot P\cdot\frac{P - P_0}{\left(2\cdot P - P_0\right)^2\cdot P_0^2}$$

$$k\cdot t = \frac{-2\cdot(R\cdot T)^2\cdot\left(P - P_0\right)\cdot P}{\left(2\cdot P - P_0\right)^2\cdot P_0^2} \ \Bigg|\begin{array}{l}\text{solve}, P\\ \text{simplify}\end{array} \rightarrow \begin{bmatrix} \frac{1}{2}\cdot\left[R^2\cdot T^2 + 2\cdot t\cdot k\cdot P_0^2 + \left[R^2\cdot T^2\cdot\left(2\cdot t\cdot k\cdot P_0^2 + R^2\cdot T^2\right)\right]^{\frac{1}{2}}\right]\cdot\frac{P_0}{2\cdot t\cdot k\cdot P_0^2 + R^2\cdot T^2} \\ \frac{1}{2}\cdot\left[R^2\cdot T^2 + 2\cdot t\cdot k\cdot P_0^2 - \left[R^2\cdot T^2\cdot\left(2\cdot t\cdot k\cdot P_0^2 + R^2\cdot T^2\right)\right]^{\frac{1}{2}}\right]\cdot\frac{P_0}{2\cdot t\cdot k\cdot P_0^2 + R^2\cdot T^2} \end{bmatrix}$$

Fig. 4.3 Derivation of the kinetic equations for butadiene dimerization

the coordinates $\ln[P_0/(2P - P_0)]$ on t; for the second reaction order one should use the coordinates $1/(2P - P_0)$ on t. Finally, the coordinates $1/(2P - P_0)^2$ on t linearize the third order reaction curve. The rate constant can be derived from the corresponding line slope in all cases.

The second part of the document (Fig. 4.4) shows the calculation of the rate constant based on this information using a built-in function `slope(x,y)`.

We want to stress out that the correlation coefficients of the linear regressions are close enough to 1 in all the three cases. This leads to ambiguity with the reaction order. The final decision can be made only by a direct comparison of the experimental data and calculated kinetic curves. Such comparison, shown in Fig. 4.4, allows to conclude that this dimerization process follows namely the second reaction order with the constant $k = 9.487 \times 10^{-3}\ \mathrm{M}^{-1}\ \mathrm{s}^{-1}$.

$P_0 := 626 \cdot \mathrm{torr} \qquad T := 400 \cdot K \qquad R \equiv 8.3144 \cdot \mathrm{joule} \cdot \mathrm{mol}^{-1} \cdot K^{-1}$

$t := \mathrm{stack}(750, 1500, 2640, 3425, 4280, 5140, 6000, 7500, 9000, 10500) \cdot s$

$P := \mathrm{stack}(579, 545, 510, 485, 465, 450, 440, 425, 410, 405) \cdot \mathrm{torr}$

$$\mathrm{corr}\left(t, \ln\left(\frac{P_0}{2 \cdot P - P_0}\right)\right) = 0.982 \qquad k_{1order} := \mathrm{slope}\left(t, \ln\left(\frac{P_0}{2 \cdot P - P_0}\right)\right) \qquad k_{1order} = 1.1011 \times 10^{-4}\,\frac{1}{s}$$

$$\mathrm{corr}\left(t, \frac{1}{2 \cdot P - P_0}\right) = 0.998 \qquad k_{2order} := \mathrm{slope}\left(t, \frac{R \cdot T}{2 \cdot P - P_0}\right) \qquad k_{2order} = 9.467 \times 10^{-3}\,\frac{\mathrm{liter}}{s \cdot \mathrm{mol}}$$

$$\mathrm{corr}\left[t, \frac{R^2 \cdot T^2}{(2 \cdot P - P_0)^2}\right] = 0.993 \qquad k_{3order} := \mathrm{slope}\left[t, \frac{R^2 \cdot T^2}{(2 \cdot P - P_0)^2}\right] \qquad k_{3order} = 1.7395\,\frac{\mathrm{liter}^2}{\mathrm{mol}^2 \cdot s}$$

$$p_{1order}(x) := \frac{1}{2} \cdot P_0 \cdot \frac{\exp(x \cdot k_{1order}) + 1}{\exp(x \cdot k_{1order})} \qquad p_{2order}(x) := \frac{1}{2} \cdot P_0 \cdot \frac{2 \cdot R \cdot T + x \cdot k_{2order} \cdot P_0}{R \cdot T + x \cdot k_{2order} \cdot P_0}$$

$$p_{3order}(x) := \frac{P_0 \left[4 \cdot x \cdot k_{3order} \cdot P_0^{\,2} + 2 \cdot R^2 \cdot T^2 + 2 \cdot \left(2 \cdot x \cdot k_{3order} \cdot P_0^{\,2} \cdot R^2 \cdot T^2 + R^4 \cdot T^4\right)^{\frac{1}{2}}\right]}{2 \cdot \left(4 \cdot x \cdot k_{3order} \cdot P_0^{\,2} + 2 \cdot R^2 \cdot T^2\right)}$$

$x := 0 \cdot s, 10 \cdot s \,..\, \max(t)$

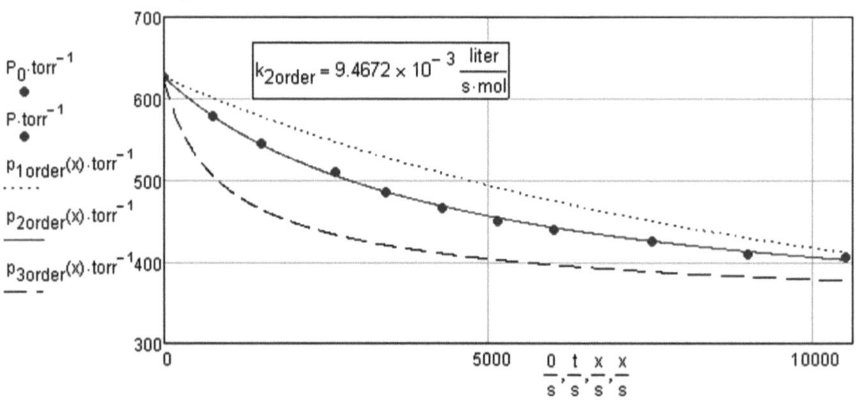

Fig. 4.4 Finding the dimerization reaction order

4.2.3 Exclusion of Time as an Independent Variable

The method of exclusion of time as an independent variable (discussed in Sect. 1.3.3) can be applied effectively for several inverse problems. For example, some compound A participates in three parallel steps of a process and each of the steps follows the first reaction order:

$$A \xrightarrow{k_1} B,\ A \xrightarrow{k_2} C,\ A \xrightarrow{k_3} D.$$

Assume the corresponding file contains information about concentrations of all the reaction components at certain moments of time. One is asked to derive the rate constants for each of the three steps given the initial concentrations of the components are (M): $A_0 = 0.844$; $B_0 = 0.071$, $C_0 = D_0 = 0$.

From the mathematical model of the reaction,

$$dA(t)/dt = -(k_1 + k_2 + k_3)A(t), \tag{4.8}$$

$$dB(t)/dt = k_1 A(t), \tag{4.9}$$

$$dC(t)/dt = k_2 A(t), \tag{4.10}$$

$$dD(t)/dt = k_3 A(t), \tag{4.11}$$

and the given initial conditions follows:

$$A(t) = A_0 e^{-(k_1+k_2+k_3)t}. \tag{4.12}$$

Division of (4.9) by (4.10) eliminates time t as an independent variable:

$$dB/dC = k_1/k_2.$$

Considering the initial concentrations of the compounds B and C, a function $B(C)$ can be derived:

$$B(C) = B_0 + \frac{k_1}{k_2} C. \tag{4.13}$$

Similarly, (4.10) and (4.11) allow defining the relationship between the concentrations of C and D:

$$C(D) = \frac{k_2}{k_3} D. \tag{4.14}$$

From (4.12)–(4.14) follows that the rate constants can be calculated by linearization of the starting kinetic data. For example, the relationship between $\ln A(t)$ and

t is linear with a slope $(k_1 + k_2 + k_3)$. The slopes of two other linear relationships, $B(C)$ and $C(D)$, are k_1/k_2 and k_2/k_3 respectively. Figure 4.5 shows a solution of this inverse problem. After computing the ratios between the constants, each of the latter is found with a GIVEN/FIND solver. Then kinetic curves are simulated using the computed constants and compared with the experimental values of concentration. The results indicate that the constants are determined with a satisfactory precision.

$$\text{DATA} := \text{READPRN}(\text{"example1.prn"}) \qquad tm := \text{DATA}^{\langle 0\rangle}\cdot min$$

$$Am := \text{DATA}^{\langle 1\rangle}\cdot\frac{mol}{liter} \qquad Bm := \text{DATA}^{\langle 2\rangle}\cdot\frac{mol}{liter} \qquad Cm := \text{DATA}^{\langle 3\rangle}\cdot\frac{mol}{liter} \qquad Dm := \text{DATA}^{\langle 4\rangle}\cdot\frac{mol}{liter}$$

$$\text{corr}(Dm, Cm) = 0.998 \qquad \text{corr}\left(tm, \ln\left(\frac{Am}{mol\cdot liter^{-1}}\right)\right) = -0.999 \qquad \text{corr}(Cm, Bm) = 0.994$$

$$(k_1\ \ k_2\ \ k_3) := (0.1\ \ 0.2\ \ 0.3)\cdot min^{-1}$$

$$\text{Given} \qquad \text{slope}(Cm, Bm) = \frac{k_1}{k_2} \qquad -\text{slope}\left(tm, \ln\left(\frac{Am}{mol\cdot liter^{-1}}\right)\right) = k_1 + k_2 + k_3$$

$$\text{slope}(Dm, Cm) = \frac{k_2}{k_3} \qquad (k_1\ \ k_2\ \ k_3) := \text{Find}(k_1, k_2, k_3)^T$$

$$\boxed{(k_1\ \ k_2\ \ k_3) = (0.0157\ \ 0.0083\ \ 0.0289)\,min^{-1}}$$

$$f(a,x) := \begin{bmatrix} -(k_1+k_2+k_3)\cdot x_0\cdot min \\ k_1\cdot x_0\cdot min \\ k_2\cdot x_0\cdot min \\ k_3\cdot x_0\cdot min \end{bmatrix} \qquad in := \begin{pmatrix} Am_0\cdot liter\cdot mol^{-1} \\ Bm_0\cdot liter\cdot mol^{-1} \\ 0 \\ 0 \end{pmatrix} \qquad S := \text{rkfixed}\left(in, 0, \max\left(\frac{tm}{min}\right), 1000, f\right)$$

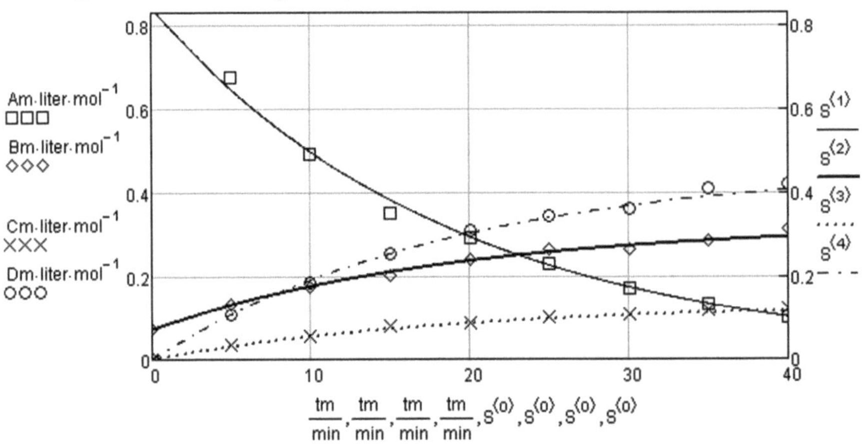

Fig. 4.5 Solution of the inverse problem for a parallel irreversible reaction

4.2.4 Linearization with Numerical Integration of Kinetic Data: Basic Hydrolysis of Diethyl Adipate

The reader might think that the linearization method is of limited usage and can be applied only to simple or formally simple reactions. We can show, however, that the problems that can be solved with thi s method are rather numerous and include complex reactions (including the ones that have a mathematical model represented as a system of non-linear ODEs).

Consider a kinetic scheme for the following two-step reaction:

$$\begin{aligned}&(1)\ A_1 + A_2 \longrightarrow A_3 + A_4\ (k_1),\\ &(2)\ A_1 + A_3 \longrightarrow A_4 + A_5\ (k_2).\end{aligned}$$

This mechanism corresponds, in particular, to the reaction of basic hydrolysis of diethyl adipate. Here A_1 is $NaOH$; A_2 is diethyl adipate; A_3 is monoethyl adipate; A_4 is ethanol; A_5 is adipic acid. The scheme is represented by the following mathematical model of a competitive reaction:

$$\begin{aligned}&dC_1/dt = -k_1C_1C_2 - k_2C_1C_3;\\ &dC_2/dt = -k_1C_1C_2;\\ &dC_3/dt = k_1C_1C_2 - k_2C_1C_3;\\ &dC_4/dt = k_1C_1C_2 + k_2C_1C_3;\\ &dC_5/dt = k_2C_1C_3.\end{aligned}$$

Bearing (4.1) in mind, one can deduce:

$$\begin{aligned}&\Delta C_1 = C_{10} - C_1 = k_1\int_0^t C_1C_2dt + k_2\int_0^t C_1C_3dt,\\ &\Delta C_2 = C_{20} - C_2 = k_1\int_0^t C_1C_2dt,\end{aligned}$$

where C_{10} and C_{20} are the initial concentrations of reagents A_1 and A_2 respectively. Obviously:

$$\Delta C_1 - \Delta C_2 = k_2\int_0^t C_1C_3dt.$$

We will focus on the last two equations because they will be used later in determination of the kinetic constants. To do this, we need information about the

Table 4.3 Starting data for the basic ester hydrolysis problem

t (min)	$C_1 \times 10^3$ (M)	$C_2 \times 10^3$ (M)	$C_3 \times 10^3$ (M)
0	9.02	4.510	0.000
2.667	8.49	4.004	0.482
12.667	7.04	2.696	1.648
22.667	6.04	1.930	2.180
22.667	5.31	1.444	2.426
22.667	4.75	1.116	2.581
22.667	4.31	0.885	2.540
22.667	3.95	0.716	2.518
22.667	3.65	0.590	2.470

current concentration of A_1, A_2, A_3 (Table 4.3). The data should be linearized in rather unusual coordinates: ΔC_2 on $\int_0^t C_1C_2dt$, and $(\Delta C_1 - \Delta C_2)$ on $\int_0^t C_1C_3dt$. Here ΔC_1 and ΔC_2 are the changes in the concentrations of A_1 and A_2 respectively, at time t. Values of the desired constants k_1 and k_2 can be deduced independently from the slopes of the two plotted lines (both of them go through the origin). To do this, the values of the corresponding integral should be first calculated numerically.

All these considerations are implemented in a document shown in Fig. 4.6. The starting data (put in the vectors c_1, c_2, c_3) is used to form the new vectors of the concentration products – $[c_1 \cdot c_2]$ and $[c_1 \cdot c_3]$. Numerical integration becomes possible after preliminary interpolation of the modified data set with built-in functions `pspline` and `interp`. Then the desired rate constants are easily deduced from the obtained linear regressions.

4.2.5 Estimation of Confidence Intervals for the Calculated Constants

It is often important not only to calculate the values of unknown constants but also to statistically evaluate the obtained results. Such estimation tells a lot about the quality of the solution. For instance, the calculated kinetic parameters are sometimes reported with the corresponding confidence intervals. A few formulas from regression analysis are enough to find these intervals. For example, one can calculate a sum of square errors *SSE* of the experimental data from calculated regression line (the so called *least squares method*):

$$SSE = \sum_{k=0}^{N-1} (Y_k - mX_k - b)^2,$$

where m and b are the values of `slope` and `intercept` respectively; N is the number of experimental data pairs from the vectors X and Y. SSE is then divided by the number of degrees of freedom, giving a mean squares error, *MSE*:

$$MSE = SSE/(N - 2).$$

$t := \text{stack}(0, 2.667, 12.667, 22.667, 32.667, 42.667, 52.667, 62.667, 72.667)$

$c_1 := \text{stack}(9.02, 8.49, 7.04, 6.04, 5.31, 4.75, 4.31, 3.95, 3.65) \cdot 10^{-3}$

$c_2 := \text{stack}(4.51, 4.004, 2.696, 1.93, 1.444, 1.116, 0.885, 0.716, 0.590) \cdot 10^{-3}$

$c_3 := \text{stack}(0, 0.482, 1.648, 2.18, 2.426, 2.581, 2.54, 2.518, 2.47) \cdot 10^{-3}$

$$\Delta c_1 = c_{1_0} - c_1 = k_1 \cdot \int_0^t c_1 \cdot c_2 \, dt + k_2 \cdot \int_0^t c_1 \cdot c_3 \, dt \qquad \Delta c_2 = c_{2_0} - c_2 = k_1 \cdot \int_0^t c_1 \cdot c_2 \, dt$$

$$\Delta c_1 - \Delta c_2 = k_2 \cdot \int_0^t c_1 \cdot c_3 \, dt \qquad [c_1 \cdot c_2] := \overrightarrow{(c_1 \cdot c_2)} \qquad [c_1 \cdot c_3] := \overrightarrow{(c_1 \cdot c_3)}$$

$y_{12} := \text{pspline}(t, [c_1 \cdot c_2]) \qquad y_{13} := \text{pspline}(t, [c_1 \cdot c_3])$

$y_{12i}(x) := \text{interp}(y_{12}, t, [c_1 \cdot c_2], x) \qquad y_{13i}(x) := \text{interp}(y_{13}, t, [c_1 \cdot c_3], x)$

$$\text{Intc}_{12}(x) := \int_0^x y_{12i}(\xi)\, d\xi \qquad \text{Intc}_{13}(x) := \int_0^x y_{13i}(\xi)\, d\xi \qquad \Delta c_1 := c_{1_0} - c_1 \quad \Delta c_2 := c_{2_0} - c_2$$

$i := 0 .. \text{rows}(t) - 1 \qquad x1_i := \text{Intc}_{12}(t_i) \qquad y1_i := \Delta c_{2_i} \qquad x2_i := \text{Intc}_{13}(t_i) \qquad y2_i := \Delta c_{1_i} - \Delta c_{2_i}$

$\text{corr}(x1, y1) = 1.000000 \qquad \text{corr}(x2, y2) = 0.99996$

$Y_1(x) := \text{intercept}(x1, y1) + \text{slope}(x1, y1) \cdot x \qquad Y_2(x) := \text{intercept}(x2, y2) + \text{slope}(x2, y2) \cdot x$

$k_1 := \text{slope}(x1, y1) \quad \boxed{k_1 = 5.132} \qquad k_2 := \text{slope}(x2, y2) \quad \boxed{k_2 = 1.828}$

$x := 0, 0.0005 .. 0.001$

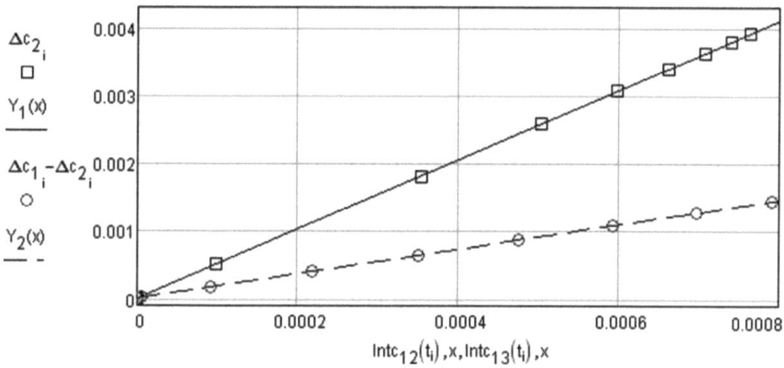

Fig. 4.6 Solution of the inverse problem using numerical integration of the starting data

In this particular case the number of degrees of freedom is $N - 2$ (a number of the experimental data pairs minus a number of unknown parameters). In the next step, the vector X is transformed into a matrix XX with the zero column consisting of unities and the first column being equal to the vector X. Standard deviations s_b and s_m are then calculated for the parameters b and m:

$$s_b = \sqrt{MSE\left[XX^T \cdot XX\right]^{-1}_{0,0}},$$

$$s_m = \sqrt{MSE\left[XX^T \cdot XX\right]^{-1}_{1,1}}.$$

To continue the calculations one needs a value of Student's t-distribution coefficient for the given number of degrees of freedom. Special tables are usually used for this; however, they are not required when working with Mathcad as the suite has a built-in function qt. For the most used confidence level of 95%, t_a equals:

$$t_\alpha = qt\left(1 - \left(1 - \frac{0,95}{2}\right), N - 2\right) = qt(0.975, N - 2).$$

Finally, the values of $m_{\min}$ and $m_{\max}$, that define the upper and the lower limits of the confidence interval for the parameter m, are calculated using the following equations:

$$m_{\min} = m - t_\alpha \cdot s_m,$$
$$m_{\max} = m + t_\alpha \cdot s_m.$$

Consequently, the limits of the confidence interval are:

$$b_{\min} = b - t_\alpha \cdot s_b,$$
$$b_{\max} = b + t_\alpha \cdot s_b.$$

Figure 4.7 shows a part of network Mathcad calculations of the rate constants for diethyl adipate hydrolysis with the corresponding confidence intervals.

4.2.6 Kinetics of α-Pinene Isomerization

It is known that specialized computer software have been created for solution of different types of kinetic problems. For example, one can find a specialized software suite `Dynafit` that is capable of solving inverse kinetic problems. Surely, the developers of such programs usually include some tests or sample files to demonstrate advantages of their products. In particular, a real kinetic problem of α-pinene isomerization is given as one of the examples in a `Dynafit` tutorial.

Pinenes are colorless liquids. Three isomers differing in the position of a double bond are known: α-, β- and δ-pinene. Pinenes have rather high reactivity. When heated to 250°C, α-pinene mainly isomerizes to limonene; at temperatures above 400°C it forms a mixture of allo-ocimene and limonene. The isomerization process can be represented by a scheme with five elementary steps and five substances:

$$Pinene \xrightarrow{k_1} Limonene,$$
$$Pinene \xrightarrow{k_2} Allo - ocimene,$$
$$Allo - ocimene \xrightarrow{k_3} Pyrolene,$$
$$Allo - ocimene \underset{k_5}{\overset{k_4}{\rightleftarrows}} Dimer.$$

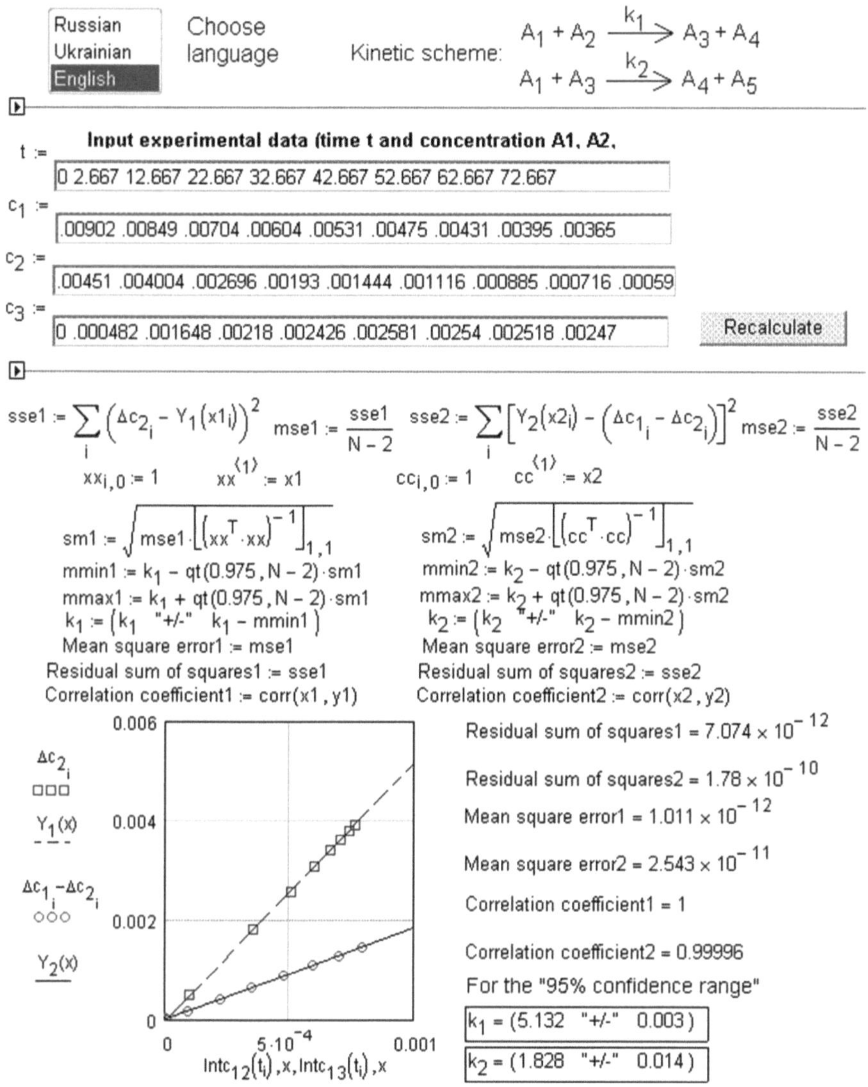

Fig. 4.7 Calculation of confidence intervals for the derived constants

It is known from the preliminary experiments that each step is of the first order. The experimental kinetic data, given by developers of `Dynafit`, is shown in Table 4.4. However, authors do not disclose the calculation sequence; consequently, we can only compare the final results. A document that solves this problem in Mathcad is shown in Fig. 4.7. First, a stoichiometric matrix and a rate vector are built in order to define the mathematical model of the process. Then a system of equations is formed based on this model and (4.1):

Table 4.4 Experimental data for α-pinene isomerization

Time (s)	Weight fraction of the reactants (%)				
	Pinene A	Limonene B	Allo-ocimene C	Pyrolene D	Dimer E
0	100	0	0	0	0
1,230	88.35	7.3	2.3	0.4	1.75
3,060	76.4	15.6	4.5	0.7	2.8
4,920	65.1	23.1	5.3	1.1	5.8
7,800	50.4	32.9	6.0	1.9	12.0
10,680	37.5	42.7	6.0	1.9	12.0
15,030	25.9	49.1	5.9	2.2	17.0
22,620	14.0	57.4	5.1	2.6	21.0
36,420	4.5	63.1	3.8	2.9	25.7

$$\underline{A} = A_0 - (k_1 + k_2)\underline{\varphi_A},$$
$$\underline{B} = k_1\underline{\varphi_A},$$
$$\underline{C} = k_2\underline{\varphi_A} - (k_3 + k_4)\underline{\varphi_C} + k_5\underline{\varphi_E},$$
$$\underline{D} = k_3\underline{\varphi_C},$$
$$\underline{E} = k_4\underline{\varphi_C} - k_5\underline{\varphi_E}.$$

Here $\underline{A}, \underline{B}, \underline{C}, \underline{D}$, and $\underline{E}$ are vectors of the current concentrations for the reaction components (experimental values); $\underline{\varphi_A}, \underline{\varphi_B}, \underline{\varphi_C}, \underline{\varphi_D}$, and $\underline{\varphi_E}$ are vectors defined by (4.2). They contain interpolated values of the reactant concentrations and were obtained by numerical integration of the kinetic data.

In fact, here we use the approach that was already discussed in the previous chapter – numerical integration of the kinetic curves. The difference is that this problem is transformed into a system of vector equations that is solved using the `Given/Minerr` block (Fig. 4.8). The initial guesses for the desired constants were set randomly.

It is known that the result produced by such functions as `MINERR`, `GENFIT`, and `MINIMIZE` depends greatly on the initial guesses of sought values. Also, the user often has to repeat the calculations using the obtained results as the new approximations for the next step. The `MINERR` function, in particular, has several options to set a certain method of solving algebraic equation systems. The desired algorithm can be set using a context menu (Fig. 4.9) that is called with a mouse in a `MINERR` keyword area.

Solutions of the α-pinene kinetics problem using different algorithms are given in Table 4.5. One can see that the linear method of least square deviations gives an unsatisfactory result. Among the non-linear algorithms, the `Conjugate Gradient` and the `Quasi-Newton` methods lead to the encouraging results: the error `ERR` is decreased after each calculation step. However, these methods converge rather slowly. The best results are obtained with the `Levenberg-Markquardt` algorithm.

▶ Input Data (array DATA)

$$\alpha := \begin{pmatrix} -1 & 1 & 0 & 0 & 0 \\ -1 & 0 & 1 & 0 & 0 \\ 0 & 0 & -1 & 1 & 0 \\ 0 & 0 & -1 & 0 & 1 \\ 0 & 0 & 1 & 0 & -1 \end{pmatrix} \quad \alpha^T \cdot \begin{pmatrix} k_1 \cdot A \\ k_2 \cdot A \\ k_3 \cdot C \\ k_4 \cdot C \\ k_5 \cdot E \end{pmatrix} \text{ collect}, A \rightarrow \begin{bmatrix} (-k_1 - k_2) \cdot A \\ k_1 \cdot A \\ k_2 \cdot A + E \cdot k_5 - C \cdot k_4 - C \cdot k_3 \\ C \cdot k_3 \\ C \cdot k_4 - E \cdot k_5 \end{bmatrix} \quad \begin{pmatrix} t & A \\ B & C \\ D & E \end{pmatrix} := \begin{pmatrix} DATA^{\langle 0 \rangle} & DATA^{\langle 1 \rangle} \\ DATA^{\langle 2 \rangle} & DATA^{\langle 3 \rangle} \\ DATA^{\langle 4 \rangle} & DATA^{\langle 5 \rangle} \end{pmatrix}$$

$SA := cspline(t, A)$ $SB := cspline(t, B)$ $SC := cspline(t, C)$ $SD := cspline(t, D)$ $SE := cspline(t, E)$

$fitA(x) := interp(SA, t, A, x)$ $fitB(x) := interp(SB, t, B, x)$ $fitC(x) := interp(SC, t, C, x)$

$fitD(x) := interp(SD, t, D, x)$ $fitE(x) := interp(SE, t, E, x)$

$$ifitA(x) := \int_0^x fitA(\xi)\, d\xi \quad ifitB(x) := \int_0^x fitB(\xi)\, d\xi \quad ifitC(x) := \int_0^x fitC(\xi)\, d\xi \quad ifitD(x) := \int_0^x fitD(\xi)\, d\xi$$

$i := 0 .. rows(DATA) - 1$

$$ifitE(x) := \int_0^x fitE(\xi)\, d\xi \quad IA_i := ifitA(t_i) \quad IB_i := ifitB(t_i) \quad ID_i := ifitD(t_i) \quad IC_i := ifitC(t_i) \quad IE_i := ifitE(t_i)$$

$A_0 := 100 \qquad (k_1\ k_2\ k_3\ k_4\ k_5) := (1\ 1\ 1\ 1\ 1)$

Given $A = A_0 - (k_1 + k_2) \cdot IA$ $\quad B = k_1 \cdot IA$ $\quad C = k_2 \cdot IA - (k_3 + k_4) \cdot IC + k_5 \cdot IE$ $\quad D = k_3 \cdot IC$ $\quad E = k_4 \cdot IC - k_5 \cdot IE$

$(k_1\ k_2\ k_3\ k_4\ k_5) := Minerr(k_1, k_2, k_3, k_4, k_5)^T$

$(k_1\ k_2\ k_3\ k_4\ k_5) = (5.936 \times 10^{-5}\ \ 2.937 \times 10^{-5}\ \ 1.978 \times 10^{-5}\ \ 3.084 \times 10^{-4}\ \ 5.146 \times 10^{-5})$

DYNAFIT calculation (www.biokin.com):

k_1=5,926 10^{-5}; k_2=2,963 10^{-5}; k_3=2,047 10^{-5}; k_4=2,745 10^{-4}; k_5=3,998 10^{-5}

▶ Direct problem solution

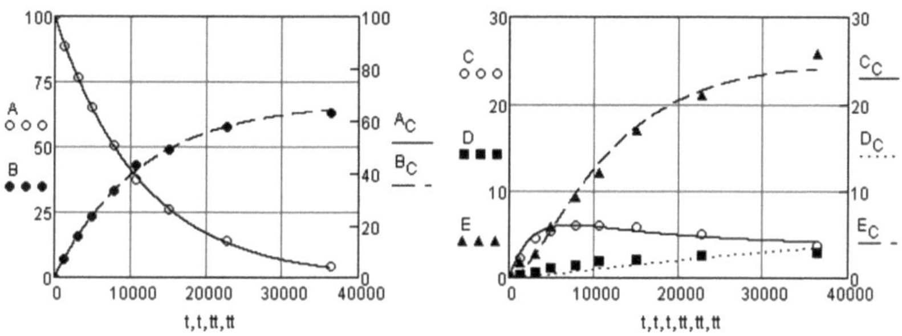

Fig. 4.8 To the problem of α-pinene isomerization

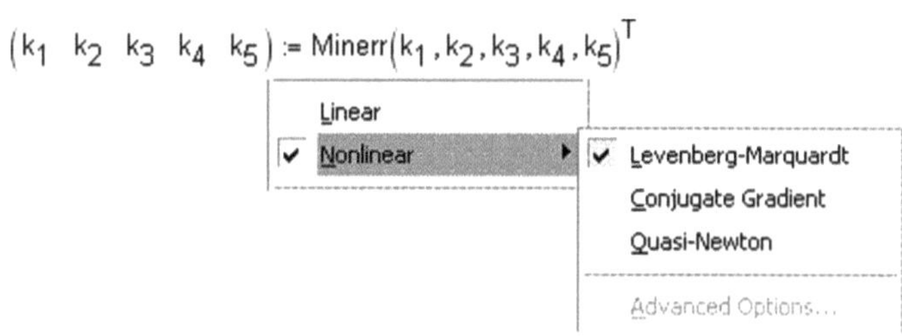

Fig. 4.9 Setting options for `Minerr` function

Table 4.5 Results of the kinetic parameters calculations using different algorithms (the initial guess values are 1 for all the constants)

Unknown rate constants					ERR
k_1	k_2	k_3	k_4	k_5	
Linear method of smallest square deviations (option `Linear`)					
1	1	1	1	1	
5.817×10^{-5}	2.300×10^{-5}	1.633×10^{-5}	2.515×10^{-4}	3.231×10^{-5}	19.1792
5.817×10^{-5}	2.300×10^{-5}	1.633×10^{-5}	2.515×10^{-4}	3.231×10^{-5}	19.1792
Non-linear methods of smallest square deviations					
1. Gradient method (option `Conjugate Gradient`)					
1	1	1	1	1	
6.303×10^{-5}	1.074×10^{-4}	6.496×10^{-4}	2.241×10^{-3}	1.201×10^{-3}	518.145
5.105×10^{-5}	4.162×10^{-5}	8.815×10^{-5}	6.921×10^{-4}	1.656×10^{-4}	39.5077
5.817×10^{-5}	2.389×10^{-5}	2.183×10^{-5}	3.391×10^{-4}	5.878×10^{-5}	20.2179
6.106×10^{-5}	2.455×10^{-5}	2.183×10^{-5}	3.221×10^{-4}	5.978×10^{-5}	12.5836
6.200×10^{-5}	2.629×10^{-5}	2.183×10^{-5}	3.222×10^{-4}	6.005×10^{-5}	9.6179
6.200×10^{-5}	2.629×10^{-5}	2.183×10^{-5}	3.222×10^{-4}	6.005×10^{-5}	9.6175
6.200×10^{-5}	2.629×10^{-5}	2.183×10^{-5}	3.222×10^{-4}	6.005×10^{-5}	9.6175
2. Newton method (option `Quasi-Newton`)					
1	1	1	1	1	
6.303×10^{-5}	1.074×10^{-4}	6.496×10^{-4}	2.241×10^{-3}	1.201×10^{-3}	518.145
5.105×10^{-5}	4.162×10^{-5}	8.815×10^{-5}	6.921×10^{-4}	1.656×10^{-4}	39.5077
5.817×10^{-5}	2.389×10^{-5}	2.183×10^{-5}	3.391×10^{-4}	5.878×10^{-5}	20.2179
6.106×10^{-5}	2.455×10^{-5}	2.183×10^{-5}	3.221×10^{-4}	5.978×10^{-5}	12.5836
6.200×10^{-5}	2.629×10^{-5}	2.183×10^{-5}	3.222×10^{-4}	6.005×10^{-5}	9.6179
6.200×10^{-5}	2.629×10^{-5}	2.183×10^{-5}	3.222×10^{-4}	6.005×10^{-5}	9.6175
6.200×10^{-5}	2.629×10^{-5}	2.183×10^{-5}	3.222×10^{-4}	6.005×10^{-5}	9.6175
3. Levenberg-Markquardt method (option `Levenberg-Markquardt`)					
1	1	1	1	1	
5.936×10^{-5}	2.937×10^{-5}	1.978×10^{-5}	3.084×10^{-4}	5.146×10^{-5}	5.5414

Figure 4.8 shows that simplification of inverse problem to a system of algebraic equations generally gives satisfactory results. However, we have to mention that even though `Dynafit` does not give the exact solution, it is close enough. For the same reason this methodic can be considered only as an estimation or approximation.

4.3 Inverse Problem and Specialized Minimization Methods

Inverse problem can also be considered as an optimization problem. If one has an analytical expression for the component concentration $C(k_1, k_2, \ldots, k_n, t)$ and a set of experimental values $C_{\exp_i}$ corresponding to time t_i, it is possible to find the unknown constants by minimization of a sum of squared differences between the experimental and calculated values.

$$SSq(k_1, k_2, \ldots, k_n) = \sum_i \left[C_{\exp_i} - C(k_1, k_2, \ldots, k_n, t_i)\right]^2 \Rightarrow \min .$$

This is a typical minimization problem for a function of n variables that can be solved using a Mathcad built-in function `MINIMIZE`. The latter implements gradient search algorithms to find the local minimum. The `SSq` function in this case is called the *target function*, and the unknown kinetic constants are the *optimization parameters*. When there are no additional limitations for the values of optimization parameters or the sought function, we have a case of the so called *unconstrained optimization*. Likewise, if the unknown parameters or the target function itself are mathematically constrained with some equalities or inequalities, then one deals with the *constrained optimization*. Such additional constrains are usually set on the basis on the physical nature of the problems (e.g. rate constants must be positive, a ratio of the direct reaction rate to that of the inverse one must equal the equilibrium constant, etc.) The constraints are sometimes added in order to speed up the computations (for example, the value of target function in the found minimum should not exceed some number `TOL`).

Usage of the Mathcad `MINIMIZE` function requires following three main rules:

- Initial guesses of the sought parameters should be set before the optimization
- The `MINIMIZE` function uses only numerical calculation algorithms and does not work with symbol mathematics
- For constrained optimization problems, the `MINIMIZE` function is applicable only as a part of solving block

4.3.1 Deriving Parameters for an Empirical Rate Equation of Phosgene Synthesis

Phosgene synthesis occurs according to the overall reaction:

$$CO + Cl_2 \Leftrightarrow COCl_2$$

An empirical equation was proposed to describe the process in terms of the product formation rate, v:

$$v = k_1[CO]^a[Cl_2]^b - k_2[Cl_2]^c[COCl_2]^d$$

The equation contains current concentrations of the reactants, $[CO],[Cl_2]$, and$[COCl_2]$, as well as the effective rate constants for the direct and inverse reactions, $k_1$andk_2. The parameters a, b, c, and d can be considered as the formal reaction orders on each reactant. An experiment gave the values of the reaction rates depending on the ratios of component concentration (Table 4.6). It is necessary to derive the unknown parameters of the empirical equation as well as the pre-exponential terms and activation energies of the direct and reverse reactions.

The problem is solved in several steps (Fig. 4.10). First, we will define the reaction orders and rate constants by minimization of the target function of six

Table 4.6 Experimental data for a phosgene photosynthesis reaction ($T = 450$ K)

Current concentrations (M)			v (mol·L^{-1} s^{-1})
$[CO]$	$[Cl_2]$	$[COCl_2]$	
1	1	1	−1
2	1	1	0
1	2	1	0
2	2	1	2.828
1	1	2	−3
2	1	2	−2
2	2	2	0
2	2	2	0
1.5	1.5	1.5	−0.992

$$[CO] := (1\ \ 2\ \ 1\ \ 2\ \ 1\ \ 2\ \ 2\ \ 2\ \ 1.5)^T \qquad [Cl_2] := (1\ \ 1\ \ 2\ \ 2\ \ 1\ \ 1\ \ 2\ \ 2\ \ 1.5)^T$$

$$[COCl_2] := (1\ \ 1\ \ 1\ \ 1\ \ 2\ \ 2\ \ 2\ \ 2\ \ 1.5)^T \qquad v := (-1\ \ 0\ \ 0\ \ 2.828\ \ -3\ \ -2\ \ 0\ \ 0\ \ -0.992)^T$$

$$SSq(k_1, k_2, a, b, c, d) := \sum \overrightarrow{\left[v - \left[k_1 \cdot ([CO])^a \cdot ([Cl_2])^b - k_2 \cdot ([Cl_2])^c \cdot ([COCl_2])^d\right]\right]^2}$$

$$(k_1\ \ k_2\ \ a\ \ b\ \ c\ \ d) := (0.73\ \ 1.68\ \ 1.17\ \ 1.65\ \ 0.48\ \ 1.14)$$

$$(k_1\ \ k_2\ \ a\ \ b\ \ c\ \ d) := \text{Minimize}(SSq, k_1, k_2, a, b, c, d)^T$$

$$(k_1\ \ k_2\ \ a\ \ b\ \ c\ \ d) = (0.98\ \ 1.99\ \ 1.01\ \ 1.53\ \ 0.52\ \ 1)$$

$$\boxed{(a\ \ b\ \ c\ \ d) := \left(1\ \ \frac{3}{2}\ \ \frac{1}{2}\ \ 1\right)}$$

$$SSq1(k_1, k_2) := \sum \overrightarrow{\left[v - \left[k_1 \cdot ([CO])^a \cdot ([Cl_2])^b - k_2 \cdot ([Cl_2])^c \cdot ([COCl_2])^d\right]\right]^2}$$

$$(k_1\ \ k_2) := (0.429\ \ 1.636)$$

$$(k_1\ \ k_2) := \text{Minimize}(SSq1, k_1, k_2) \qquad \boxed{(k_1\ \ k_2) = (1.002\ \ 2.007)}$$

$$R := 8.3144 \qquad T := 450$$

$$SSq2(k_{01}, k_{02}, E_1, E_2) := \sum \overrightarrow{\left[v - \left[k_{01} \cdot e^{\frac{-E_1}{R \cdot T}} \cdot ([CO])^a \cdot ([Cl_2])^b - k_{02} \cdot e^{\frac{-E_2}{R \cdot T}} \cdot ([Cl_2])^c \cdot ([COCl_2])^d\right]\right]^2}$$

$$(k_{01}\ \ k_{02}\ \ E_1\ \ E_2) := (8.553 \times 10^9\ \ 1.447 \times 10^{10}\ \ 8.556 \times 10^4\ \ 8.493 \times 10^4)$$

$$(k_{01}\ \ k_{02}\ \ E_1\ \ E_2) := \text{Minimize}(SSq2, k_{01}, k_{02}, E_1, E_2)$$

$$\boxed{(k_{01}\ \ k_{02}\ \ E_1\ \ E_2) = (8.553 \times 10^9\ \ 1.447 \times 10^{10}\ \ 8.556 \times 10^4\ \ 8.493 \times 10^4)}$$

Fig. 4.10 Derivation of the equation parameters for phosgene synthesis arguments: $SSq(k_1, k_2, a, b, c, d)$. One can see that the calculated values allow rewriting the empirical equation in the following form:

$$v = k_1[CO][Cl_2]^{3/2} - k_2[Cl_2]^{1/2}[COCl_2].$$

We want to note that exactly this form of the equation was proposed in classical *Bodenstein* and *Christiansen* works. In addition, the same equation can be derived by analysis of a multi-step kinetic scheme of the chain mechanism for thermal phosgene synthesis in a quasi-stationary approximation (see Chap. 2).

Then the computed orders are fixed and the target function is reset as a function of two arguments in order to calculate the rate constants more precisely. In the last step, a target function of four arguments is minimized and the values of the pre-exponential terms and effective activation energies are found.

Initial guesses for the optimized parameters were chosen arbitrarily in each step, and the results the `MINIMIZE` function calculations were reused as the initial guesses. This procedure was repeated until the initial guesses did not influence the final result.

We want to stress out one syntax detail that is important for solving inverse kinetic problems in Mathcad. It concerns the way to define the target function. For example, the document in Fig. 4.10 shows a function that contains a vectorization operator. At the same time, the target function can be defined with index variables. This approach is not recommended – constructions with index variables cal lead to exceedingly long computation time, especially when experimental data sets contain many elements.

4.3.1.1 Kinetics of *Trans*-Stilbene Isomerization

Stilbene (1,2-diphenylethylene) obtained by catalytic oxidation of toluene exists as a pair of isomers. The *trans*-isomer is converted into the *cis*-isomer upon UV irradiation:

$$trans - stilbene \underset{K_{-1}}{\overset{k_1}{\rightleftharpoons}} cis - stilbene.$$

Kinetic data for such conversion at some temperature is given in Table 4.7.

Given that each reaction is of the first order, find the rate constants for the reversible isomerization.

In principle, this problem can be solved using data linearization but we want to discuss an approach based on a `MINIMIZE` function. The corresponding document is shown in Fig. 4.11. An expression for the minimized functional `SSE`

Table 4.7 Kinetic data for *trans*-stilbene isomerization

Time (min)	*0*	*20*	*50*	*80*	*120*	*170*	2,000
Cis-isomer content (%)	100.0	92.5	82.3	73.6	63.7	53.8	17.1

$t := \begin{pmatrix} 0 & 20 & 50 & 80 & 120 & 170 & 2\cdot10^3 \end{pmatrix}^T \qquad CIS := \begin{pmatrix} 100 & 92.5 & 82.3 & 73.6 & 63.7 & 53.8 & 17.1 \end{pmatrix}^T$

$TRANS := 100 - CIS$

$$\frac{d}{dt}x(t) = k_{\_1}\cdot(C_0 - x(t)) - k_1\cdot x(t)$$

$$\int_0^{Trans} \frac{1}{k_{\_1}\cdot(C_0 - x) - k_1\cdot x}\,dx = \int_0^{\tau} 1\,dy \text{ solve}, Trans \rightarrow k_{\_1}\cdot C_0\cdot\frac{e^{\tau\cdot k_{\_1}+\tau\cdot k_1} - 1}{e^{\tau\cdot k_{\_1}+\tau\cdot k_1}\cdot(k_{\_1} + k_1)}$$

$C_0 := 100 \qquad Trans(k_1, k_{\_1}, x) := k_{\_1}\cdot C_0\cdot\frac{\exp(x\cdot k_{\_1} + x\cdot k_1) - 1}{\exp(x\cdot k_{\_1} + x\cdot k_1)\cdot(k_{\_1} + k_1)} \qquad i := 0 .. \text{rows}(t) - 1$

$$SSE(k_1, k_{\_1}) := \sum_i (Trans(k_1, k_{\_1}, t_i) - TRANS_i)^2$$

$H := \text{CreateMesh}(SSE, 1.01\cdot10^{-6}, 0.002, 0.002, 0.0055)$

$H1 := \text{CreateMesh}(SSE, 8\cdot10^{-4}, 8.4\cdot10^{-4}, 3.95\cdot10^{-3}, 4.\cdot10^{-3}) \qquad k_1 := 0.001 \qquad k_{\_1} := 0.004$

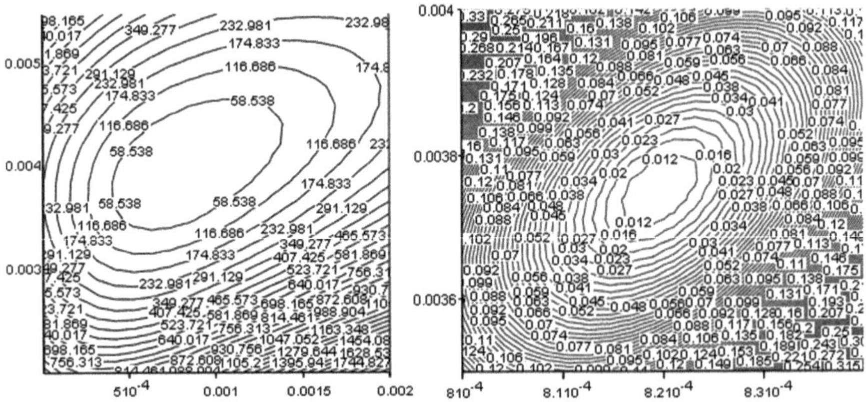

$\begin{pmatrix} k_1 \\ k_{\_1} \end{pmatrix} := \text{Minimize}(SSE, k_1, k_{\_1}) \qquad \begin{pmatrix} k_1 \\ k_{\_1} \end{pmatrix} = \begin{pmatrix} 8.196\times10^{-4} \\ 3.975\times10^{-3} \end{pmatrix} \qquad SSE(k_1, k_{\_1}) = 8.414\times10^{-3}$

$K := \frac{k_1}{k_{\_1}} \qquad K = 0.206$

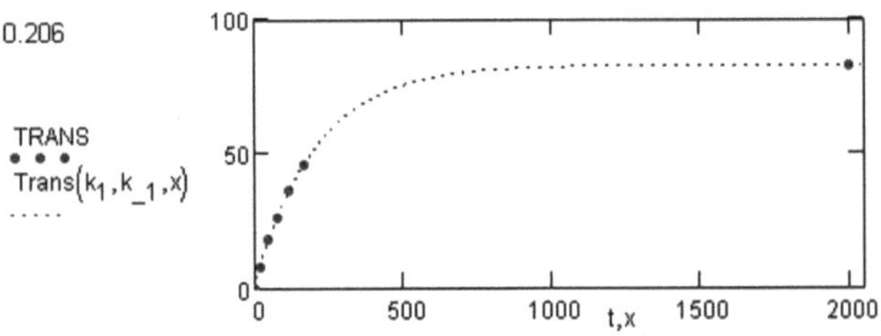

Fig. 4.11 Calculation of the rate constants for reversible stilbene isomerization

(k_1, k_{-1}) includes a function for the *trans*-isomer concentration, `Trans(k1, k-1)`, that was obtained using a symbol processor.

Graphical areas created with a function `CreateMesh` are of particular importance in this document. It has five required arguments (a target function name and ranges for both constants) that provide data for calculation of a contour plot of the target function. One can clearly see an oval region on such contour plot where the sum of square errors has the minimal value inside the chosen range. Apparently, the coordinates of any point inside this oval region can be used as the initial guesses. The second graphical region of the document basically duplicates the first one; it is a magnified version used to show the contour plot minimum more clearly. Consequently, `CreateMesh` provides tools for conscious setting of the initial guesses for the optimized rate constants. The calculation ends with a plot of the calculated kinetic curve for the *trans*-isomer, which shows a satisfactory agreement with the experiment.

It is apparent that the graphical function `CreateMesh` can be used effectively only in a case of two optimization parameters.

4.3.2 *Kinetics of Stepwise Ligand Exchange in Chrome Complexes*

One might think that solution of a reverse problem with two unknown constants should not have any difficulties related to the initial guess setting. We want to show that even though `CreateMesh` gives some advantages, one can face more complex situations due to the features of target function.

An experimenter investigated kinetics of a stepwise ligand exchange reaction in chrome complexes:

$$\underset{A}{Cr(en)_2L_2{}^{3+}} \xrightarrow{k_1} \underset{B}{Cr(en)_2LS^{3+}} \xrightarrow{k_2} \underset{P}{Cr(en)_2S_2{}^{3+}}.$$

Here *en* is ethylenediamine, *L* is dimethylacetamide, and *S* is dimethylsulfoxide. Optical density *D* of the solution was measured at the wavelength $\lambda = 546$ nm in a 1 cm cuvette. Molar extinction coefficients for the discussed complexes were 146, 119, and 100 M^{-1} cm^{-1} respectively. An initial concentration of the $Cr(en)_2L_2{}^{3+}$ complex was 9.6×10^{-3} M. Experimental data is given in Table 4.8. Find the rate constants k_1 and k_2.

This problem is interesting for an experimenter because it demonstrates an important feature of research in kinetics: in many cases it is not necessary to register concentrations of the reaction system components. It is enough to register some parameter (a physical property of the system) that is quantitatively proportional

Table 4.8 Optical density dynamics in a system of chrome complexes

t, s	*0*	*100*	*200*	*300*	*400*	*500*	*600*	800
D	1.402	1.381	1.362	1.343	1.326	1.309	1.295	1.266
t. s	1,000	1,200	1,400	1,600	2,000	2,500	3,000	3,500
D	1.242	1.220	1.183	1.166	1.151	1.121	1.097	1.077
t.c	4,000	4,500	5,000	5,500	6,000	6,500	7,000	7,500
D	1.060	1.047	1.035	1.026	1.017	1.010	1.004	0.988

to the concentration. This principle lies in the basis of many experimental methods of chemical kinetics. In particular, a powerful experimental tool is spectrophotometry – a method based on measuring optical densities of compounds. In the discussed reaction, each component is optically active and adds to the total (and in this case – changing over time) optical density (light absorption) of the system D:

$$D(t) = D_A(t) + D_B(t) + D_P(t).$$

At the same time, the optical density depends on the current component concentrations as represented by the Beer–Lambert law:

$$D(t) = [\varepsilon_A C_A(t) + \varepsilon_B C_B(t) + \varepsilon_P C_P(t)]l.$$

Here ε is a molar extinction coefficient; l is an optical path length (cuvette length). Behavior of the concentrations for the components A, B, and P over time is known from the direct kinetic problem solution (see Chap. 1). After some substitution and simplifications, one can obtain the following equation for optical density in time:

$$D(t) = C_{A_0} l \left[\varepsilon_P + \left(\varepsilon_A + \frac{\varepsilon_B k_1 - \varepsilon_P k_2}{k_2 - k_1} \right) e^{-k_1 7} + \frac{k_1(\varepsilon_P - \varepsilon_B)}{k_2 - k_1} e^{-k_2 t} \right]. \quad (4.15)$$

A target function *SSError* should be defined now. This function represents a sum of squared errors of the experimental optical densities $D_{\exp_m}$ from their theoretical values (obtained from (4.15)):

$$SSError(k_1, k_2, t_m) = \sum_m \left[D(k_1, k_2, t_m) - D_{\exp_m} \right]^2.$$

Here m is a position number in the vector of experimental data.

One way to solve the problem is shown in Fig. 4.12. The data from Table 4.8 – time and optical density values – is put into the vectors `time` and `density` respectively. These initial vectors, together with the constants (the initial concentration of the complex A, the molar extinction coefficients), are hidden to make the document clearer. We do not have any clue as for the initial guesses and so we use the graphical function `CreateMesh` again. This function gives a plot of the sums of squared deviations in a defined range for each rate constant. The limits of the range can be defined through the logarithms of the constants instead of their absolute values; it allows covering a wider range. A graphical zone in Fig. 4.12 shows the plot of the sum of squared deviations in a range $e^{-8} \le k_1 \le e^{-6}$ and $e^{-10} \le k_2 \le e^{0}$. One can clearly see two wells on the plot where the values of `SSError` function are below 0.01. Two random points are chosen inside the wells (the coordinates of the points represent logarithms of the constants k_1 and k_2) and each of them is used as an initial guess for calculations. Two independent

▶ Input Data and constants

$$D(k_1,k_2,t) := A_0\cdot\left[\varepsilon_P + \left(\varepsilon_A + \frac{\varepsilon_B\cdot k_1 - \varepsilon_P\cdot k_2}{k_2 - k_1}\right)\cdot e^{-k_1\cdot t} + k_1\cdot\frac{\varepsilon_P - \varepsilon_B}{k_2 - k_1}\cdot e^{-k_2\cdot t}\right]$$

$$SSE(lnk_1, lnk_2) := \sum \overrightarrow{\left[\left(D\left(e^{lnk_1}, e^{lnk_2}, time\right) - density\right)^2\right]}$$

H := CreateMesh(SSE,−8,−6,−10,0,100,50)

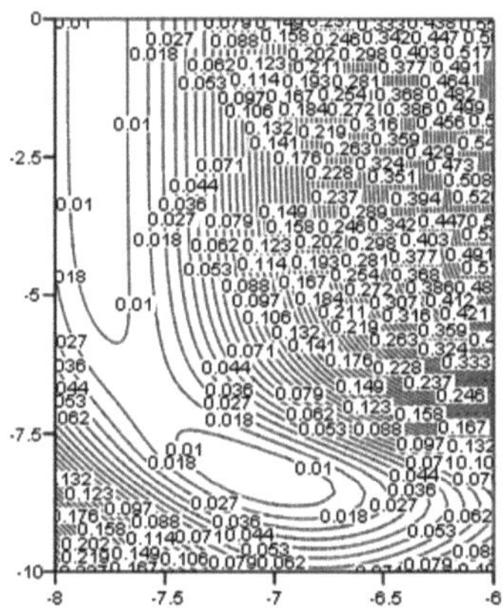

H

$$\begin{pmatrix} k_1 \\ k_2 \end{pmatrix} := \begin{pmatrix} e^{-7.8} \\ e^{-2} \end{pmatrix} \qquad \begin{pmatrix} k_1 \\ k_2 \end{pmatrix} = \begin{pmatrix} 4.097\times 10^{-4} \\ 0.135 \end{pmatrix}$$

$$\begin{pmatrix} k_1 \\ k_2 \end{pmatrix} := \text{Minimize}(SSE, k_1, k_2) \qquad \begin{pmatrix} k_1 \\ k_2 \end{pmatrix} = \begin{pmatrix} 4.097\times 10^{-4} \\ 0.135 \end{pmatrix}$$

$$\boxed{SSE(\ln(k_1), \ln(k_2)) = 4.63\times 10^{-3}}$$

$$\begin{pmatrix} k_{1\_2} \\ k_{2\_2} \end{pmatrix} := \begin{pmatrix} e^{-7} \\ e^{-8} \end{pmatrix} \qquad \begin{pmatrix} k_{1\_2} \\ k_{2\_2} \end{pmatrix} := \begin{pmatrix} 8.562\times 10^{-4} \\ 2.587\times 10^{-4} \end{pmatrix}$$

$$\begin{pmatrix} k_{1\_2} \\ k_{2\_2} \end{pmatrix} := \text{Minimize}(SSE, k_{1\_2}, k_{2\_2}) \qquad \begin{pmatrix} k_{1\_2} \\ k_{2\_2} \end{pmatrix} = \begin{pmatrix} 8.562\times 10^{-4} \\ 2.587\times 10^{-4} \end{pmatrix}$$

$$\boxed{SSE(\ln(k_{1\_2}), \ln(k_{2\_2})) = 5.335\times 10^{-4}}$$

$$t := 0, \frac{\max(time)}{100} .. \max(time)$$

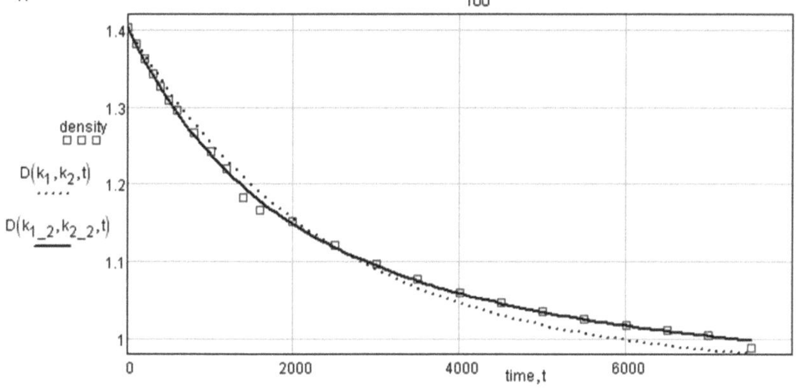

Fig. 4.12 Calculation of rate constant for stepwise ligand exchange

calculations of the rate constants are then performed with different initial guesses. The results, shown in Fig. 4.12, indicate that choosing the first well (top left region of the graphic zone) was a mistake: the minimum, which has been found inside the well, is a local one. The last plot in the document confirms this as well: it shows a change of the optical density over time (lines) overlapped with the experiment data (points).

This example has shown that an incorrect choice of initial guesses can lead to an erroneous result. The MINIMIZE function uses gradient algorithms to find a minimum. The main disadvantage of these algorithms is that they can lead to a set of the sought parameters that correspond to a local minimum and not the global one. In this case, the built-in function `CreateMesh` has helped us to find several minima but, obviously, such assistance is possible only in a case when the minimization is performed on two parameters. An increase in dimensionality of the optimization problem (i.e. an increase in the number of the unknown rate constants) may lead to serious issues in the solution process.

In any case, one might need more sophisticated algorithms for correct solution of inverse problems than the ones built-in into Mathcad (e.g. `MINIMIZE`).

In summary, the standard algorithms of gradient descent start with some initial guess chosen either arbitrarily or using some auxiliary method. During each local descent step, the current solution is changed to the neighboring one with a smaller target function value until a local minimum is reached. That is why the local descent algorithms do not allow finding the global minimum of the problem.

Optimization methods recently received a powerful incentive from essentially new algorithmic schemes that mimic life and nature. This has changed the global conception of optimization algorithm creation; now it is not dominated by the requirement of the monotonous improvement of a target function. Here we speak about the so called *metaheuristics* – probabilistic algorithm schemes for that are applicable to any optimization problem. Analogies with natural processes lead to somewhat exotic names of these algorithms. For example, metaheuristics include *simulated annealing*, *tabu search*, *genetic* algorithms, *neural networks*, probabilistic *greedy* algorithms, etc. Let us briefly discuss some of them.

Tabu search: The algorithm is based on the idea that the computation does not stop in a local optimum point but goes from one optimum to another in search of the global optimum. The main tool that allows passing a local minimum is the so called *Tabu* list. It is created from the preceding results of the calculation and "prohibits" a part of the current solution vicinity. The tabu list depends on the problem and excludes the solution "fragments" (graph nodes, coordinates of a vector) that were changed during the preceding l steps of the algorithm. The parameter l defines the "memory" of the algorithm. In a case of "short" memory ($l = 0$) the algorithm degenerates into the standard gradient descent.

Simulated annealing algorithm: The exotic name of this algorithm is related to imitative modeling methods of statistical physics that are based on Monte Carlo method. Investigation of crystal lattices and atom behavior during a slow cooling process has led to the creation of probabilistic algorithms that are effective in combinatorial optimization. Simulated annealing belongs to the class of *threshold algorithms*. On each step of this algorithm, a new solution is chosen in a vicinity of the current one. If the difference between the values of the target function of the new and the current solutions is below the defined threshold, than the new solution substitutes the current one. Otherwise, a new neighboring solution is chosen. This is a type of search where an arbitrary worsening of the target function is allowed. However, it has been proven that the possibility of such transition is inversely proportional to the worsening magnitude.

Mathcad has a function `anneal` that is based on simulated annealing. This function is a part of the `Numerical Recipes` extension pack. The website of the Mathcad 12 guide demonstrates a solution of classical optimization problem – the travelling salesman problem.

The idea of *genetic* algorithms is borrowed from the living nature. In this approach, an evolution process is organized, the ultimate goal of which is to find the optimal solution of a complex combinatorial problem. A developer of the algorithm plays a role of the "Creator" who needs to set the correct rules of evolution in order to achieve the goal as soon as possible. Genetic algorithm starts with the formation of starting population – a finite set of possible problem solutions. It is noteworthy that a choice of the starting population is not of ultimate importance for the process convergence – it defines only the computation time. During each evolution step, two "parent" solutions are chosen using a probabilistic selection operator. A crossover operator converts these solutions into a new one. The latter is then modified slightly in a process of mutation. Finally, this solution is added to the population, whereas the solution with the smallest value of the target function is deleted from the population.

One of the genetic algorithm modifications is implemented in Mathcad as a user function MGA (*Sabanin V. R. et al.*). Its description can be found in literature. A real-life demonstration of the function is available online on Mathcad Calculation Server (http://twt.mpei.ac.ru/MAS/Worksheets/Minimum.mcd).

Figure 4.13 shows a document that calculates rate constants for stepwise ligand exchange using a genetic algorithm. One can see that the vectors of rate constant vicinities ($\ln k_{in}$, $\ln k_{fin}$) are defined instead of the vectors of initial guesses. A parameter μ defines the size of the initial population. We have already seen that there are both the local and global minima in this vicinity. Nonetheless, the results of calculation indicate that the MGA function was able to find namely the global minimum.

4.3.3 Computing Kinetic Parameters Using Non-Linear Approximation Tools

If a kinetic curve equation is described by some analytical expression that cannot be transformed into a linear form, than the kinetic parameters of this equation can be found using a genfit function of Mathcad.

This function should be discussed in detail. First, it has proven itself as a valuable and powerful tool starting from the early versions of Mathcad (interestingly enough, such a versatile suite as Maple did not have similar function for non-linear approximation, at least until the Maple 10 version). Additionally, the genfit function has changed a lot with the release of Mathcad 13.

Figure 4.14 shows a solution of inverse kinetic problem for a two-step consecutive reaction in Mathcad 11. The solution uses the known concentrations of an intermediate at different moments of time (this data is read from the corresponding data file and placed into vectors Be and te respectively). The analytical expression

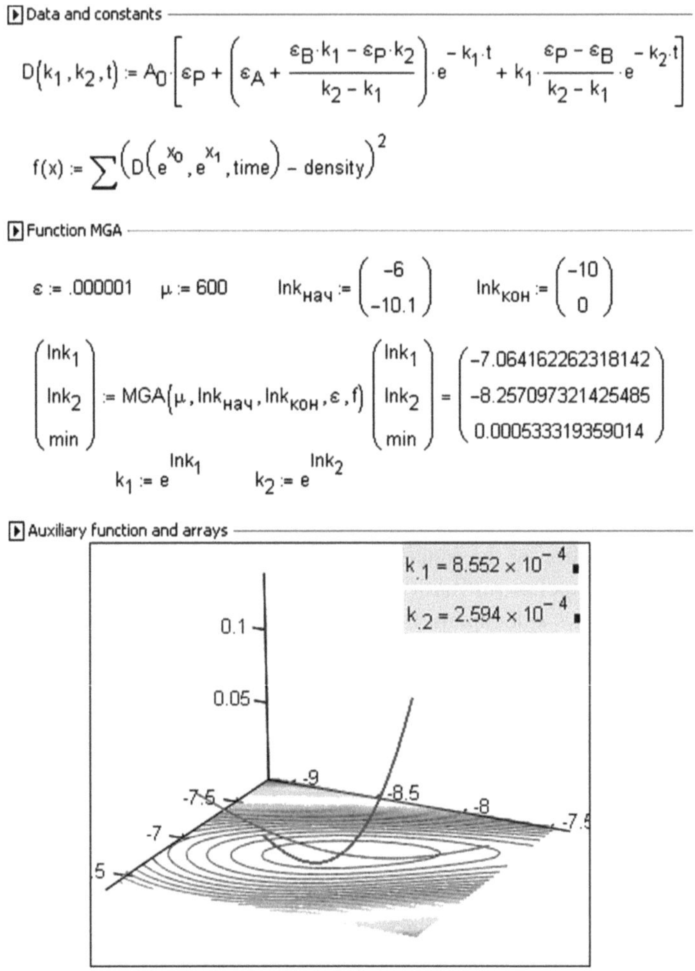

Fig. 4.13 Rate constant calculation using a genetic algorithm

for the concentration of compound B is known from the solution of direct kinetic problem:

$$A \xrightarrow{k_1} B \xrightarrow{k_2} P,$$

$$C_B(t) = C_{A_0} \frac{k_1}{k_2 - k_1} \left(e^{-k_1 t} - e^{-k_2 t} \right). \tag{4.16}$$

It is also known that the initial concentration of the compound A was $C_{A_0} = 0.241$ M.

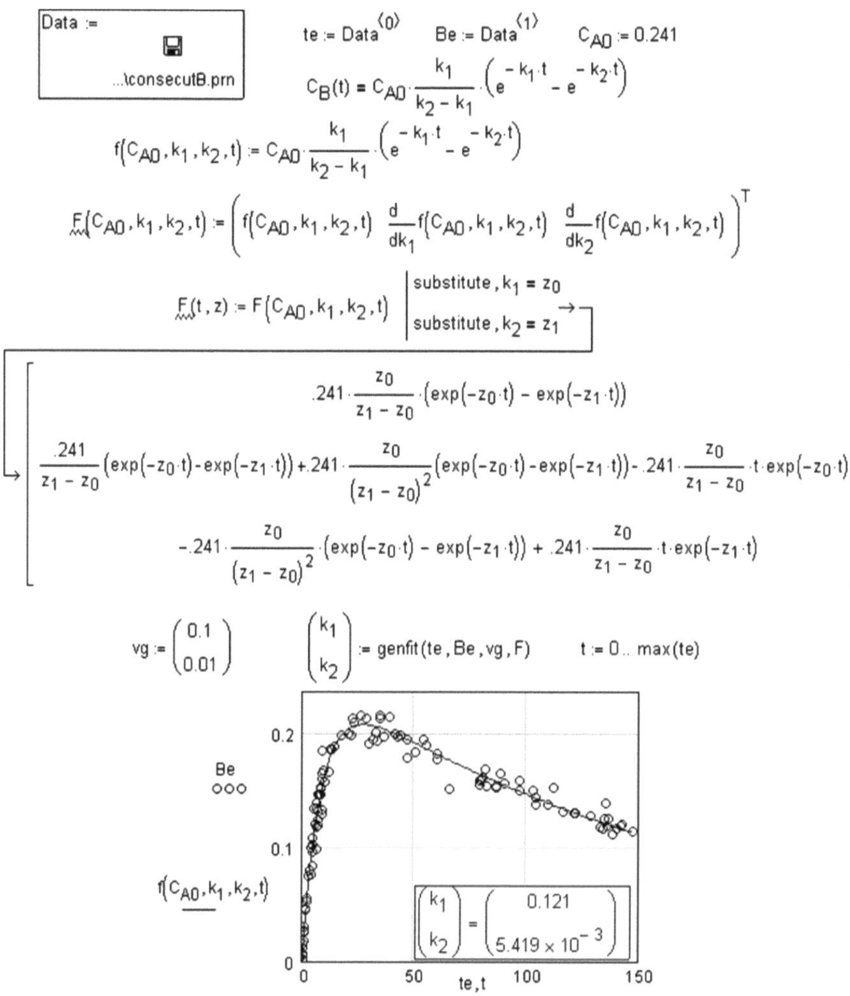

Fig. 4.14 Solution of inverse kinetic problem in Mathcad 11 using `genfit`

Mathcad 11 has successfully solved the problem as can be seen from comparison of the experimental data with the calculated kinetic curve. However, application of `genfit` required rather intensive preliminary work, since it was necessary to find the analytical expressions for partial derivatives of the function (4.16). Only two parameters are searched in this case and it required quite bulky constructions to be created. In Mathcad 12 and older versions, a user had to explicitly define a function vector containing the approximation expression as well as its partial derivatives on the approximation parameters. Additionally, this function had to have two arguments: the first one being the independent variable, and the second one being the vector of sought parameters. However, the main inconvenience was in the fact that if the approximation function `f(t,z)` was defined according to

Mathcad syntax, it could not be used for the formation of the function vector `F(x, z)` – the Mathcad suite is not capable of index variable differentiation.

The authors of this book were able to somewhat improve the function `genfit` in its network version on `MCS`. Figure 4.15 shows a part of the document that is

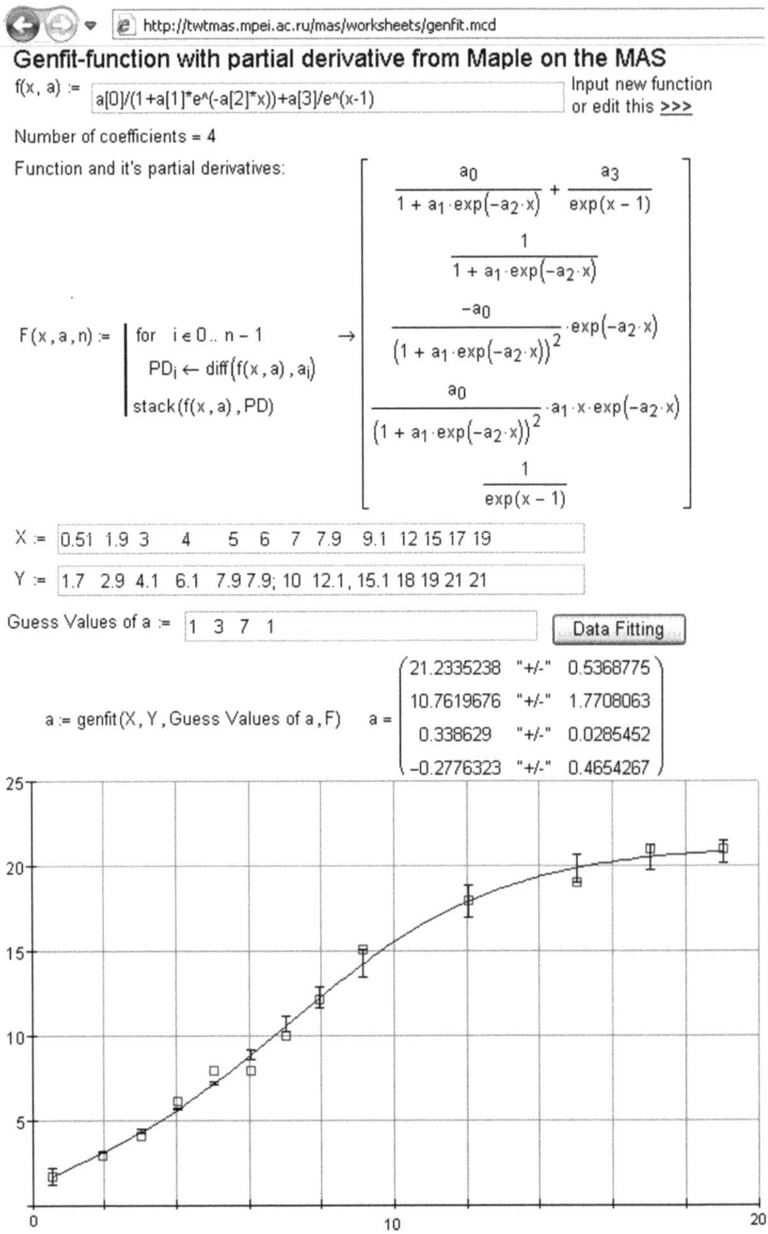

Fig. 4.15 Network version of `genfit` function

available online on http://twt.mpei.ac.ru/mas/woksheets/genfit.mcd. The user needs only to input the fitting function, the experimental data and the initial guesses for unknown parameters into the corresponding fields. The function vector, required for `genfit`, is generated automatically. In addition, the document calculates confidence intervals for the found parameters. We will not discuss the essence of these improvements as they are based on undocumented features of the Mathcad 11.2 symbol engine. It is more important that the document works and can solve many kinetic problems quite efficiently.

In Mathcad 13, the function `genfit` has been significantly changed by developers. The user does not need to routinely (and tediously) create the function vector anymore. An expression for the approximation function is enough for computation of the sought parameters.

The reader can compare the documents in Figs. 4.14 and 4.16 that solve the same kinetic problem.

The calculation algorithm, used by `genfit`, was also changed significantly (Fig. 4.17). One can now choose between options `LevenbergMarquardt` and `Optimized Levenberg Marquardt` (the latter is recommended when the function vector is absent).

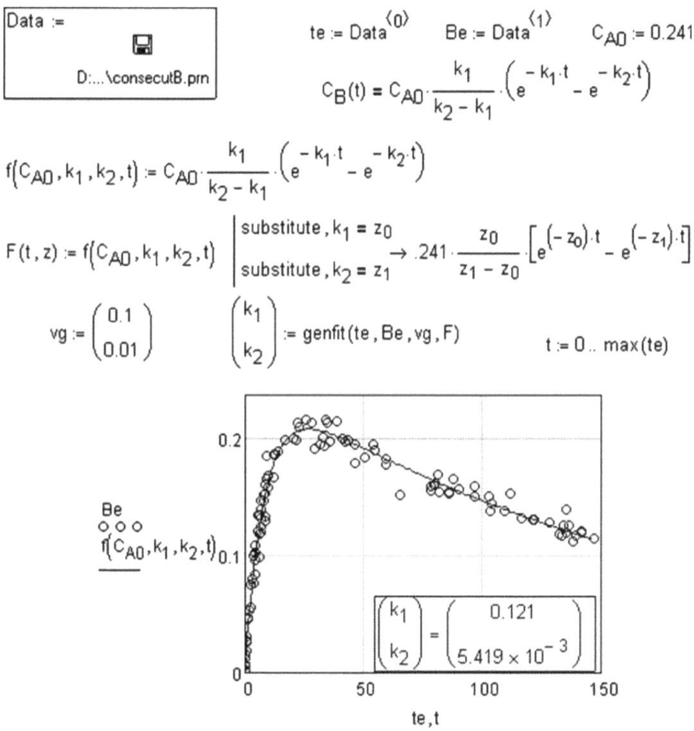

Fig. 4.16 Solution of the inverse kinetic problem in Mathcad 13 using `genfit` function

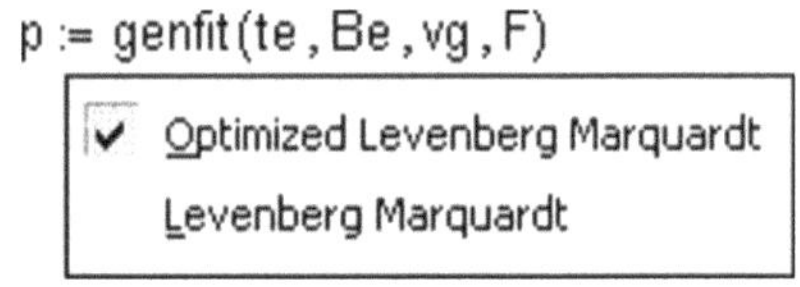

Fig. 4.17 Algorithms of `genfit` function implemented in Mathcad 13/14

An extension pack `DAEP (Data Analysis Extension Pack)` includes some functions that might be useful for solving inverse problems requiring non-linear fitting. This pack is automatically installed with the latest Mathcad versions. In Chap. 1, we obtained a solution for direct kinetic problem of the following consecutive second-order reaction:

$$A + B \xrightarrow{k_1} C,$$
$$C \xrightarrow{k_2} P.$$

It has been shown that the concentration of the final product changes over time according to a quite complex equation:

$$C_p(t) = C_{A_0}C_{B_0}(C_{A_0} - C_{B_0})k_1k_2 \int_0^t \int_0^u \frac{e^{\left[k_2+k_1\left(C_{A_0}-C_{B_0}\right)\right]u}}{\left[C_{A_0}e^{k_1\left(C_{A_0}-C_{B_0}\right)u} - C_{B_0}\right]^2} du\, e^{-k_2u} du.$$

Figure 4.18 shows a solution for a problem of finding the kinetic constants k_1 and k_2 from the experiment data contained in the vectors t and P (initial concentrations of the reagents and their combinations are represented numerically in this document).

The problem is solved using the function `genfit` as well as a DEAP function `LeastSquareFit`. One can see (Fig. 4.18) that `genfit` returns an approximation coefficient vector, whereas `LeastSquareFit` gives a matrix containing the sought parameters in the zero column, the lower limit of a confidence interval in the first column, and the upper limit in the second column.

Certainly, a problem of setting the correct initial guess values still remains and directly defines the overall success of the calculation.

A built-in function for non-linear fit, `rationalfit`, (from the `DAEP` pack) might be useful for solving some inverse problems. Assume it is necessary to define the rate constant and initial reagent concentration for some second-order reaction described by the following kinetic equation:

$$C_A(t) = \frac{C_0}{1 + kC_0t}.$$

Substitute $\beta_0 = C_0$ and $\beta_1 = kC_0$; it gives a fractionally rational function:

$$C_A(t) = \frac{\beta_0}{1 + \beta_1 t}.$$

Experiment (vectors t, P)

t := stack(0,.12,.13,.14,.23,.45,.65,1.1,1.5,2.4,3,4.02,4.40,5,5.90,6.12,7.43,8.99,9.15,10.00)

P := stack(0,2.1,2.9,3.4,8.6,29,55,128,202,377,500,667,726,813,928,953,1063,1199,1211,1259)$\cdot 10^{-3}$

Experiment (vectors t, P)

Fitting Function

$$f(x,z) := \frac{3}{4}\cdot z_0\cdot z_1\cdot\int_0^x\int_0^u \frac{e^{\left(\frac{-z_0}{2}+z_1\right)\cdot u}}{\left(\frac{3}{2}\cdot e^{\frac{-z_0}{2}\cdot u}-2\right)^2}\,du\cdot e^{-z_1\cdot u}\,du$$

$$guess := \begin{pmatrix}0.3\\0.2\end{pmatrix}$$

S := genfit(t, P, guess, f)

Res := LeastSquaresFit(t, P, f, guess, 0.95)

$y(x) := f(x,S) \quad fit(x) := f\left(x, Res^{\langle 0\rangle}\right)$

$f1(x) := f\left(x, Res^{\langle 1\rangle}\right) \quad f2(x) := f\left(x, Res^{\langle 2\rangle}\right)$

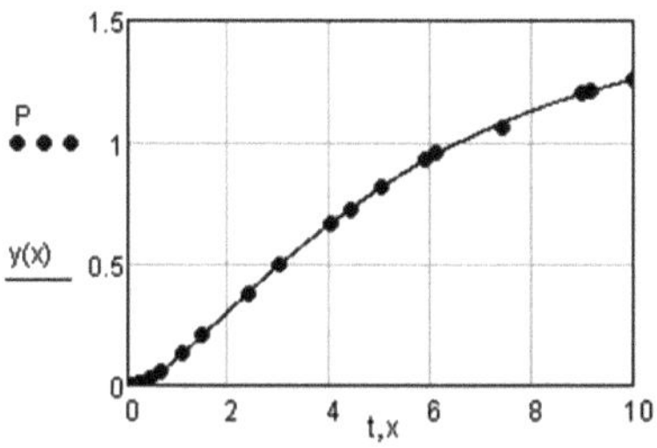

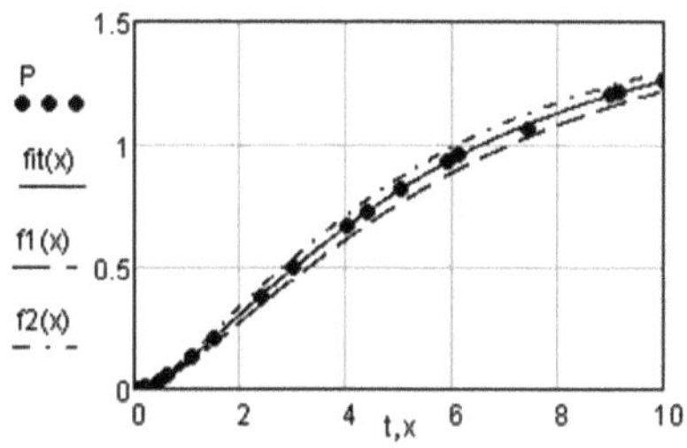

$$S = \begin{pmatrix}0.558\\0.237\end{pmatrix} \qquad Res = \begin{pmatrix}0.558 & 0.487 & 0.63\\0.237 & 0.225 & 0.25\end{pmatrix}$$

Fig. 4.18 Usage of `genfit` and `LeatSquaresFit` functions (Mathcad 14)

t := stack(1,2,3,5,7,10,15,20,25)

C := stack(0.00920,0.008520,.0070,0.0070,.00550,0.00464,.00363,0.0029,0.00254)

$m := 0 \quad n := 1 \qquad Y(x,\beta) := \frac{\beta_0}{1+\beta_1\cdot x}$ Res := rationalfit(t, C, 0.95, m, n)

$$fit(x) := Y\left(x, Res^{\langle 0\rangle}\right) \qquad Res = \begin{pmatrix}0.01034 & 0.01031 & 0.01036\\0.12186 & 0.12108 & 0.12263\end{pmatrix}$$

$$x := 0, \frac{max(t)}{100} .. max(t) \quad corr\left(\overrightarrow{fit(t)}, C\right) = 0.9923 \quad \sum\left(C-\overrightarrow{fit(t)}\right)^2 = 7.198\times 10^{-7}$$

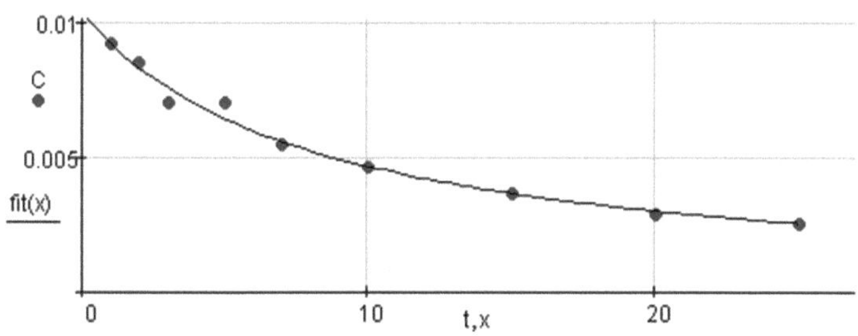

Fig. 4.19 Usage of `rationalfit` function from the DAEP pack

The coefficients β_0 and β_1 are found using a function `rationalfit(vx, vy,P,f,m,n)`, where `vx` and `vy` are the experimental data vectors, `P` is the confidence level, and `m` and `n` are the highest powers in the nominator and denominator of the approximation function `f`, correspondingly. A part of the solution is shown in Fig. 4.19. An obvious advantage of this approach is that it does not require setting the initial guesses for the sought parameters. Then, with the β_0 and β_1 values in hand, one can easily calculate the rate constant k and the initial reagent concentration. As with the `LeastSquaresFit` function, the upper and the lower limits of the confidence interval are found for each parameter using the given confidence level.

4.4 Universal Approaches to Inverse Chemical Kinetics Problem

4.4.1 *Reversible Reaction with Dimerization of an Intermediate*

We admit that all the examples, discussed in this chapter, represent only particular cases of inverse problems in chemical kinetics. One needs some universal approach that could be applied to the majority of the inverse problems, if not all of them. Obviously, such approach has to be solely based on numerical computations as the mathematical models for the majority of reaction mechanisms cannot be analytically integrated.

Consider the problem of finding the kinetic parameters for a kinetic scheme where the final product D is formed by reversible dimerization of the intermediate C.

$$A + B \underset{k_2}{\overset{k_1}{\rightleftarrows}} C; \quad C + C \underset{k_4}{\overset{k_3}{\rightleftarrows}} D.$$

The reaction mixture initially consists of the reagents A and B with the concentrations 2 and 3 M respectively. Concentrations of the compounds C and D were experimentally registered at the specific moments of time t (Table 4.9). It is necessary to define the rate constants for each step – k_1, k_2, k_3, and k_4.

This problem can be found in a manual for the `Mlab` software suite developed by Civilized, Inc (should not be confused with `MATLAB`). The problem is really

Table 4.9 Kinetic data for a reaction with reversible dimerization of an intermediate

t, min	0	7	14	21	28	35
C_C, M	0	1.0650	1.3830	0.9793	1.1070	9.7289
C_D, M	0	0.0058	0.2203	0.4019	0.3638	0.4650
t, min	42	49	56	63	70	–
C_C, M	0.7236	0.4674	0.6031	0.6149	0.3369	–
C_D, M	0.5014	0.7150	0.4723	0.7219	0.7294	–

interesting due to several reasons. First, information about the current concentrations of the components A and B is absent. Second, the mathematical model for the reaction is a system of non-linear differential equations that cannot be solved analytically. Third, the experimental data vary significantly. Fourth, there is no any information that would quantitatively describe equilibrium of the system and give a clue about the initial guesses for sought parameters (at least – the ratios of constants). Finally, the interest is attracted by a claim of the suite developers who say, "Only MLAB can solve a kinetics-modeling problem like this so easily!"

This reaction scheme was previously discussed in Sect. 2.1 and we proved that its mathematical model is defined by a system of two differential equations,

$$\begin{cases} dC_C/dt = -k_1 C_A C_B - k_2 C_C - 2k_3 {C_C}^2 + 2k_4 C_D \\ dC_D/dt = k_3 {C_C}^2 - k_4 C_D \end{cases},$$

in which the concentrations of A and B can be expressed from the current concentrations of the intermediate C and the product D:

$$\begin{cases} dC_C/dt = -k_1 (C_{A_0} - C_C - 2C_D)(C_{B_0} - C_C - 2C_D) - k_2 C_C - 2k_3 {C_C}^2 + 2k_4 C_D \\ dC_D/dt = k_3 {C_C}^2 - k_4 C_D \end{cases}.$$

Let's solve the problem in the Mathcad environment. One of the solutions for kinetic parameters is shown in Fig. 4.20. In fact, the beginning of the calculations corresponds to a solution of direct kinetic problem with arbitrary values of the constants (more precisely, logarithms of the constants) as initial guesses. Note an interesting approach implemented here to form the vector of the "right parts" of the differential equation system. Rate constants are used as *the functions on time*; therefore, the vector includes four null elements (according to the number of constants) that represent the right parts of four differential equations:

$$\frac{dk_1(t)}{dt} = \frac{dk_2(t)}{dt} = \frac{dk_3(t)}{dt} = \frac{dk_4(t)}{dt} = 0.$$

The new system of differential equations is then solved numerically with a built-in integrating function `Rkadapt`. This calculation step serves the two goals. First, the initial guesses are obtained that allow for more or less good fit of the calculated kinetic curves and experimental data (first figure). Second, the computed result is assigned to a user function Z ($\mathtt{lnk}_1$, $\mathtt{lnk}_2$, $\mathtt{lnk}_3$, $\mathtt{lnk}_4$). This function is used to calculate the sum of square errors, S ($\mathtt{lnk}_1$, $\mathtt{lnk}_2$, $\mathtt{lnk}_3$, $\mathtt{lnk}_4$), which is then minimized (`Minimize` with a `Conjugate Gradient` option). Such way of defining the target function forces Mathcad to solve both the direct and inverse problems during minimization, thus giving the optimal values of rate constants. The initial guesses vector is updated with the new values until the sum of square deviations stop changing significantly. In the end, the direct problem solution provides calculated kinetic curves and compares them to experimental data.

DATA

$$f(t,x) := \begin{bmatrix} e^{x_2}\cdot(A_0 - x_0 - 2\cdot x_1)(B_0 - x_0 - 2\,x_1) - e^{x_3}\cdot x_0 - 2\left[e^{x_4}\cdot(x_0)^2 - x_1\cdot e^{x_5}\right] \\ e^{x_4}(x_0)^2 - e^{x_5}\,x_1 \\ 0 \\ 0 \\ 0 \\ 0 \end{bmatrix}$$

$lnk1 := \ln(0.01)$
$lnk2 := \ln(0.002)$
$lnk3 := \ln(0.02)$
$lnk4 := \ln(0.002)$

$$Z(lnk1, lnk2, lnk3, lnk4) := Rkadapt\left[(0\ \ 0\ \ lnk1\ \ lnk2\ \ lnk3\ \ lnk4)^T, 0, 70, 10, f\right]$$

$tt := Z(lnk1, lnk2, lnk3, lnk4)^{\langle 0\rangle}$ $\quad C_n := Z(lnk1, lnk2, lnk3, lnk4)^{\langle 1\rangle}$ $\quad D_n := Z(lnk1, lnk2, lnk3, lnk4)^{\langle 2\rangle}$

$i := 0 .. rows(t) - 1$

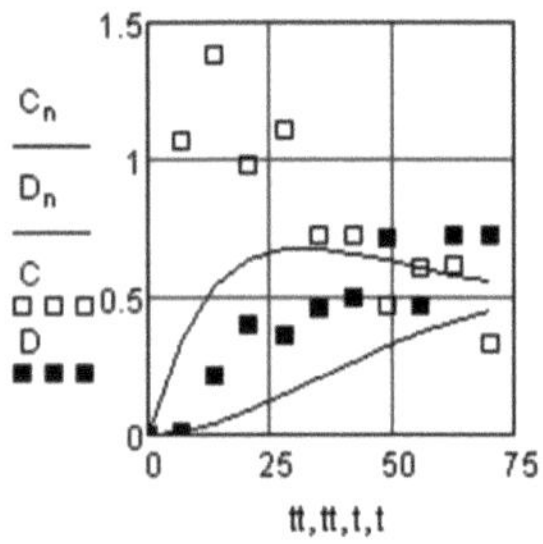

$$S(lnk1, lnk2, lnk3, lnk4) := \left[\sum_i \left(Z(lnk1, lnk2, lnk3, lnk4)_{i,1} - C_i\right)^2 \ldots + \sum_i \left(Z(lnk1, lnk2, lnk3, lnk4)_{i,2} - D_i\right)^2\right]$$

$S(lnk1, lnk2, lnk3, lnk4) = 2.247$ $\quad start := time(0)\cdot sec$

$$\begin{pmatrix} lnk1 \\ lnk2 \\ lnk3 \\ lnk4 \end{pmatrix} := \begin{pmatrix} -2.66845844490474 \\ -4.59747526639913 \\ -4.25044061783608 \\ -24.7488636284988 \end{pmatrix}$$

Conjugate Gradient

$$\begin{pmatrix} lnk1 \\ lnk2 \\ lnk3 \\ lnk4 \end{pmatrix} := Minimize(S, lnk1, lnk2, lnk3, lnk4) \qquad \begin{pmatrix} lnk1 \\ lnk2 \\ lnk3 \\ lnk4 \end{pmatrix} = \begin{pmatrix} -2.668458438663634 \\ -4.597475444192439 \\ -4.250440755259043 \\ -24.748863628570135 \end{pmatrix}$$

$S(lnk1, lnk2, lnk3, lnk4) = 0.22244915033108$

$k_1 := e^{lnk1}$ $\quad k_2 := e^{lnk2}$ $\quad k_3 := e^{lnk3}$ $\quad k_4 := e^{lnk4}$ $\quad$ $S(lnk1, lnk2, lnk3, lnk4) = 0.22244915033108$

$finish := time(0)\cdot sec$ $\quad Calc\_time := finish - start$ $\quad Calc\_time = 4.703\cdot s$

$k_1 = 6.93591\times 10^{-2}$ $\quad k_2 = 1.00772\times 10^{-2}$ $\quad k_3 = 1.42579\times 10^{-2}$ $\quad k_4 = 1.78527\times 10^{-11}$

$$Z1 := Rkadapt\left[(0\ \ 0\ \ lnk1\ \ lnk2\ \ lnk3\ \ lnk4)^T, 0, 70, 1000, f\right]$$

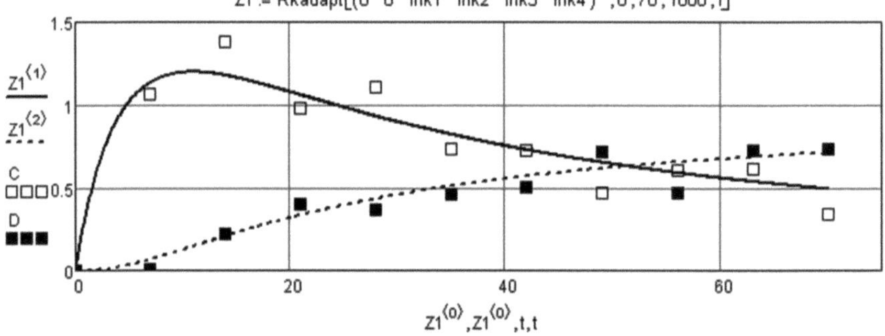

Fig. 4.20 Solution of inverse problem using `Rkadapt` and `Minimize` functions

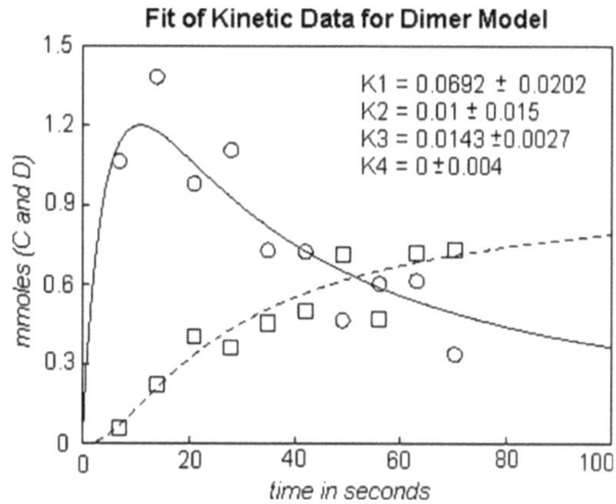

Fig. 4.21 Solution of the problem in MLAB environment

The rate constants calculated this way in Mathcad are inside the confidence interval of the `Mlab` solution. According to Fig. 4.21, $k_1 = 0.0692 \pm 0.0202$; $k_2 = 0.01 \pm 0.015$, $k_3 = 0.0143 \pm 0.027$, $k_4 = 3.458 \times 10^{-20} \pm 3.96 \times 10^{-3}$. It allows us to say that the discussed approach gives a satisfactory result. However, we have to mention that this algorithm can take a while (up to several hours) when a multi-step process with many reactants is modeled, particularly when experimental data set is large.

The calculation time can be decreased significantly if the target function S(lnk$_1$, lnk$_2$, lnk$_3$, lnk$_4$) is not created at all and the user function Z(lnk$_1$, lnk$_2$, lnk$_3$, lnk$_4$) is used inside the solving block `Given/Minerr`. This approach is shown in Fig. 4.22. We can see that the computation took 40 times less – without any decrease in precision.

`Given/Minerr` solving block.

Figure 4.23 contains a solution of another kinetic problem – defining three rate constants for a consecutive first-order reaction with a reversible second step. The unknown constants are calculated analogously to the previous example. However, `GIVEN/ODESOLVE` solving block is used instead of the built-in integrating function. This method has an advantage: it allows to quickly change an integration method by setting the corresponding option of `ODESOLVE` without affecting the document itself. However, one cannot use the vector of ODE system right parts in this approach, and this is not always convenient. Consider, for example, a model with a stoichiometric matrix and a rate constant vector.

4.4.1.1 Thermal Decomposition of Disilane

It is interesting to check Mathcad ability to solve inverse kinetic problems with stoichiometric matrices of high dimensionality. Consider the following problem.

DATA

$$f(t,x) := \begin{bmatrix} e^{x_2}(A_0 - x_0 - 2\cdot x_1)(B_0 - x_0 - 2\,x_1) - e^{x_3}\cdot x_0 - 2\left[e^{x_4}\cdot(x_0)^2 - x_1\cdot e^{x_5}\right] \\ e^{x_4}(x_0)^2 - e^{x_5}\,x_1 \\ 0 \\ 0 \\ 0 \\ 0 \end{bmatrix} \qquad \begin{pmatrix} lnk1 \\ lnk2 \\ lnk3 \\ lnk4 \end{pmatrix} := \begin{pmatrix} -2.668 \\ -4.597 \\ -4.25 \\ -6154 \end{pmatrix}$$

$$Z(lnk1, lnk2, lnk3, lnk4) := Rkadapt\left[(0\ \ 0\ \ lnk1\ \ lnk2\ \ lnk3\ \ lnk4)^T, 0, 70, 10, f\right]$$

$$tt := Z(lnk1, lnk2, lnk3, lnk4)^{\langle 0\rangle} \qquad C_n := Z(lnk1, lnk2, lnk3, lnk4)^{\langle 1\rangle} \qquad D_n := Z(lnk1, lnk2, lnk3, lnk4)^{\langle 2\rangle}$$

$i := 0 .. rows(t) - 1 \qquad start := time(0)\cdot sec$

Don't create SSE's in order to minimize them. It's slow and inaccurate.

Given

$$\left|\begin{array}{l} S \leftarrow Z(lnk1, lnk2, lnk3, lnk4) \\ augment\left(S^{\langle 1\rangle}, S^{\langle 2\rangle}\right) \end{array}\right. = augment(C, D)$$

$$\begin{pmatrix} lnk1 \\ lnk2 \\ lnk3 \\ lnk4 \end{pmatrix} := Minerr(lnk1, lnk2, lnk3, lnk4) = \begin{pmatrix} -2.6684897050930974 \\ -4.597530723022669 \\ -4.250440622796833 \\ -6.153892679652555\times 10^3 \end{pmatrix}$$

$ERR^2 = 0.22244914994428 \qquad k_1 := e^{lnk1} \qquad k_2 := e^{lnk2} \qquad k_3 := e^{lnk3} \qquad k_4 := e^{lnk4}$

$finish := time(0)\cdot sec \qquad Calc\_time := finish - start \qquad Calc\_time = 0.501\cdot s$

$k_1 = 6.93569\times 10^{-2}$ $k_2 = 1.00767\times 10^{-2}$ $k_3 = 1.4258\times 10^{-2}$ $k_4 = 0$

$$Z1 := Rkadapt\left[(0\ \ 0\ \ lnk1\ \ lnk2\ \ lnk3\ \ lnk4)^T, 0, 70, 1000, f\right]$$

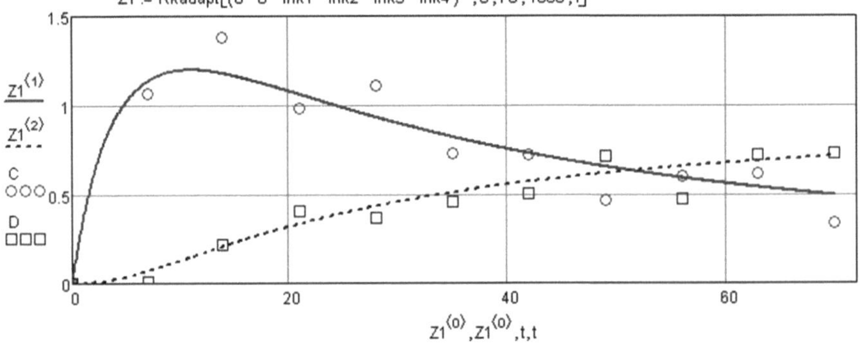

Fig. 4.22 Calculation of the rate constants using the `Rkadapt` function

Mick et al. investigated thermal decomposition of disilane Si_2H_6. The reaction is the following:

$$Si_2H_6 \longrightarrow SiH_4 + SiH_2.$$

The reagent was rapidly (less than in 1 μs) heated to 1,200 K using a special experimental methodic. This initiated disilane decomposition. Concentration of SiH_2 was measured as a function of time from the initiation ($t = 0$) to $t = 100$ μs.

	0	1	2	3
0	0	1	0	0
1	0.020543	0.8175	0.1578	0.0247
2	0.039978	0.6727	0.246	0.0813
3	0.059347	0.5538	0.2984	0.1478
4	0.078943	0.4509	0.3312	0.2179
5	0.099077	0.3695	0.3362	0.2943
6	0.12023	0.3002	0.3306	0.3692
7	0.14139	0.2363	0.3304	0.4333
8	0.1643	0.1906	0.317	0.4924
9	0.18788	0.1533	0.3074	0.5393
10	0.21281	0.1194	0.2954	0.5852
11	0.23698	0.0931	0.2892	0.6177
12	0.26292	0.0728	0.2842	0.643
13	0.29031	0.0562	0.28	0.6638
14	0.31787	0.0432	0.27	0.6868
15	0.34602	0.0343	0.2636	0.7021
16	0.37686	0.0254	0.2628	0.7118
17	0.40634	0.019	0.2622	0.7188
18	0.43691	0.0137	0.254	0.7323
19	0.47015	$9.8\cdot10^{-3}$	0.2518	0.7384
20	0.50899	$6.7\cdot10^{-3}$	0.2514	0.7419

$te := Data^{\langle 0\rangle}$ $\quad v(f,t) := \overrightarrow{f(t)}$

$Ae := Data^{\langle 1\rangle}$ $\quad A_0 := 1$

$Be := Data^{\langle 2\rangle}$ $\quad B_0 := Be_0$ $\quad \begin{pmatrix} k_1 \\ k_2 \\ k_3 \end{pmatrix} := \begin{pmatrix} 1 \\ 1 \\ 1 \end{pmatrix}$

$Ce := Data^{\langle 3\rangle}$ $\quad C_0 := Ce_0$

Given $\quad A'(t) = -k_1\cdot A(t) \quad C'(t) = k_2\cdot B(t) - k_3\cdot C(t)$

$B'(t) = k_1\cdot A(t) - k_2\cdot B(t) + k_3\cdot C(t)$

$A(0) = A_0 \quad B(0) = B_0 \quad C(0) = C_0$

$$S(k_1,k_2,k_3) := \text{Odesolve}\left[\begin{pmatrix} A \\ B \\ C \end{pmatrix}, t, \max(te), 1000\right]$$

Given

$$\begin{pmatrix} A \\ B \\ C \end{pmatrix} \leftarrow S(k_1,k_2,k_3) \qquad = \text{augment}(Ae, Be, Ce)$$

$$\text{augment}(v(A,te), v(B,te), v(C,te))$$

$$\begin{pmatrix} k_1 \\ k_2 \\ k_3 \end{pmatrix} := \text{Minerr}(k_1,k_2,k_3) = \begin{pmatrix} 10.029 \\ 15.021 \\ 5.053 \end{pmatrix}$$

$$\begin{pmatrix} A \\ B \\ C \end{pmatrix} := S(k_1,k_2,k_3) \qquad ERR^2 = 6.439\times10^{-4}$$

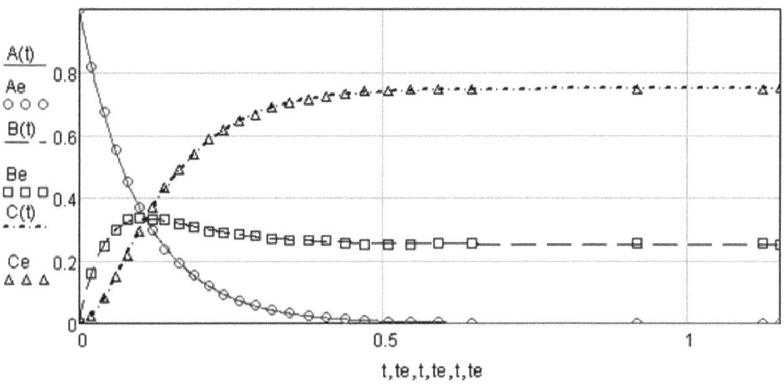

Fig. 4.23 Solution of inverse problem using `Given/Odesolve` and `Given/Minerr`

Argon atmosphere was used to eliminate possible side reactions of Si_2H_6. Obtained experiment data is shown in Fig. 4.24.

The authors explained the obtained data with a mechanism shown in Table 4.10. The rate constants for direct reactions and equilibrium constants for reversible reactions are calculated for $T = 1{,}200$ K and total argon concentration 4.0×10^{-6} mol cm^{-3} (0.40 bar at 1,200 K).

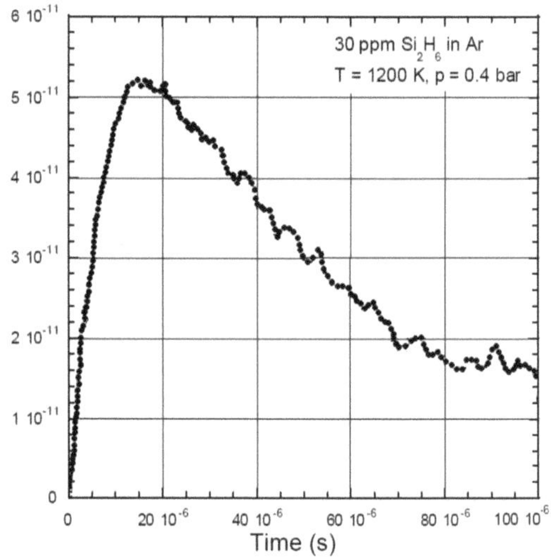

Fig. 4.24 Concentration of SiH_2 over time (data from H. Mick et al)

Table 4.10 Mechanism of thermal decomposition of Si_2H_6

Reaction	Rate constant of forward reaction	Equilibrium constant
(1) $Si_2H_6 \Leftrightarrow SiH_4 + SiH_2$	Unknown	3.6×10^{-9} mol cm^{-3}
(2) $SiH_4 \Leftrightarrow SiH_2 + H_2$	1.3×10^2 s^{-1}	8.0×10^{-9} mol cm^{-3}
(3) $SiH_2 + Ar \Leftrightarrow Si + H_2 + Ar$	3.1×10^8 cm^{-3} mol^{-1} s^{-1}	2.5×10^{-8} mol cm^{-3}
(4) $2SiH_2 \Leftrightarrow Si_2H_2 + H_2$	Unknown	4.2×10^4
(5) $SiH_4 + SiH_2 \Leftrightarrow H_3SiSiH + H_2$	1.3×10^{13} cm^{-3}mol^{-1}s^{-1}	8.1×10^{-1}
(6) $Si_3H_8 \Leftrightarrow SiH_4 + H_3SiSiH$	1.8×10^6 s^{-1}	5.2×10^{-9} mol cm^{-3}
(7) $Si_3H_8 \Leftrightarrow SiH_2 + Si_2H_6$	4.1×10^5 s^{-1}	1.4×10^{-8} mol cm^{-3}
(8) $H_3SiSiH \Leftrightarrow H_2SiSiH_2$	9.5×10^{10} s^{-1}	1.4×10^1

The overall process, as seen from Table 4.10, includes 9 compounds and 16 elementary steps. Rate constants for some of these reactions are known. In the original paper, the rate constant for the first step is determined. To complicate the problem, assume the k_4 value is also unknown.

The general solving protocol matches that in Sect. 4.5.1. The corresponding Mathcad listings are shown in Figs. 4.25–4.27. The first two figures contain the mathematical model in form of a vector of right parts of the ODE system. Computations of the unknown constants using a built-in `Radau` function and `Given/ Minerr` solving block is given in Fig. 4.27. The construction at the beginning of the document is of particular interest. Application of a `linterp` function after the user-created one allows solving the inverse problem even when experiment data is not evenly distributed.

The kinetic curve, calculated using the found k_1 and k_4 values, apparently fits the experiment quite well.

▸ Problem Statement

▾ Mathematical model

t0 := time(1)

$\mathrm{lnk}_1 = x_0 \qquad \mathrm{lnk}_2 = x_1$

$[Si_2H_6] = x_2 \qquad [SiH_4] = x_3$

$[SiH_2] = x_4 \quad (Si) = x_5 \quad [H_2] = x_6$

$[Si_2H_2] = x_7 \qquad [H_3SiSiH] = x_8$

$[Si_3H_8] = x_9 \qquad [H_2SiSiH_2] = x_{10}$

StM :=

"Reagents/Steps"	"(1)"	"(2)"	"(3)"	"(4)"	"(5)"	"(6)"	"(7)"	"(8)"
"Si2H6"	−1	0	0	0	0	0	1	0
"SiH4"	1	−1	0	0	−1	1	0	0
"SiH2"	1	1	−1	−2	−1	0	1	0
"Si"	0	0	1	0	0	0	0	0
"H2"	0	1	1	1	1	0	0	0
"Si2H2"	0	0	0	1	0	0	0	0
"H3SiSiH"	0	0	0	0	1	1	0	−1
"Si3H8"	0	0	0	0	0	−1	−1	0
"H2SiSiH2"	0	0	0	0	0	0	0	1

$$r_1 = e^{x_0}\cdot\left(x_2 - \frac{x_3\cdot x_4}{K_1}\right) \qquad r_2 = k_2\cdot\left(x_3 - \frac{x_4\cdot x_5}{K_2}\right) \qquad r_3 = k_3\cdot Ar\cdot\left(x_4 - \frac{x_5\cdot x_6}{K_3}\right)$$

$$\alpha := \mathrm{submatrix}(StM, 1, \mathrm{rows}(StM) - 1, 1, \mathrm{cols}(StM) - 1)$$

$$r_4 = e^{x_1}\cdot\left[(x_4)^2 - \frac{x_6\cdot x_7}{K_4}\right] \quad r_5 = k_5\cdot\left(x_3\cdot x_4 - \frac{x_6\cdot x_7}{K_5}\right) \quad r_6 = k_6\cdot\left(x_9 - \frac{x_3\cdot x_8}{K_6}\right) \quad r_7 = k_7\cdot\left(x_9 - \frac{x_2\cdot x_4}{K_7}\right) \quad r_8 = k_8\cdot\left(x_8 - \frac{x_{10}}{K_8}\right)$$

Fig. 4.25 Stoichiometric matrix for thermal decomposition of disilane

$$\alpha\cdot\begin{bmatrix} e^{x_0}\cdot\left(x_2 - \frac{x_3\cdot x_4}{K_1}\right) \\ k_2\cdot\left(x_3 - \frac{x_4\cdot x_5}{K_2}\right) \\ k_3\cdot Ar\cdot\left(x_4 - \frac{x_5\cdot x_6}{K_3}\right) \\ e^{x_1}\cdot\left[(x_4)^2 - \frac{x_6\cdot x_7}{K_4}\right] \\ k_5\cdot\left(x_3\cdot x_4 - \frac{x_6\cdot x_7}{K_5}\right) \\ k_6\cdot\left(x_9 - \frac{x_3\cdot x_8}{K_6}\right) \\ k_7\cdot\left(x_9 - \frac{x_2\cdot x_4}{K_7}\right) \\ k_8\cdot\left(x_8 - \frac{x_{10}}{K_8}\right) \end{bmatrix} \rightarrow \begin{bmatrix} -\exp(x_0)\cdot\left(x_2 - x_3\cdot\frac{x_4}{K_1}\right) + k_7\cdot\left(x_9 - x_2\cdot\frac{x_4}{K_7}\right) \\ \exp(x_0)\cdot\left(x_2 - x_3\cdot\frac{x_4}{K_1}\right) - k_2\cdot\left(x_3 - x_4\cdot\frac{x_5}{K_2}\right) - k_5\cdot\left(x_3\cdot x_4 - x_6\cdot\frac{x_7}{K_5}\right) + k_6\cdot\left(x_9 - x_3\cdot\frac{x_8}{K_6}\right) \\ \exp(x_0)\cdot\left(x_2 - x_3\cdot\frac{x_4}{K_1}\right) + k_2\cdot\left(x_3 - x_4\cdot\frac{x_5}{K_2}\right) - k_3\cdot Ar\cdot\left(x_4 - x_5\cdot\frac{x_6}{K_3}\right) - 2\cdot\exp(x_1)\cdot\left[(x_4)^2 - x_6\cdot\frac{x_7}{K_4}\right] - k_5\cdot\left(x_3\cdot x_4 - x_6\cdot\frac{x_7}{K_5}\right) + k_7\cdot\left(x_9 - x_2\cdot\frac{x_4}{K_7}\right) \\ k_3\cdot Ar\cdot\left(x_4 - x_5\cdot\frac{x_6}{K_3}\right) \\ k_2\cdot\left(x_3 - x_4\cdot\frac{x_5}{K_2}\right) + k_3\cdot Ar\cdot\left(x_4 - x_5\cdot\frac{x_6}{K_3}\right) + \exp(x_1)\cdot\left[(x_4)^2 - x_6\cdot\frac{x_7}{K_4}\right] + k_5\cdot\left(x_3\cdot x_4 - x_6\cdot\frac{x_7}{K_5}\right) \\ \exp(x_1)\cdot\left[(x_4)^2 - x_6\cdot\frac{x_7}{K_4}\right] \\ k_5\cdot\left(x_3\cdot x_4 - x_6\cdot\frac{x_7}{K_5}\right) + k_6\cdot\left(x_9 - x_3\cdot\frac{x_8}{K_6}\right) - k_8\cdot\left(x_8 - \frac{x_{10}}{K_8}\right) \\ -k_6\cdot\left(x_9 - x_3\cdot\frac{x_8}{K_6}\right) - k_7\cdot\left(x_9 - x_2\cdot\frac{x_4}{K_7}\right) \\ k_8\cdot\left(x_8 - \frac{x_{10}}{K_8}\right) \end{bmatrix}$$

$K_1 := 3.6\cdot10^{-9}\quad K_2 := 8.0\cdot10^{-9}\quad K_3 := 2.5\cdot10^{-8}\quad K_4 := 4.2\cdot10^{4}\quad K_5 := 8.1\cdot10^{-1}\quad K_6 := 5.2\cdot10^{-9}\quad K_8 := 1.4\cdot10^{1}$

$k_2 := 1.3\cdot10^{2}\quad k_3 := 3.1\cdot10^{8}\quad k_5 := 1.3\cdot10^{13}\quad k_6 := 1.8\cdot10^{6}\quad k_7 := 4.1\cdot10^{5}\quad K_7 := 1.4\cdot10^{-8}\quad k_8 := 9.5\cdot10^{10}\quad Ar := 4\cdot10^{-6}$

Fig. 4.26 Mathematical model for decomposition of Si_2H_6

In the end, we want to point out that all the examples of inverse kinetic problem given here, required calculations of kinetic parameters for the *known* kinetic scheme of the reaction. Real-life research often includes a much more complicated

$$Si2H6_0 := 4 \cdot 10^{-6} \cdot \frac{30}{1 \cdot 10^6} \qquad TOL := 10^{-15} \qquad \begin{pmatrix} lnk_1 \\ lnk_4 \end{pmatrix} := \begin{pmatrix} 12 \\ 35 \end{pmatrix}$$

$$SS(lnk_1, lnk_4) := Radau\left(\left(lnk_1 \;\; lnk_4 \;\; Si2H6_0 \;\; 0 \;\; 0 \;\; 0 \;\; 0 \;\; 0 \;\; 0 \;\; 0 \;\; 0\right)^T, 0, 0.0001, 1000, f\right)$$

$$tt := data^{\langle 0 \rangle} \qquad \left[SiH_{2\,measured}\right] := data^{\langle 1 \rangle} \qquad i := 0 .. rows(tt) - 1$$

$$fint(lnk_1, lnk_4, xx) := linterp\left(SS(lnk_1, lnk_4)^{\langle 0 \rangle}, SS(lnk_1, lnk_4)^{\langle 5 \rangle}, xx\right)$$

$$SSE(lnk_1, lnk_4) := \sum_i \left[fint(lnk_1, lnk_4, tt_i) \cdot 10^{11} - \left(\left[SiH_{2\,measured}\right]\right)_i \cdot 10^{11}\right]^2$$

$$SS_c := SS(lnk_1, lnk_4) \qquad \left[SiH_2\right] := SS_c^{\langle 5 \rangle} \qquad SSE(lnk_1, lnk_4) = 513.030601634941$$

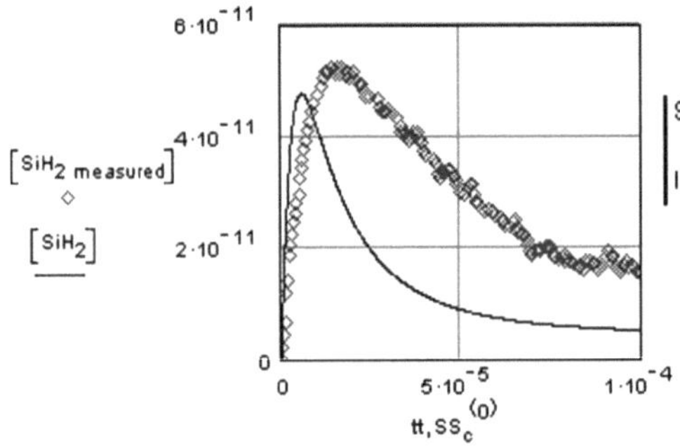

Given

$$\left| \begin{matrix} SS_c \leftarrow SS(lnk_1, lnk_4) \\ linterp\left(SS_c^{\langle 0 \rangle}, SS_c^{\langle 5 \rangle}, tt\right) \cdot 10^{11} \end{matrix} \right. = \left[SiH_{2\,measured}\right] \cdot 10^{11}$$

$$\begin{pmatrix} lnk_1 \\ lnk_4 \end{pmatrix} := Minerr(lnk_1, lnk_4)$$

$$\begin{pmatrix} lnk_1 \\ lnk_4 \end{pmatrix} = \begin{pmatrix} 11.0579680081634 \\ 33.7332685138631 \end{pmatrix}$$

$$ERR^2 = 2.08210278702073 \qquad SSE(lnk_1, lnk_4) = 2.08210278702073$$

$$SS_c := SS(lnk_1, lnk_4) \qquad \left[SiH_2\right] := SS_c^{\langle 5 \rangle}$$

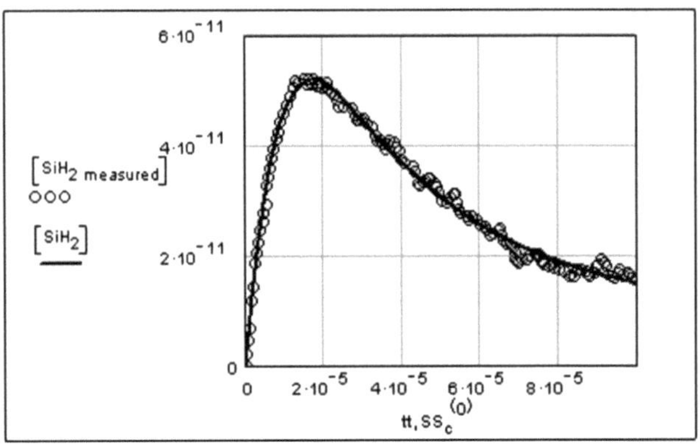

$$k_1 := e^{lnk_1}$$

$$k_1 = 6.344750 \times 10^4$$

$$k_4 := e^{lnk_4}$$

$$k_4 = 4.468609 \times 10^{14}$$

$$time(1) - t0 = 2.860000$$

Fig. 4.27 Final stage of rate constant calculations for decomposition of disilane

stage: defining the kinetic scheme for a complex chemical transformation, finding the nature of each step, and proving that the reaction is theoretically possible. Investigation of kinetics in not always enough for this.

Chapter 5
Introduction into Electrochemical Kinetics

5.1 General Features of Electrode Processes

A special kind of chemical reactions are the redox (reduction–oxidation) processes in electrochemical circuits – the systems consisting of two metal electrodes in solution of electrolyte (internal circuit) that are connected using an electronic conductor (external circuit). *Electrolytic* and *Galvanic cells* are the examples of electrochemical circuits. The reactions occurring in such cells are called *electrochemical reactions* to stress out their essential difference from ordinary chemical processes.

An elementary act of a traditional homogeneous redox reaction can be considered as a collision of the reacting particles with a simultaneous electron transfer from the reductant to the oxidant. This approach does not work with electrochemical processes. Here, the reactant and the oxidant are separated in space and the electron transfer occurs through the external circuit. Thus, an electrochemical reaction is accompanied by electron flow. The directed electron flow might be induced by a spontaneous reaction in the circuit. In this case, one deals with a *Galvanic cell*. On the contrary, in an *electrolytic cell*, the electric current from an external source may cause non-spontaneous chemical transformations.

The overall reaction in electrochemical circuit consists of two half-reactions: reduction and oxidation. Each of them takes place on a separate electrode – a metallic conductor immersed into electrolyte solution. The electrode where oxidation occurs is called *anode*. Another electrode, where the reduction takes place, is called *cathode*.

The *Faraday laws of electrolysis* are quantitative relationships between the amount of substance chemically transformed on an electrode and the amount of electricity that passed through the electrochemical circuit. For example, the same amount of electricity – 96,485 C – is required for a reaction of 1 mEq of any compound. The following equation is commonly used in calculations:

$$m = \frac{M}{nF} q. \tag{5.1}$$

V.I. Korobov and V.F. Ochkov, *Chemical Kinetics with Mathcad and Maple*,
DOI 10.1007/978-3-7091-0531-3_5,

Here, m is the mass (in grams) of the compound that has reacted of formed during the electrolysis; M is its molar mass, g mol^{-1}; n is the number of electrons required for the transition from the oxidized into reduced form; q is the amount of electricity, C; F is the *Faraday constant*, C mol eq^{-1}. In real life experiments, one commonly measures the electric current I that can change over time due to different reasons. Therefore, (5.1) can be rewritten as:

$$m(t) = \frac{M}{nF}\int_0^t I(t)dt.$$

If the current is constant, then:

$$m(t) = \frac{M}{nF}It.$$

Introduce the *current density*, i, which is the electric current per unit area of cross section S. Then,

$$m(t) = \frac{M}{nF}iSt.$$

Differentiation leads to the following expressions:

$$\frac{dm}{dt} = \frac{M}{nF}iS,$$

and

$$\frac{dm}{MSdt} = \frac{i}{nF}.$$

The left part of the latter equation contains the change of the reactant amount over time on unit area, i.e. the rate of this process, r. Consequently, the Faraday laws also represent an extremely important kinetic relationship:

$$i = nFr. \tag{5.2}$$

This equation shows that the current density can serve as a rate unit for an electrode process, i.e. electrochemical reaction.

Strictly speaking, the reaction occurs only at the interface between the electrode and solution. Therefore, it is always heterogenic. This distinguishes electrochemical processes from other redox reactions and implies their second most important feature – *electrochemical reactions always have several steps*. The reacting particle has to approach the electrode surface (mass transfer step), react (actual electrochemical step), and leave the reaction area into bulk solution (second mass transfer). The mass transfer step, as well as the electrochemical one, are always present in any electrochemical transformation. Importantly, the electrochemical step is always accompanied by *transfer of a charged particle through the interface*. That is why this step is called *the transfer step* or *the discharge-ionization step*. Other complications are also possible. They are related to the formation of a new phase on the electrode (surface diffusion of adatoms, recombination of adatoms, formation of crystals or gas bubbles, etc.). The transfer step may be accompanied by different chemical reactions, both in bulk and on the electrode surface. A set of all the possible transformations is called *the electrode process*. Electrochemical kinetics works with the general description of electrode processes over time. While related to chemical kinetics, electrochemical kinetics has several important features. They are specific to the certain processes, in particular – the discharge-ionization step. Determination of a possible step order and the slowest (rate-determining) step is crucial for the description of the specific electrode process.

Mathematical description of electrochemical kinetics is based on a functional relationships between the process rate i and the electrode potential φ. At the equilibrium (i.e. when there is no current), the electrode potential is equal to its equilibrium value φ_{eq} defined by the *Nernst equation:*

$$\varphi_{eq} = \varphi^0 + \frac{RT}{nF} \ln \frac{C^0_{Ox}}{C^0_{\text{Red}}},$$

where φ^0 is the standard electrode potential of the electrochemical system, and C_i^0 is the concentration (activity) of the oxidized and reduced forms in solution. The current that passes through the circuit changes the electrode potential. The value of this shift is called *electrode polarization* $\Delta\varphi = |\varphi - \varphi_p|$. The extent of polarization is determined by the decrease in the rate of some process step. If the origin of polarization is known (in other words, one knows, which step is responsible for the shift of the potential from its equilibrium value), then it is called *overpotential*. For example, if the polarization is caused by slow diffusion of a reactant to the electrode surface, then one deals with concentrational polarization; its value equals to the diffusion overpotential. If the electrode process has several slow steps, the overall polarization is a sum of polarizations for each step.[1]

The functional relationships between the current density and electrode polarization (so called *polarization properties of electrode*) usually depend on the nature of the limiting step. Experimental determination of polarization properties allows to hypothesize about the character of the investigated process, i.e. about the kinetically limiting step.

[1]Here we use the generally accepted overpotential conception of K. Fetter

5.2 Kinetics of the Slow Discharge-Ionization Step

Consider a single reversible electrode process with the reduced and oxidized forms of the same reactant:

$$Ox + ne \leftrightarrow \mathrm{Red}.$$

Assume all accompanying reactions are fast and this step is the limiting one. Then the overall rate is determined by the difference between the direct and reverse reactions:

$$r = \overrightarrow{r} - \overleftarrow{r}.$$

Given (5.2), we obtain $i = \overrightarrow{i} - \overleftarrow{i}$ (since $nFr = nF\overrightarrow{r} - nF\overleftarrow{r}$). If each step is of the first order, then

$$i = \overrightarrow{i} - \overleftarrow{i} = nFC_{Ox}\overrightarrow{A}e^{-\overrightarrow{E_a}/RT} - nFC_{\mathrm{Red}}\overleftarrow{A}e^{-\overleftarrow{E_a}/RT},$$

where $\overrightarrow{A}$, $\overleftarrow{A}$ and $\overrightarrow{E_a}$, $\overleftarrow{E_a}$ are the pre-exponential terms and energies of activation of the direct and reverse reactions, respectively. Then, we make an important assumption that the energies of activation of the reduction (discharge) and oxidation (ionization) steps depend on the electrode potential:

$$\overrightarrow{E_a} = \overrightarrow{E_{a_0}} + \alpha nF\varphi,$$

$$\overleftarrow{E_a} = \overleftarrow{E_{a_0}} - \beta nF\varphi.$$

Here $\overrightarrow{E_{a_0}}$ and $\overleftarrow{E_{a_0}}$ are the energies of activation for the discharge and ionization reactions when $\varphi = 0$; α and β are the so called *transfer coefficients* ($0 < \alpha < 1$ and $\alpha + \beta = 1$). A decrease in the electrode potential (a shift of the latter into negative values) causes an increase in the discharge rate and a decrease in the ionization rate.
This gives:

$$\overleftarrow{i} = nFC_{Ox}\overrightarrow{A}e^{-\overrightarrow{E_{a_0}}/RT}e^{-\alpha nF\varphi/RT} = nFC_{Ox}\overrightarrow{k_0}e^{-\alpha nF\varphi/RT}, \tag{5.3}$$

$$\overleftarrow{i} = nFC_{\mathrm{Red}}\overleftarrow{A}e^{-\overleftarrow{E_{a_0}}/RT}e^{\beta nF\varphi/RT} = nFC_{\mathrm{Red}}\overleftarrow{k_0}e^{(1-\alpha)nF\varphi/RT}, \tag{5.4}$$

where $\overrightarrow{k_0}$ and $\overleftarrow{k_0}$ are the effective rate constants.

At equilibrium, when $\varphi = \varphi_p$, the resulting current through the electrode equals zero. This means that the rates of both reactions are equal:

$$i_0 = \overleftarrow{i} = \overleftarrow{i} = nFC_{Ox}\overrightarrow{k_0}e^{-\alpha nF\varphi_p/RT} = nFC_{\mathrm{Red}}\overleftarrow{k_0}e^{(1-\alpha)nF\varphi_p/RT}. \tag{5.5}$$

The number i_0 is called *exchange current density* and represents the discharge-ionization rate at the equilibrium. One can easily prove that:

$$i = i_0\left(\frac{\overrightarrow{i}}{i_0} - \frac{\overleftarrow{i}}{i_0}\right). \tag{5.6}$$

Considering (5.3)–(5.5), we can rewrite (5.6) as

$$i = i_0\left\{e^{\frac{\alpha nF\eta}{RT}} - e^{\frac{-(1-\alpha)nF\eta}{RT}}\right\}, \tag{5.7}$$

where η is the overpotential of discharge-ionization step, $\eta = \varphi - \varphi_p$. Equation (5.7) is called *the main equation of the slow discharge theory*.

The assumed definition of η means that the overpotential is positive when the potential is shifted to the negative values (into cathode region). Similarly, $\eta < 0$ in anode region.

Figure 5.1 shows the resulting relationship between the rate of the slow discharge-ionization step and overpotential calculated using (5.7).

One can see that a shift in potential from its equilibrium value leads to different changes in the anode and cathode process rates. The dotted line represents partial polarization characteristics for the discharge and ionization steps. The resulting reaction rate depends on the overpotential exponentially.

Analysis of the main equation of slow discharge theory shows that two cases are possible. For example, when deviations from the equilibrium are small (i.e. when

$R := 8.3144 \quad n := 1 \quad T := 298 \quad F := 96485 \quad i_0 := 1\cdot 10^{-5} \quad \alpha := 0.5$

$i(\eta) := i_0\cdot\left[e^{\frac{\alpha\cdot n\cdot F}{R\cdot T}\cdot\eta} - e^{\frac{-(1-\alpha)\cdot n\cdot F}{R\cdot T}\cdot\eta}\right] \quad i_k(\eta) := i_0\cdot\left(e^{\frac{\alpha\cdot n\cdot F}{R\cdot T}\cdot\eta}\right) \quad i_a(\eta) := -i_0\cdot e^{\frac{-(1-\alpha)\cdot n\cdot F}{R\cdot T}\cdot\eta}$

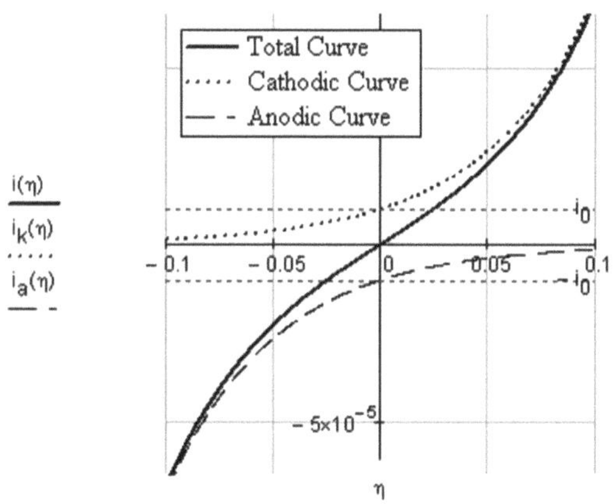

Fig. 5.1 Polarization curves for slow discharge-ionization step

$|\eta| \ll RT/F$), the exponents can be expanded into a series. Using the first two terms of the series, one can obtain a linear relationship between the overpotential η and current density in the low overpotential region *(Butler equation)*:

$$i = i_0 \frac{nF}{RT} \eta. \tag{5.8}$$

Figure 5.2 shows the derivation of the Butler equation. The plot demonstrates the region where this relationship is applicable.

The situation is different for large overpotentials. Namely, if the electrode potential is shifted into the high overpotential region (i.e. if $\eta \gg RT/F$), the second term of the series becomes negligible. In this case, we obtain so called *Tafel equation* (Fig. 5.3):

$$\eta = -\frac{RT}{\alpha nF} \ln i_0 + \frac{RT}{\alpha nF} \ln i. \tag{5.9}$$

Equation (5.9) indicates that there is a linear relationship between η and $\ln i$. A part of the polarization curve where (5.9) is valid is called the *Tafel region.*

The Tafel equation is widely used in practical electrochemistry to determine the exchange current densities and transport numbers.

We want to note that we obtained the main equation of the slow discharge theory based only on principles of formal kinetics and without any models of the elementary

$$i_0 \cdot \left[e^{\frac{\alpha \cdot n \cdot F}{R \cdot T} \cdot \eta} - e^{\frac{-(1-\alpha) \cdot n \cdot F}{R \cdot T} \cdot \eta} \right] \begin{vmatrix} \text{series}, \eta, 1 \\ \text{simplify} \end{vmatrix} \rightarrow \frac{F \cdot \eta \cdot i_0 \cdot n}{R \cdot T}$$

$R := 8.3144 \quad n := 1 \quad T := 298 \quad F := 96485 \quad i_0 := 1 \cdot 10^{-5} \quad \alpha := 0.5$

$$i_1(\eta) := i_0 \cdot \left[e^{\frac{\alpha \cdot n \cdot F}{R \cdot T} \cdot \eta} - e^{\frac{-(1-\alpha) \cdot n \cdot F}{R \cdot T} \cdot \eta} \right] \qquad i_2(\eta) := i_0 \cdot \frac{n \cdot F}{R \cdot T} \cdot \eta$$

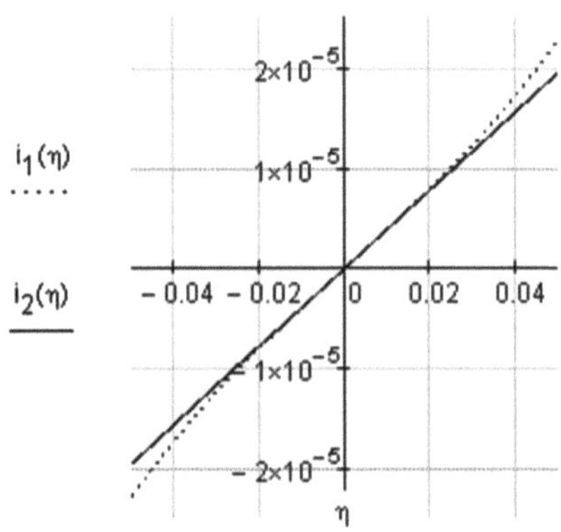

Fig. 5.2 Function of the electrode current over overpotential near equilibrium

$$j := 0..90 \qquad \eta_j := 0.003 \cdot j$$

$$\alpha := 0.5 \qquad i_j := i_0 \cdot \left[e^{\frac{\alpha \cdot n \cdot F}{R \cdot T} \cdot \eta_j} - e^{\frac{-(1-\alpha) \cdot n \cdot F}{R \cdot T} \cdot \eta_j} \right]$$

$$\eta_1(i_1) := \frac{-R \cdot T}{\alpha \cdot n \cdot F} \cdot \ln(i_0) + \frac{R \cdot T}{\alpha \cdot n \cdot F} \cdot \ln(i_1) \qquad i_1 := 10^{-7}, 5 \cdot 10^{-7} .. 10^{-4}$$

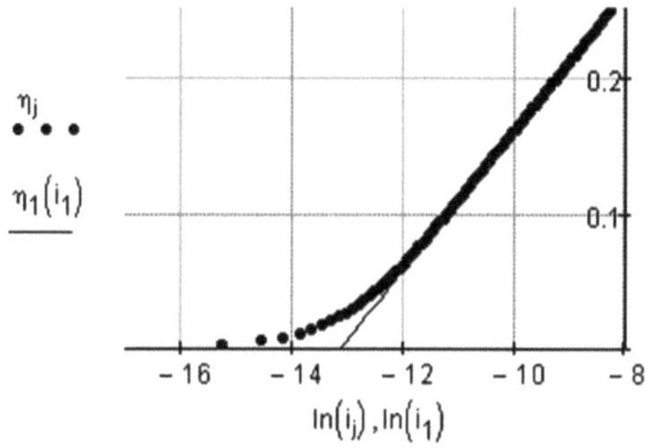

$$R \equiv 8.3144 \quad T \equiv 298 \quad n \equiv 1 \quad i_0 \equiv 2 \cdot 10^{-6} \quad F \equiv 96485$$

Fig. 5.3 Polarization curve (cathode region) in the Tafel coordinates

step mechanism. In addition, we implemented significant simplifications, ignoring the structure of double layer at the interface. More sophisticated approaches to the subject can be found in specialized literature.

5.3 Electrochemical Reactions with Stepwise Electron Transfer

Simple calculations show that when three electrons are transferred during the slow electrochemical step and the transfer number is 0.5, a slope of the Tafel correlation between η and $\lg i$ must equal $2.3RT/(\alpha nF) = 39$ mV. However, completely different slopes can be measured experimentally. Data for the same electrochemical system can heavily depend on the electrolyte composition. This leads to a conclusion that electrooxidation or electroreduction of charged particles can proceed through a stepwise electron transfer mechanism.

Consider with Maple an analysis of the following kinetic scheme that represents stepwise ionization of iron:

$$\text{(I)}\ Fe - e \underset{k_{-1}}{\overset{k_1}{\rightleftharpoons}} Fe^{+},$$

$$\text{(II)}\ Fe^{+} - e \underset{k_{-2}}{\overset{k_2}{\rightleftharpoons}} Fe^{2+},$$

$$\text{(III)}\ Fe^{2+} - e \xrightarrow{k_3} Fe^{3+}$$

```
> restart;
```

Write the kinetic equations for each step according to the mass action law.

```
> step1:=v1=k[1]-k[-1]*C[Fe1]; step2:=v2=k[2]*C[Fe1]-k[-
2]*C[Fe2];    step3:=v3=k[3]*C[Fe2];
```

$$step1 := v1 = k_1 - k_{-1}\, C_{Fe1},$$
$$step2 := v2 = k_2\, C_{Fe1} - k_{-2}\, C_{Fe2},$$
$$step3 := v3 = k_3\, C_{Fe2}$$

Substitute the concentration of Fe^{+} in the second equation with the expression from the first one. Then express the concentration of Fe^{2+} from the second equation and put it into the third expression.

```
> C[Fe1]:=solve(step1,C[Fe1]);
```

$$C_{Fe1} := -\frac{v1 - k_1}{k_{-1}},$$

```
> C[Fe2]:=solve(step2,C[Fe2]);
```

$$C_{Fe2} := -\frac{v2 k_{-1} + k_2 v1 - k_2 k_1}{k_{-1} k_{-2}}$$

In a stationary mode, rates of all the reactions are identical.

```
> subs(step3,v1=v,v2=v,v3=v);
```

$$-\frac{k_3(v k_{-1} + k_2 v - k_2 k_1)}{k_{-1} k_{-2}} = v$$

Consequently, the overall process rate can be expressed in terms of the rate constants of each step:

```
> v:=solve(%,v);
```

$$v := \frac{k_3\, k_2\, k_1}{k_3\, k_{-1} + k_3\, k_2 + k_{-1}\, k_{-2}}$$

Assume that each step follow kinetic laws of slow discharge. Also, assume for simplicity that the transfer coefficients is 0.5.

```
> k[1]:=k1[0]*exp(F*eta/(2*R*T));
k[2]:=k2[0]*exp(F*eta/(2*R*T));
  k[3]:=k3[0]*exp(F*eta/(2*R*T)); k[-1]:=km1[0]*exp(-
F*eta/(2*R*T));
  k[-2]:=km2[0]*exp(-F*eta/(2*R*T));
```

$$k_1 := k1_0 e^{\left(\frac{F\eta}{2RT}\right)} \quad k_2 := k2_0 e^{\left(\frac{F\eta}{2RT}\right)} \quad k_3 := k3_0 e^{\left(\frac{F\eta}{2RT}\right)} \quad k_{-1} := km1_0 e^{\left(-\frac{F\eta}{2RT}\right)}$$

$$k_{-2} : km2_0 e^{\left(-\frac{F\eta}{2RT}\right)}.$$

Rewrite the rate equation with the obtained rate constants:

```
> v;
```

$$\frac{k3_0\left(e^{\left(\frac{F\eta}{2RT}\right)}\right)^3 k2_0\,k1_0}{k3_0 e^{\left(\frac{F\eta}{2RT}\right)}\,km1_0 e^{\left(-\frac{F\eta}{2RT}\right)} + k3_0\left(e^{\left(\frac{F\eta}{2RT}\right)}\right)^2 k2_0 + km1_0\left(e^{\left(-\frac{F\eta}{2RT}\right)}\right)^2 km2_0}.$$

The current that passes through the electrode depends on the overpotential according to the following equation:

```
> eq_end:=i=simplify(%);
```

$$eq\_end := i = \frac{k3_0 e^{\left(\frac{3F\eta}{2RT}\right)} k2_0\ k1_0}{k3_0\,km1_0 + k3_0 e^{\left(\frac{F\eta}{RT}\right)} k2_0 + km1_0 e^{\left(-\frac{F\eta}{RT}\right)} km2_0}.$$

When the first step of the process is slow, one can drop the $km1_0$ term in denominator. For a slow second step, the terms $k2_0$ and $km2_0$ can be dropped. Finally, the $k3_0$ term is negligibly small in the case when the third step is slow. This leads to three kinetic equations describing the overall process kinetics:

```
> d:=denom(rhs(eq_end)): n:=numer(rhs(eq_end)):
> i[rds1]:=simplify(n/op(2,d));
  i[rds2]:=simplify(n/op(1,d));
  i[rds3]:=simplify(n/op(3,d));
```

$$i_{rds1} := k1_0 e^{\left(\frac{F\eta}{2RT}\right)}\ i_{rds2} := \frac{e^{\left(\frac{3F\eta}{2RT}\right)} k2_0\,k1_0}{km1_0}\ i_{rds3} := \frac{k3_0\,k2_0\,k1_0 e^{\left(\frac{5F\eta}{2RT}\right)}}{km1_0\,km2_0}.$$

It can be shown that each of the expressions can be presented in a Tafel equation form:

```
>tafel1:=lni=ln(i[rds1]):eta1:=expand(solve(tafel1,eta));
```

$$\eta 1 := \frac{2RT\ \ln\left(\frac{1}{k1_0}\right)}{F} + \frac{2RT\ \ln i}{F},$$

```
>tafel2:=lni[rds2]=ln(i[rds2]):eq2:=eta2=expand(solve(taf
el2,eta));
```

$$eq2 := \eta 2 = \frac{2}{3}\frac{RT\ \ln\left(\frac{km1_0}{k2_0 k1_0}\right)}{F} + \frac{2}{3}\frac{RT\ \ \ln i_{rds2}}{F},$$

```
> tafel3: =lni=ln(i[rds3]):eq2: = eta3 =
expand(solve(tafel3,eta));
```

$$eq2 := \eta 3 = \frac{2}{5}\frac{RT\ \ln\left(\frac{km1_0 km2_0}{k3_0 k2_0 k1_0}\right)}{F} + \frac{2\ RT\ \ln i}{5F}.$$

The slopes of the Tafel equations are the following:

```
> b1:=2.303*2*R*T/F;b2:=2.303*2/3*R*T/F;
b3:=2.303*2/5*R*T/F;
```

$$b1 := \frac{4.606\ RT}{F},$$

$$b2 := \frac{1.535333333\ RT}{F},$$

$$b3 := \frac{0.9212000000\ RT}{F}.$$

```
> R:=8.3144: T:=298.15: F:=96485: >
evalf([[b1],[b2],[b3]]);
```

$$[[0.1183395355],\ [0.03944651183],\ [0.02366790710]],$$

> # End of the Maple worksheet.

These calculations lead to an important conclusion. If the limiting step of the process is the first electron transfer then a Tafel equation slope equals 118 mV; for the second electron transfer – 39 mV; and for the third one – 24 mV. One can see that the slope decreases with an increase in the slow step number.

5.4 Electrode Processes Under Slow Diffusion Conditions

Any electrode process has several stages and the electrochemical step is just one of them. For a consecutive series of transformations with a slow mass transfer step, the overall rate depends solely on the latter. Therefore, kinetic description of such process is determined by mass transfer of the starting material from bulk solution to the electrode surface and/or departure of the products from the reaction zone. Such processes are investigated by *diffusion kinetics* and called *diffusion processes.* Diffusion kinetics has three main equations. The first one represents a relationship between the process rate and distribution of the reactant concentration near the surface of an electrode. For a linear diffusion, the reactant flux j is perpendicular to the interface. If this direction is taken as the x axis, the equation looks this way:

$$i = nFD_j \left(\frac{\partial C_j}{dx} \right)_{x=0}, \quad (5.10)$$

where D is the diffusion coefficient of the reactant.

The second equation correlates the electrode potential to the surface concentrations of reactants, C^S_{Ox} and C^S_{Red}:

$$\varphi = \varphi_0 + \frac{RT}{nF} \ln \frac{C^S_{Ox}}{C^S_{\text{Red}}}. \quad (5.11)$$

We want to note that these concentrations are not the equilibrium ones, in contrast to the Nernst equation. Similarly, the potential φ is not the equilibrium potential φ_{eq}.

These two equations allow to define a relationship between the process rate i and the electrode potential φ. However, one needs to know the concentration distribution in the electrode vicinity. This can be found from the *Fick's second law*. For linear diffusion:

$$\frac{\partial c_j}{\partial t} = D_j \frac{\partial^2 c_j}{\partial x^2}. \quad (5.12)$$

Equation (5.12) is the third main equation of diffusion kinetics.

Thus, a general approach to the problem solving in diffusion kinetics is the following:

1. A system of differential equations of the second Fick's law, (5.12), is solved for the given initial and boundary conditions in order to determine the functions that describe the oxidant and reductant concentrations, $C_j(x,t)$, as the functions of time and distance from electrode.
2. Partial derivatives $\partial C_j(x,t)/\partial x$ are calculated. Their values at $x = 0$ are put into the main (5.10). This gives a relationship for the overall process rate.
3. The electrode potential is found from the compound concentrations at the surface of electrode, $C_j(0,t)$.
4. Based on the data from entries 2 and 3, the desired function of the current density i and electrode potential is found (the polarization curve equation). Sometimes other relationships are investigated. For example, in the case of potentiostatic electrolysis one determines the rate as a function of time (chronoamperometry), and the potential does not change. Under galvanostatic conditions *($i = const$)*, kinetics of the process is described by a potential–time relationship (chronopotentiometry).

It is important to note that the described methods are applicable only when mass transport into and from the reaction space is caused by molecular diffusion.

5.4.1 Relationship Between Rate and Potential Under Stationary Diffusion

Consider the following situation. The oxidized form of a reactant consists of metal ions in solution, and the electrode material (metal) is the reduced form. Such electrochemical system is called *class 1 electrode*. An example may be a copper plate immersed into a solution of copper sulfate. If some external source of current is used to create a potential on such an electrode and the potential is more negative that the equilibrium one, then copper ions will be reduced on the cathode. Ions near the electrode will be reduced first. This will lead to the decrease in concentration of the discharging particles and new portions of the oxidized form will diffuse from the bulk of the electrode.

Consider a situation when this stage is rate determining.

Assume the diffusion is *stationary*, i.e. the reagent concentration does not change over time and we can put $\partial C/dt = 0$ into the *Fick's law*. Then, obviously, $\partial^2 C/dx^2 = 0$ and $\partial C/\partial x = \text{const}$.

The concentration of the reductant near surface, C^S, is smaller than that in the bulk, C^0. Assume the concentration linearly changes inside some specific layer δ (called *diffusion layer*):

$$\frac{\partial C}{\partial x} = \frac{C^0 - C^S}{\delta}.$$

Combining this with the main equation, we obtain:

$$i = nFD\frac{C^0 - C^S}{\delta}.$$

As we can see, the process rate cannot exceed some maximal value. This value corresponds to a situation when the reagent is reduced immediately after approaching the surface and, consequently, its surface concentration is zero. For $C^S = 0$:

$$i = i_d = nFD\frac{C^0}{\delta}.$$

The maximal current i_d is called *limiting diffusion current.*

The last two equations give:

$$C^S/C^0 = 1 - i/i_d.$$

On the other hand, for a class 1 electrode the concentration polarization equals:

$$\Delta\varphi = \varphi^0 + \frac{RT}{nF}\ln C^S - \varphi^0 - \frac{RT}{nF}\ln C^0 = \frac{RT}{nF}\ln\frac{C^S}{C^0}.$$

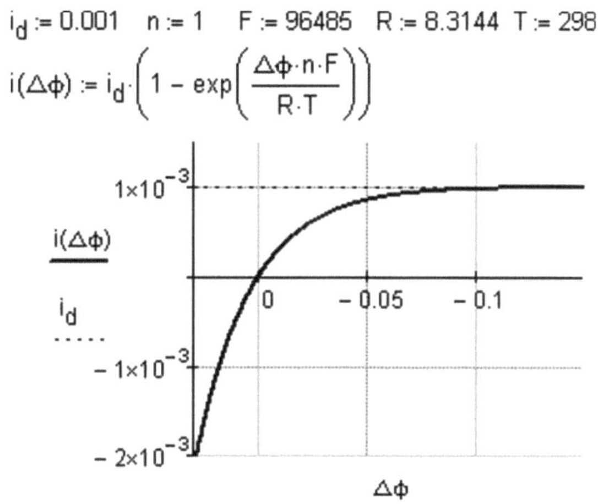

Fig. 5.4 Polarization curves under stationary diffusion conditions

Therefore,

$$\Delta\varphi = \frac{RT}{nF}\ln\left(1 - \frac{i}{i_d}\right).$$

This leads to the equation of a polarization curve under stationary diffusion conditions:

$$i = i_d\left[1 - \exp\left(\frac{nF}{RT}\Delta\varphi\right)\right].$$

We want to note cathode polarizations are considered negative. The corresponding polarization curve is shown in Fig. 5.4. A negative shift of the potential leads to an increase of the process rate up to some maximal value. This value is determined by the limiting diffusion current. A further increase in rate is impossible, since the reagent cannot diffuse into the reaction area faster.

Now let's discuss more complex problems.

5.4.1.1 Nonstationary Linear Diffusion to a Planar Electrode Under Electrostatic Conditions

Consider an electrode reaction occurring on an infinite planar surface. Assume electrons or molecules of the electrochemically active compound are transported into the reaction zone (i.e. the electrode surface) only by means of molecular diffusion. The compound is transferred to the surface only tangentially, in other words – along only one axis. In this case, we have a nonlinear diffusion. It is also *nonstationary*, $dC_{Ox}(x,t)/dt \neq 0$. Let's find the function of concentration distribution as well as the current–time relationship.

First, we have to set the initial and boundary condition. The initial condition defines the concentration distribution of the oxidized form at $t = 0$. At this moment, the reactant concentration is the same at any part of the solution. We will label this concentration as $C^0_{Ox} : C_{Ox}(x, 0) = C^0_{Ox}$.

The first boundary condition describes how the concentration on an electrode surface changes during the electrolysis. Assume such potential is applied to the electrode that the concentration of Ox on the surface immediately drops to zero. This potential is kept stationary during the electrolysis; therefore, $C_{Ox}(0, t) = 0$ for any $t > 0$. So, the first boundary condition characterizes the electrolysis regime. For other regime, this condition would be different, as would the solution.

Also, assume that, for any regime, the concentration of oxidized form changes only on the electrode surface or nearby. At some distant point this concentration will remain C_{Ox_0} even after extended electrolysis. This can be written as:

$$C_{Ox}(\infty, t) = C^0_{Ox}.$$

Now all the information required for solving the Fick's law is known. We will use an operator approach. When a Laplace transform is applied to a linear differential equation, it gives an operator algebraic equation. Similarly, a linear differential equation is obtained from the partial differential equation (5.10):

$$D_{Ox}\frac{d^2LC(x)}{dx^2} - s \cdot LC(x) + C^0_{Ox} = 0. \tag{5.13}$$

Here, $LC(x)$ is a Laplace representation of the original function $C_{Ox}(x,t)$ (Fig. 5.5).

Note that the initial condition $C_{Ox}(x, 0) = C^0_{Ox}$ is used during the transformation. Equation (5.13) has the following general solution (see Chap. 2):

$$LC(x) = \frac{C_{Ox_0}}{s} + C_1 \exp\left(-\sqrt{\frac{s}{D_{Ox}}}x\right) + C_2 \exp\left(\sqrt{\frac{s}{D_{Ox}}}x\right), \tag{5.14}$$

Here C_1 and C_2 are the integration constants. It is necessary to determine them prior to finding the specific solution. It can be done based on the information

$$\frac{d}{dt}C_{Ox}(x,t) \text{ laplace}, t \rightarrow s\cdot\text{laplace}(C_{Ox}(x,t), t, s) - C_{Ox}(x,0)$$

$$D_{Ox}\cdot\frac{d^2}{dx^2}C_{Ox}(x,t) \text{ laplace}, t \rightarrow D_{Ox}\cdot\left(\frac{d}{dx}\frac{d}{dx}\text{laplace}(C_{Ox}(x,t), t, s)\right)$$

$$s\cdot\text{laplace}(C(x,t), t, s) - C_{Ox}(x,0) \text{ substitute}, C_{Ox}(x,0) = C_{Ox0} \rightarrow s\cdot\text{laplace}(C(x,t), t, s) - C_{Ox0}$$

$$\boxed{s\cdot\text{laplace}(C(x,t), t, s) - C_{Ox0} = D_{Ox}\cdot\left(\frac{d}{dx}\frac{d}{dx}\text{laplace}(C(x,t), t, s)\right)}$$

Fig. 5.5 Transformation of the second Fick's law into an operator form provided by the boundary conditions. For example, the condition $C_{Ox}(\infty, t) = C^0_{Ox}$ is transformed into $LC(\infty) = C^0_{Ox}/s$. In other words, the representation approaches some limiting value when $x \to \infty$. At the same time, analysis of (5.14) indicates that the right-hand part of this equation increases infinitely with an increase in x due to the positive exponent in the last term. This will not contradict the physical model only if $C_2 = 0$.

From $C_{Ox}(0, t) = LC(0) = 0$ we obtain:

$$\frac{C^0_{Ox}}{s} + C_1 \exp\left(-\sqrt{\frac{s}{D_{Ox}}}x\right)_{x=0} = 0.$$

This leads to an intermediate conclusion that C_1 equals $-C^0_{Ox}/s$. Therefore, the solution is

$$LC(x) = \frac{C^0_{Ox}}{s}\left[1 - \exp\left(-\sqrt{\frac{s}{D}}x\right)\right].$$

Now let's do the reverse transformation to the original function using Mathcad symbol editor (Fig. 5.6).

So, in potentiostatic electrolysis, the concentration of oxidized form depends on the distance from electrode and electrolysis time according to the following law:

$$C_{Ox}(x, t) = C^0_{Ox} erf\left(\frac{x}{2\sqrt{D_{Ox}t}}\right). \tag{5.15}$$

The concentration curves obtained from (5.15) give a pictorial representation of the concentration distribution near the electrode surface over time (Fig. 5.7).

Note a somewhat unusual (for chemists) function erf(λ) in the obtained equation. This is a so called *error function* that can be expressed as an integral with a variable upper limit:

$$\mathrm{erf}(\lambda) = \frac{2}{\sqrt{\pi}}\int_0^{\lambda} e^{-z^2}dz,$$

$$\frac{C_{Ox0}}{s}\cdot\left(1 - \exp\left(-\sqrt{\frac{s}{D_{Ox}}}\cdot x\right)\right) \left|\begin{array}{l}\text{assume}, D_{Ox} > 0\\ \text{assume}, x > 0\\ \text{invlaplace}, s\end{array}\right. \to C_{Ox0}\cdot \mathrm{erf}\left[\frac{1}{2\cdot(t\cdot D_{Ox})^{\frac{1}{2}}}\cdot x\right]$$

Fig. 5.6 Inverse Laplace transformation in the planar electrode diffusion problem

$D_{ox} := 1 \cdot 10^{-5}$ $\quad Cox_0 := 1 \cdot 10^{-5}$

$$C(x,t) := Cox_0 \cdot \mathrm{erf}\left(\frac{x}{2 \cdot D_{ox}^{\frac{1}{2}} \cdot t^{\frac{1}{2}}}\right) \qquad x := 0, 0.01 .. 2$$

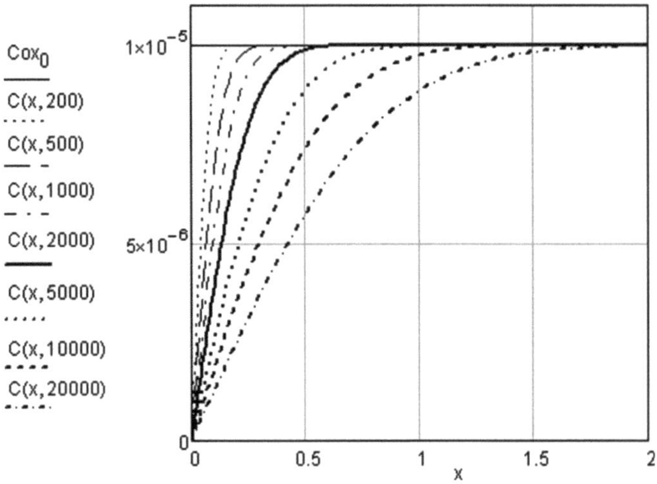

Fig. 5.7 Concentration distribution near electrode surface during potentiostatic electrolysis

where z is some auxiliary variable. The function plot is shown in Fig. 5.8. A reader can see that $\mathrm{erf}(\lambda) = 0$ when $\lambda = 0$, and $\mathrm{erf}(\lambda) \rightarrow 1$ even for small deviations of the argument. The function $\mathrm{erf}(\lambda)$ is related to a function $\mathrm{erfc}(\lambda)$:

$$\mathrm{erfc}(\lambda) = 1 - \mathrm{erf}(\lambda).$$

Recall that finding an analytical expression for the function $C_{Ox}(x,t)$ is the important but not the last step in the overall solution. Now we need to derive a relationship between the process rate (current density) and time using the first equation of diffusion kinetics, (5.10). For this, we need to find a derivative of $C_{Ox}(x,t)$ with respect to the distance x and calculate its value at $x = 0$. These operations are shown in a document in Fig. 5.9.

Thus, the current density in this case is described with the following expression:

$$i(t) = nFD_{Ox}\frac{C_{Ox}^0}{\sqrt{\pi D_{Ox} t}}. \tag{5.16}$$

We can see that the process monotonously slows down during electrolysis and its rate asymptotically approaches zero. Therefore, the constant rate mode cannot be reached on planar electrode under given conditions. The obtained equation may be called the chronoamperogram equation of a diffusion process. Compare it to the relationship that describes the maximal rate of an electrode process under stationary linear diffusion:

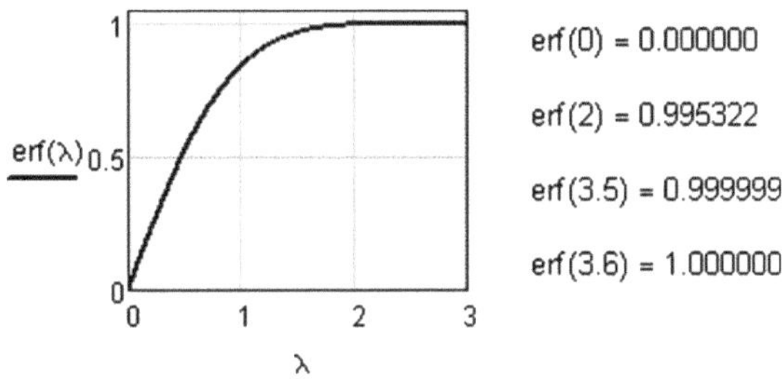

Fig. 5.8 A plot of the error function

$$Cox_0 \cdot \left(1 - \text{erf}\left(\frac{x}{2\cdot\sqrt{D_{Ox}\cdot t}}\right)\right) \text{ by differentiation, yields } -\frac{Cox_0 \cdot e^{-\frac{x^2}{4\cdot D_{Ox}\cdot t}}}{\sqrt{\pi}\cdot\sqrt{D_{Ox}\cdot t}}$$

$$-\frac{Cox_0 \cdot e^{-\frac{x^2}{4\cdot D_{Ox}\cdot t}}}{\sqrt{\pi}\cdot\sqrt{D_{Ox}\cdot t}} \text{ substitute}, x = 0 \rightarrow -\frac{Cox_0}{\sqrt{\pi}\cdot\sqrt{D_{Ox}\cdot t}}$$

$Cox_0 := 1\cdot 10^{-5}$ $\quad n := 1$ $\quad F := 96485$ $\quad D_{ox} := 10^{-5}$

$$i(t) := \frac{n\cdot F\cdot D_{ox}\cdot Cox_0}{\sqrt{\pi\cdot D_{ox}\cdot t}} \qquad t := 0.001, 0.002 .. 0.2$$

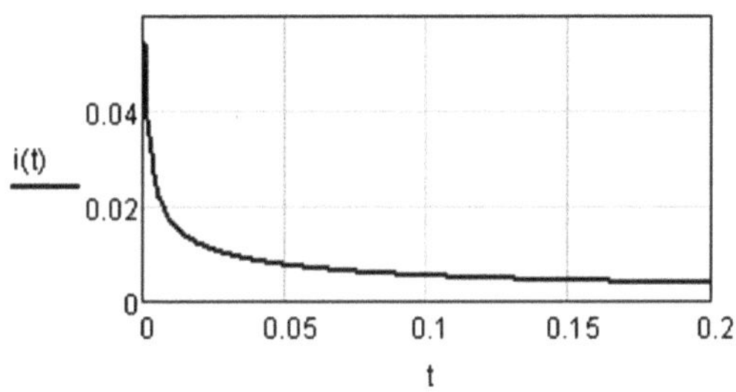

Fig. 5.9 Relationship between the process rate and time for nonstationary diffusion to a planar electrode (potentiostatic electrolysis)

$$i_d = nFD_{Ox}\frac{C^0_{Ox}}{\delta}.$$

Such a comparison shows that the delimiter in (5.16) is actually an effective thickness of the diffusion layer and it increases in proportion to a square root of time:

$$\delta_{eff} = \sqrt{\pi Dt}.$$

5.4.2 *Nonstationary Diffusion to a Spherical Electrode Under Potentiostatic Conditions*

Diffusion to a spherical electrode is of particular importance, since electrochemical experiments are often performed with an electrode consisting of mercury drops falling down from a capillary (dropping mercury electrode). In case of symmetrical spherical diffusion, the active component is transferred to the electrode surface along the lines that are tangential to the surface and end up in the center of drop.

Assume the origin is in the center of a spherical electrode with radius r_0. To simplify the problem, also assume that the radius does not change over time (i.e. the electrode is a hanging mercury drop). This case can be more easily described using a spherical coordinate system rather than the Cartesian one. The second Fick's law has the following representation in spherical coordinates:

$$\frac{\partial C_{Ox}(r,t)}{\partial t} = D_{Ox}\left[\frac{\partial^2 C(r,t)}{\partial r^2} + \frac{2}{r}\frac{\partial C(r,t)}{\partial r}\right], \tag{5.17}$$

where r is the distance from the center of electrode.

The initial and boundary conditions have the same physical meaning as the ones previously used for nonstationary diffusion to a planar electrode. Initially (when $t = 0$), the concentration of oxidized form at any distance from the electrode equals its bulk value C^0_{Ox} (initial condition):

$$C_{Ox}(r,0) = C_{Ox}^{\;0}.$$

The first boundary condition: under potentiostatic conditions the electrode remains under a potential that corresponds to the limiting diffusion current density, i.e. the concentration of oxidized form on the surface ($r = r_0$) equals zero at any moment of electrolysis ($t > 0$):

$$C_{Ox}(r_0,t) = 0.$$

The second boundary condition defines the size of diffusion area: the concentration of reagent equals its bulk value at some distance from the electrode ($r \gg r_0$):

$$C_{Ox}(\infty,t) = C_{Ox}^{\;0}.$$

The diffusion problem is completely formulated. Now we can solve it using Laplace transform. The solution is some function $C_{Ox}(r,t)$ that corresponds to the initial and boundary conditions and its derivatives fit into the second Fick's law, (5.17).

Since the solution is performed using the Maple symbol editor, we want to give a complete worksheet listing with detailed comments (the worksheet itself can be found on the book's website).

5.5 Solution of a Problem of Nonstationary Spherical Diffusion Under Potentiostatic Conditions

```
> restart; with(inttrans): #linking a library of integral transforms
```

Writing the starting equation.

```
>LawPhick:=diff(C(r,t),t)=D[Ox]*(diff(C(r,t),r$2)+2/r*dif
f(C(r,t),r));
```

$$LawPhick := \frac{\partial}{\partial t}C(r,t) = D_{Ox}\left[\left(\frac{\partial^2}{\partial r^2}C(r,t)\right) + \frac{2\left(\frac{\partial}{\partial r}C(r,t)\right)}{r}\right].$$

Subsequent operations can be significantly simplified substituting $B(r,t) = r \cdot C(r,t)$:

$B(r,t) = r.C(r,t)$:

```
> Substitution:=C(r,t)=1/r*B(r,t);
```

$$Substitution := C(r,t) = \frac{B(r,t)}{r}.$$

Rewriting the initial equation.

```
> LawPhick_Form1:=subs(Substitution, LawPhick);
```

$$Lawphick\_Form1 := \frac{\partial}{\partial t}\frac{B(r,t)}{r} = D_{Ox}\left[\left(\frac{\partial^2}{\partial r^2}\frac{B(r,t)}{r}\right) + \frac{2\left(\frac{\partial}{\partial r}\frac{B(r,t)}{r}\right)}{r}\right].$$

The obtained expression can be simplified:

```
> LawPhick_Form2:=lhs(%)*r=factor(rhs(%)*r);
```

$$LawPhick\_Form2 := \frac{\partial}{\partial t}B(r,t) = D_{Ox}\left(\frac{\partial^2}{\partial r^2}B(r,t)\right).$$

Applying the Laplace transform to the original functions:

```
> Operator_Form:=laplace(LawPhick_Form2,t,s);
```

$$Operator\_Form := s\,\mathrm{laplace}(B(r,t),\,t,\,s) - B(r,0) = D_{Ox}\left(\frac{\partial^2}{\partial r^2}\mathrm{laplace}(B(r,t),t,s)\right).$$

The value of $B(r,0)$ can be easily found. Since $B(r,0) = rC_{ox}(r,0)$, then $B(r,0) = rC_{Ox}{}^{0}$.

```
> in_cond:=B(r,0)=r*Cox[0];
```

$$in\_cond := \mathrm{B}(r,0) = rCox_0,$$

```
> Operator_Form1:=subs(in_cond, Operator_Form);
```

$$Operator\_Form1 := s\,\mathrm{laplace}(\mathrm{B}(r,\,t),\ t,\ s) - r\ Cox_0 = \mathrm{D}_{Ox}\left(\frac{\partial^2}{\partial r^2}\mathrm{laplace}\ (\mathrm{B}(r,\,t),\ t,\ s)\right).$$

Bulky expressions can be simplified substituting `laplace(B(r,t),t,s)` as `lB(r)`.

```
> vsp := laplace(B(r,t), t, s) = lB(r);
```

$$vsp := \mathrm{laplace}(\mathrm{B}(r,t),t,s) = \mathrm{lB}(r).$$

The intermediate goal of symbol calculations is almost achieved: we have obtained an ODE instead of the equation in partial derivatives. This ODE relates the transform function to its derivative on *r*.

```
> Op_Form2:=subs(vsp,Operator_Form1);
```

$$Op\_Form2 := s\,\mathrm{lB}\,(r) - r\,Cox_0 = \mathrm{D}_{Ox}\left(\frac{\partial^2}{\partial r^2}\,\mathrm{lB}(r)\right).$$

First, find a general solution.

```
> Op_sol_B:=dsolve(Op_Form2,lB(r));
```

$$Op\_sol\_\mathrm{B} := \mathrm{lB}(r) = \mathbf{e}^{\left(\frac{\sqrt{s}\,r}{\sqrt{D_{Ox}}}\right)}\ \_C2 + \mathbf{e}^{\left(-\frac{\sqrt{s}\,r}{\sqrt{D_{Ox}}}\right)}\ \_C1 + \frac{Cox_0\ r}{s}.$$

To find the specific solution, we have to define the integration constants _*C1* and _*C2*. The physical essence of the problem gives $\_C2 = 0$. Otherwise, the *Op_sol_B* term would approach infinity when $r \to \infty$. This would contradict the boundary condition $C_{Ox}(\infty,t) = C_{Ox}{}^{0}$. Thus, eq. *Op_sol_B* is simplified.

```
> Op_sol_B1:=subs(_C2=0,Op_sol_B);
```

$$Op\_sol\_\mathrm{B1} := \mathrm{lB}(r) = e^{\left(-\frac{\sqrt{s,r}}{\sqrt{D_{Ox}}}\right)}\ \_C1 + \frac{Cox_0\,r}{s}.$$

Now we can find the constant *_C1* using the first boundary condition:

```
> vsp:=subs({1B(r)=0,r=r0},Op_sol_B1);
```

$$vsp := 0 = \mathbf{e}^{\left(-\frac{\sqrt{s}\,r0}{\sqrt{D_{Ox}}}\right)}\,\_C1 + \frac{Cox_0\,r0}{s},$$

```
> const:=solve(vsp,_C1);
```

$$const := -\frac{Cox_0\,r0}{\mathbf{e}^{\left(-\frac{\sqrt{s}\,r0}{\sqrt{D_{Ox}}}\right)}\,s}.$$

The reverse substitution of the variable and obtained constants gives the `laplace(C(r,t),t,s)` transform:

```
> Op_sol_B_end:=subs({1B(r)=laplace(C(r,t),t,s)*r,
_C1=const, _C2=0}, Op_sol_B);
```

$$Op\_sol\_B\_end := \mathrm{laplace}(\mathrm{C}(r,\,t),\ t,\ s)\ r = -\frac{e^{\left(-\frac{\sqrt{s}\,r}{\sqrt{D_{Ox}}}\right)}\,Cox_0\,r0}{e^{\left(-\frac{\sqrt{s}\,r0}{\sqrt{D_{Ox}}}\right)}\,s} + \frac{Cox_0\,r}{s}.$$

It can be rewritten as:

```
> End:=laplace(C(r,t),t,s)=rhs(Op_sol_B_end)/r ;
```

$$End := \mathrm{laplace}\,(\mathrm{C}(r,\,t),\ t,\ s) = \frac{\dfrac{\mathbf{e}^{\left(-\frac{\sqrt{s}\,r}{\sqrt{D_{Ox}}}\right)}\,Cox_0\,r0}{\mathbf{e}^{\left(-\frac{\sqrt{s}\,r0}{\sqrt{D_{Ox}}}\right)}\,s} + \dfrac{Cox_0\,r}{s}}{r}.$$

Now we have all the information required for the inverse Laplace transform.

```
> Orig_Sol:=invlaplace(End,s,t);
```

$Orig\_Sol :=$

$$\mathrm{C}(r,\,t) = \frac{-Cox_0\,r0\left(\lim_{\_U1\to 0+}\mathrm{erf}\left(\frac{1}{2}\frac{r-r0}{\sqrt{D_{Ox}}\sqrt{\_U1}}\right) - \mathrm{erf}\left(\frac{1}{2}\frac{r-r0}{\sqrt{D_{Ox}}\sqrt{t}}\right)\right) + r\,Cox_0}{r}.$$

The right-hand limit can be found from the properties of the error function erf (λ): erf(λ) $\rightarrow$ 1 when $\lambda \rightarrow \infty$; consequently, the limit equals erfc(λ) (where erfc (λ) = 1 − erfc(λ)).

```
>subs(limit(erf(1/2/D[Ox]^(1/2)*(r-r0)/_U1^(1/2))-
erf(1/2/D[Ox]^(1/2)*(r-r0)   /t^(1/2)),_U1 = 0, right) =
erfc(1/2/D[Ox]^(1/2)*(r-r0)/t^(1/2)), Orig_Sol);
```

$$\mathrm{C}(r,\,t) = \frac{-Cox_0\, r0\, \mathrm{erfc}\left(\frac{1}{2}\,\frac{r-r0}{\sqrt{\mathrm{D}_{Ox}}\sqrt{t}}\right) + r\, Cox_0}{r}.$$

Thus, the concentration distribution of the oxidized form can be expressed as a function of time and distance from the electrode:

```
> C:=unapply(rhs(%),(r,t));
```

$$C := (r,\,t) \to \frac{-Cox_0\, r0\, \mathrm{erfc}\left(\frac{1}{2}\,\frac{r-r0}{\sqrt{\mathrm{D}_{Ox}}\sqrt{t}}\right) + r\, Cox_0}{r}.$$

Again, note that the most important term is the partial derivative of this function over the distance r. Its value when $r = r_0$ determines the process rate as described by the first equation of diffusion kinetics. Let's find this derivative:

```
> diff(C(r,t), r);
```

$$-\frac{-Cox_0\, r0\, \mathrm{erfc}\left(\frac{1}{2}\,\frac{r-r0}{\sqrt{\mathrm{D}_{Ox}}\sqrt{t}}\right) + r\, Cox_0}{r^2} + \frac{\frac{Cox_0\, r0\, \mathbf{e}^{\left(-1/4\frac{(r-r0)^2}{\mathrm{D}_{Ox} t}\right)}}{\sqrt{\pi}\sqrt{\mathrm{D}_{Ox}}\sqrt{t}} + Cox_0}{r}.$$

For $r = r_0$:

```
> eval(%, r=r0);
```

$$\frac{\frac{Cox_0\, r0}{\sqrt{\pi}\sqrt{\mathrm{D}_{Ox}}\sqrt{t}} + Cox_0}{r0}.$$

Using the main equation of diffusion kinetics, we can write the final expression for the rate of electrode process and this particular type of diffusion:

```
> b:=collect(%,r0): i := n*F*D[Ox]*b;
```

$$i := n\, F\, \mathrm{D}_{Ox}\left(\frac{Cox_0}{\sqrt{\pi}\sqrt{\mathrm{D}_{Ox}}\sqrt{t}} + \frac{Cox_0}{r0}\right).$$

```
> #End of the Maple worksheet
```

The final equation can be rewritten as:

$$i\,(t) = n\; FD_{Ox}\,\frac{C^0_{Ox}}{r0} + nFD_{Ox}\frac{C^0_{Ox}}{\sqrt{\pi D t}}. \tag{5.18}$$

$Cox_0 := 1 \cdot 10^{-5} \qquad n := 1 \qquad F := 96485 \qquad D_{ox} := 10^{-5} \qquad r_0 := 0.01$

$$i_1(t) := \frac{n \cdot F \cdot D_{ox} \cdot Cox_0}{\sqrt{\pi \cdot D_{ox} \cdot t}} \qquad i_2(t) := \frac{n \cdot F \cdot D_{ox} \cdot Cox_0}{\sqrt{\pi \cdot D_{ox} \cdot t}} + n \cdot F \cdot D_{ox} \cdot \frac{Cox_0}{r_0}$$

$$i_{st} := n \cdot F \cdot D_{ox} \cdot \frac{Cox_0}{r_0} \qquad t := 0.1, 0.2 .. 5000$$

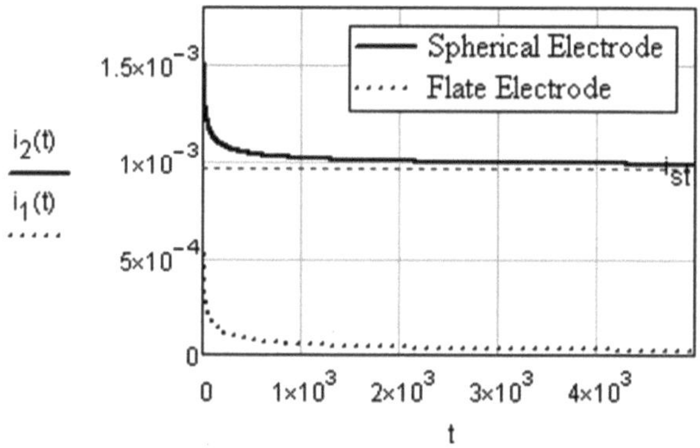

Fig. 5.10 Rate of diffusion process on a spherical electrode as a function of time

Equation (5.18), in contrast to (5.16), contains a term that does not depend on time – a stationary component of the limiting diffusion current. Therefore, potentiostatic electrolysis on a spherical electrode differs from that on a planar electrode and ends up with a stationary regime. The process rate does not change over time in this mode (Fig. 5.10).

5.5.1 *Nonstationary Diffusion Under Galvanostatic Conditions*

Consider a circuit under *galvanostatic* conditions: the current passing through the working electrode does not change during electrolysis (or, it changes according to the predefined law).

First, define the initial and boundary conditions. Initially, just like under potentiostatic conditions, $C_{Ox}(x,0) = C^0_{Ox}$. But the relationship describing the electrolysis regime is entirely different. Under potentiostatic electrolysis, this boundary condition defined the surface concentration of the oxidized form of depolarizator. Here, it sets the constant concentration gradient of the oxidized form for $x = 0$. This can be demonstrated using the main equation of diffusion kinetics:

$$i = nFD_{Ox}\left(\partial C_{Ox}(x,t)/\partial x\right)_{x=0}.$$

Given $i = const$:

$$(\partial C_{Ox}(x,t)/\partial x)_{x=0} = i/nFD_{Ox} \tag{5.19}$$

Equation (5.19) represents the boundary condition for this problem. Another condition sets the limits of diffusion area: $C_{Ox}(\infty, t) = C^0_{Ox}$. Let's define the function $C_{Ox}(x,t)$.

It has been previously shown that the second Fick's law can be transformed into an ODE with a general solution (5.14). The integration constants can be defined similarly. One of the boundary conditions requires $C_2 = 0$; otherwise, we would have an infinitely large right-hand part of (5.14) when $x \to \infty$. Therefore:

$$LC(x) = \frac{C_{Ox_0}}{s} + C_1 \exp\left(-\sqrt{\frac{s}{D_{Ox}}}x\right). \tag{5.20}$$

The boundary condition (5.19) is transformed into

$$dLC(x)/dx = i/(nFD_{Ox}s).$$

On the other hand, differentiation of (5.20) on x gives:

$$dLC(x)/dx = -C_1\sqrt{\frac{s}{D_{Ox}}}\exp\left(-\sqrt{\frac{s}{D_{Ox}}}x\right).$$

Given $x = 0$:

$$\frac{i}{nFD_{Ox}s} = -C_1\sqrt{\frac{s}{D_{Ox}}}.$$

Consequently,

$$C_1 = -\frac{i}{nFD_{Ox}{}^{1/2}s^{3/2}}.$$

The final solution in operational form can be written as:

$$LC(x) = \frac{C_{Ox_0}}{s} - \frac{i}{nFD_{Ox}{}^{1/2}s^{3/2}}\exp\left(-\sqrt{\frac{s}{D_{Ox}}}x\right).$$

Inverse Laplace transform gives the sought function $C_{Ox}(x,t)$ (Fig. 5.11). Consequently,

$$C_{Ox}(x,t) = C^0_{Ox} - \frac{2i}{nF}\sqrt{\frac{t}{\pi D_{Ox}}}\exp\left(-\frac{x^2}{4D_{Ox}t}\right) + \frac{ix}{nFD_{Ox}}erfc\left(\frac{x}{2\sqrt{D_{Ox}t}}\right).$$

$$\frac{Cox_0}{s} - \frac{i}{\left(Dox^{\frac{1}{2}} \cdot s^{\frac{3}{2}} \cdot n \cdot F\right)} \cdot \exp\left(\frac{-s^{\frac{1}{2}} \cdot x}{Dox^{\frac{1}{2}}}\right) \quad \begin{vmatrix} \text{assume}, Dox > 0 \\ \text{assume}, x = \text{RealRange}(0, \infty) \\ \text{invlaplace}, s \\ \text{simplify} \end{vmatrix} \rightarrow$$

$$\frac{Cox_0 \cdot Dox \cdot n \cdot F \cdot \pi^{\frac{1}{2}} + i \cdot x \cdot \mathrm{erfc}\left(\frac{1}{2} \cdot \frac{x}{Dox^{\frac{1}{2}} \cdot t^{\frac{1}{2}}}\right) \cdot \pi^{\frac{1}{2}} - 2 \cdot i \cdot t^{\frac{1}{2}} \cdot Dox^{\frac{1}{2}} \cdot \exp\left(\frac{-1}{4} \cdot \frac{x^2}{Dox \cdot t}\right)}{Dox \cdot n \cdot F \cdot \pi^{\frac{1}{2}}}$$

Fig. 5.11 Solution of a nonstationary diffusion problem for galvanostatic conditions

$n := 1$ electron number

$Dox := 1e{-5}$ diffusion coefficient, cm^2/s

$i := 0.1$ current density, A/cm^2

$Cox_0 := 1e{-03}$ concentration of the oxidized form, mol/cm^3

$x := 0, 0.0001 .. 0.03$ range of distances from a surface, cm

Faraday constant, C/mol $F \equiv 96485$

$$C(x,t) := Cox_0 - \frac{2 \cdot i}{n \cdot F} \cdot \sqrt{\frac{t}{\pi \cdot Dox}} \cdot \exp\left(\frac{-x^2}{4 \cdot Dox \cdot t}\right) + \frac{i \cdot x}{n \cdot F \cdot Dox} \cdot \mathrm{erfc}\left(\frac{x}{2 \cdot \sqrt{Dox \cdot t}}\right)$$

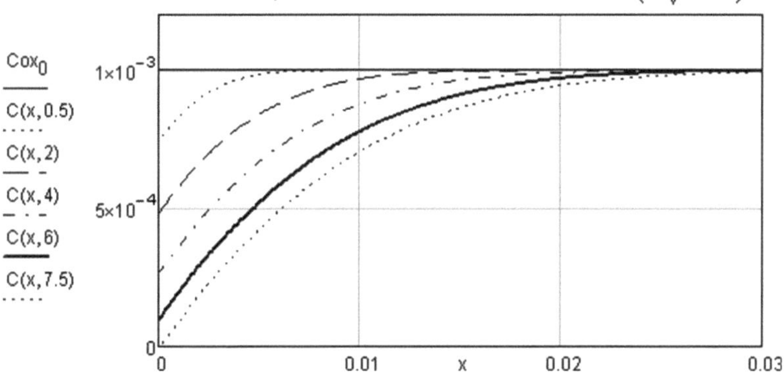

Fig. 5.12 Concentration distribution of the oxidized form under galvanostatic conditions

The last equation can be easily understood if presented graphically. Let's set reasonable parameter values and plot the concentrations of oxidized form as the functions of distance from the electrode, x, at different time. The results are shown in Fig. 5.12.

The reader can see that the surface concentration of oxidant decreases over time. The slopes at the beginning of plot remain constant. This is a consequence of the nonchanging gradient of surface concentrations (boundary condition (5.19). After some time, $C_{Ox}(0, t)$ reaches its minimal value. This period is called *transition time*, τ.

Chapter 6
Interface of Mathcad 15 and Mathcad Prime

6.1 Input/Displaying of Data

This chapter also describes a new Mathcad version, Mathcad Prime, and compares it with the current one, Mathcad 15, which will remain in-demand for a long time among those who make engineering calculations, particularly in the field of chemical technologies. We will refer to the conventional Mathcad versions (5–15) as Mathcad below, and call Mathcad Prime its full name.

When Mathcad appeared on the market of computational tools, it was advertised as, and was called a *super calculator*: if one clicks = (the evaluation equal sign) after an *expression* Mathcad shows the result (Fig. 6.1).

Comment. *The equal sign is the usual "thin" equal sign located on the upper-right corner of the keyboard. There is also a "bold" equal sign of equality, Boolean equality, inserted by clicking the keystroke < Ctrl>+<=> or by the corresponding button from the* ***Boolean*** *toolbar. A novice user often confuses these operators and presses <Ctrl> + <=> instead of <=> or vice versa, that, in turn, scrambles the calculation. There was no such confusion in the DOS-versions of Mathcad. Keystroke <Ctrl>+<=> displayed not a "bold" equal sign but a wavy line; a half of an approximate equal sign. By the way, that corresponds to the essence of the solved problem, for example, in finding the roots of a system of equations in a* `Given/Find` *solve block. The roots of such system returned by the function* `Find` *called numerically and not symbolically, make the left part equal to the right one* `approximately` *rather than* `exactly`.

There are no Math and Calculator Palettes in Mathcad Prime. Instead, there is a Math Ribbon with menus of operators, see Fig. 6.2

Comparing with traditional Mathcad versions the changes in Mathcad Prime interface follow the general trend of Windows development.

V.I. Korobov and V.F. Ochkov, *Chemical Kinetics with Mathcad and Maple*,
DOI 10.1007/978-3-7091-0531-3_6, © Springer-Verlag/Wien 2011

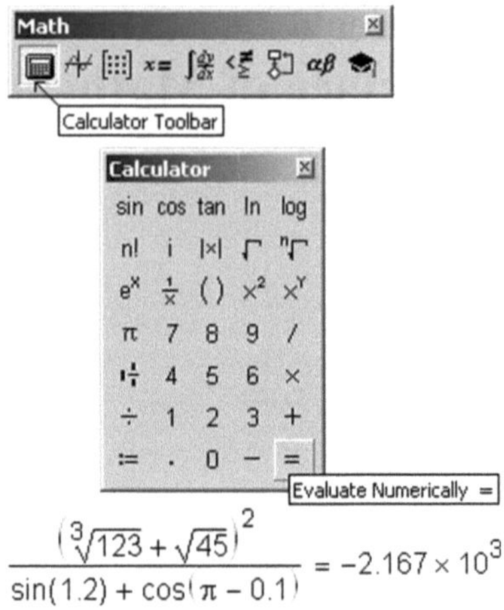

Fig. 6.1 Mathcad as a supercalculator

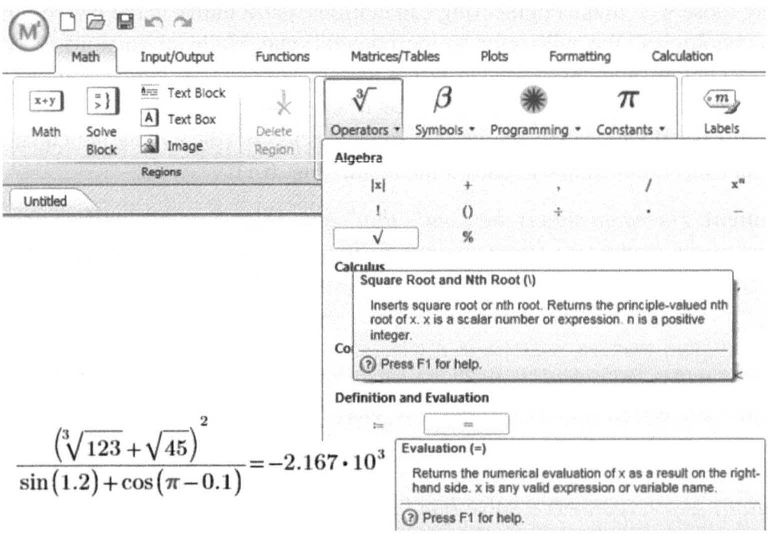

Fig. 6.2 Mathcad Prime as a supercalculator

The term “super calculator” is retained in Mathcad Prime in the name of the toolbar[1] **Calculator** (see Figs. 6.1). It contains basic mathematical functions and operators as well as the operator = (the latter showing as **Evaluate Numerically** when the mouse pointer hovers over the button). This toolbar actually copies the

[1]Another name for a toolbar is “palette” where a mouse cursor, like a brush, takes the “paint”, the necessary instrument to solve a problem.

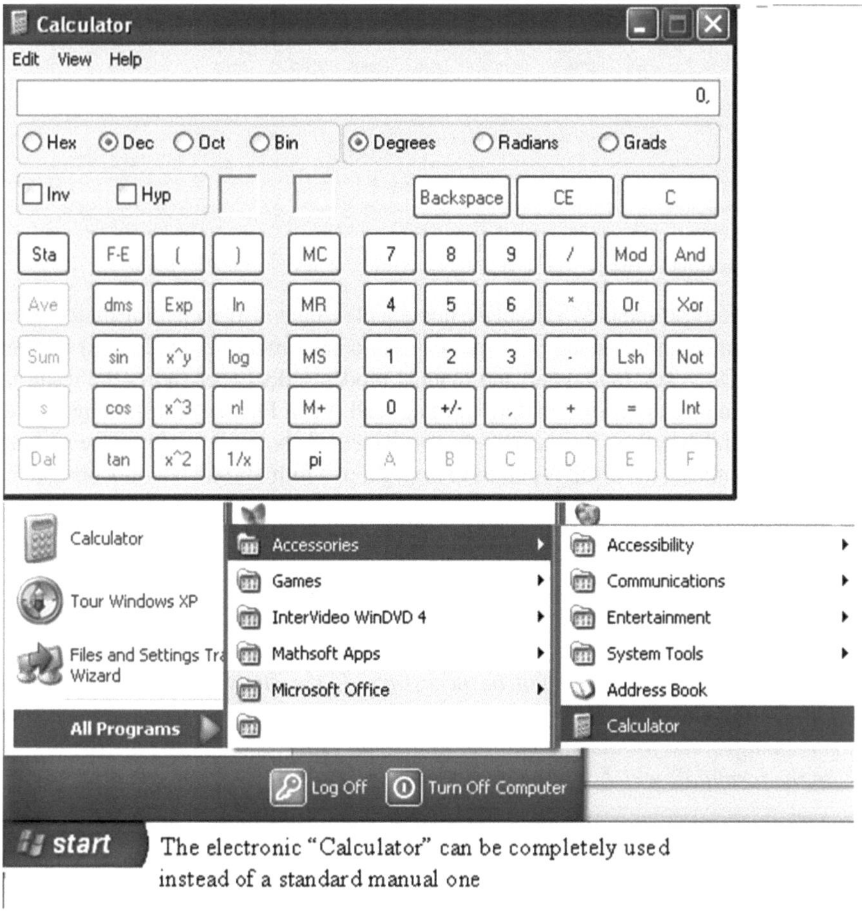

Fig. 6.3 Windows scientific calculator

keyboard of the "scientific" calculator,[2] the real one or virtual, for example, such as contained in Windows. Figure 6.3 shows the so-called scientific calculator, and the commands to call it in one Windows version.

The equal sign = works the same as that in another popular calculation software, the electronic worksheet MS Excel.[3] However, there it is placed not at the end of a mathematical expression as in Mathcad (see Fig. 6.1), but in the beginning (Fig. 6.4).

---

[2]Talking about the numeric interface, note that digital keyboards of calculators (computers) and phones are antipodes: the first have numbers one, two, and three on the bottom, the second on the top. The arrangement probably cannot be changed now, although these two devices come to meet each other (Pocket PC and SmartPhone).

[3]The Mathcad developers often note, that the environment "originated" from the electronic worksheets but did not adopt their main disadvantage: the formulae are close for analysis and modernization.

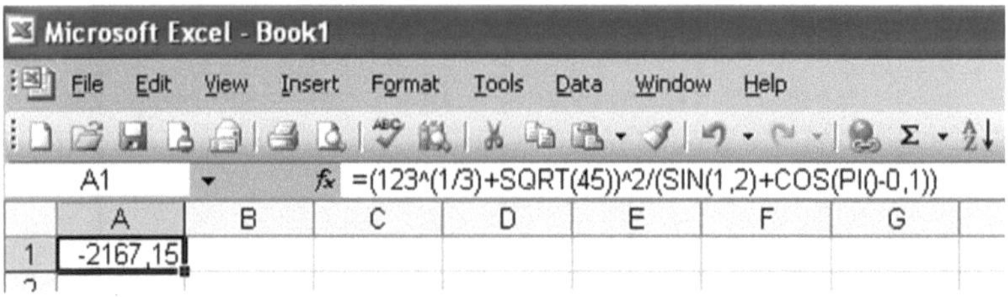

Fig. 6.4 Calculating in Excel

Both in Mathcad and in MS Excel the numerical result is displayed immediately after typing a formula and pressing the Enter key (automatic mode set by default) or after pressing < F9 > key (automatic and manual modes). MS Excel shows the result in the cell containing the corresponding formula (cell `A1` in Fig. 6.4) substituting for the formula itself. Therefore, we cannot see *simultaneously* both *all* the formulae and the results. In Mathcad, the result is placed to the right of a formula and does not obscure it.

Comment. *Mathcad has a mode, which shows the result of an analytic conversion in place of the initial expression. Whether to show simultaneously both the formula and its result or to do it sequentially depends first, on the saving of spare place on a screen and in print out and, second, on the purpose of calculation, its direction. If the purpose of a calculation is mere practice, processing new data and showing a result, the formula would not be necessary. If the study of the calculation is required for educational purposes, additional checking of the result, or making some changes, the formulae would be desirable. Mathcad has the functionality to hide the formulas, which will be discussed later.*

Let us say right away that the calculator in Mathcad (Fig. 6.1) differs from other similar calculation systems (see Figs. 6.3 and 6.4). It can work both with *numbers* and with *physical quantities*. Figure 6.5 shows the Mathcad calculation of the power of a human heart having the following parameters: it pumps 70 ml of blood per second (mean statistic of a man at rest), the pressure increases from 80 to 120 torr, and efficiency of this life pump is 70%.

In Mathcad Prime units are entered via ribbon **Math** and drop-down menu **Units**, Fig. 6.6.

Mathcad has built-in units for basic physical quantities as well as fundamental mathematical constants, such as `e` (the root of natural logarithm) or π (ratio of circumference to its diameter), etc. Mathcad also has a quite thorough reference book containing basic mathematical, physical, and chemical constants, which can be copied into a worksheet with corresponding units. For example, Fig. 6.7 shows the well-known Einstein formula $\mathtt{E} = \mathtt{mc}^2$ which could be calculated in Mathcad, taking velocity of light in a vacuum from the Reference book.[4]

[4]The velocity of light in a vacuum in Mathcad 15 is transferred (copied) from the Reference book into Mathcad itself. Now variable `c` retains this physical constant. Earlier Mathcad versions included only one constant like that, g – acceleration of gravity.

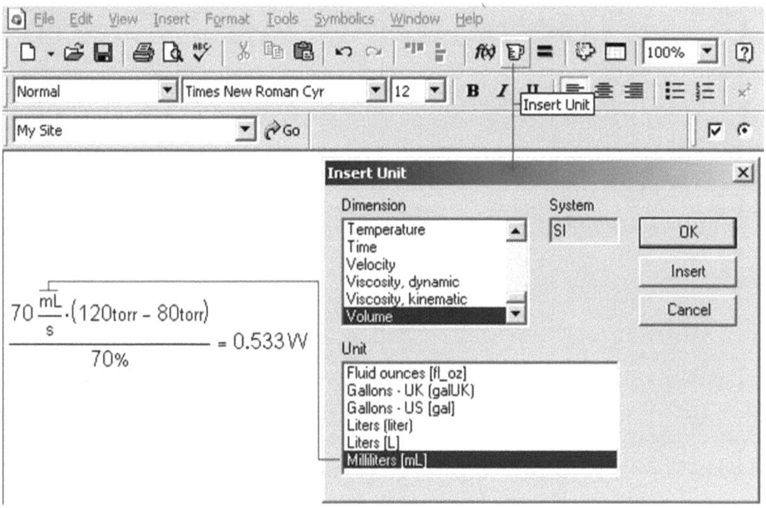

Fig. 6.5 Calculation with physical quantities in Mathcad

In Mathcad Prime physical constants are entered via ribbon **Math** and drop-down menu **Constants**, Fig. 6.8. The number of constants in Mathcad Prime, is much lower than in the Resource Center of traditional Mathcad (see Fig. 6.7). Only 15 constants have been left in Mathcad Prime but all of them are built-in now: to call them one need not open Resource Center, but just type the corresponding name on a worksheet: c (sound velocity in a vacuum – besides, it was built-in in traditional Mathcad, too), e.c. (Elementary Charge – this constant and constants below are new in Mathcad Prime), h (Planck's constant), etc. Such a reduction of constant list is due to the fact that today almost all computers used by engineers and students are Internet-connected and it is always possible to find required constant there.

There are two other principal differences between Mathcad and Excel.

The formulae in Excel are typed in *a single string* and the mathematical operators are written in "programmer" form: for example, x ^ 2 instead of x^2 etc. This notation has such a number of parentheses that only a computer but not a human[5] can read long equations. For example, it is sometimes difficult to understand where the nominator or the denominator is. A formula in Mathcad follows a "multilevel" natural format.

[5]The latest versions of Excel mark pairs of opening and closing parenthesis with different colors at scanning the formula by the pointer when editing a formula.

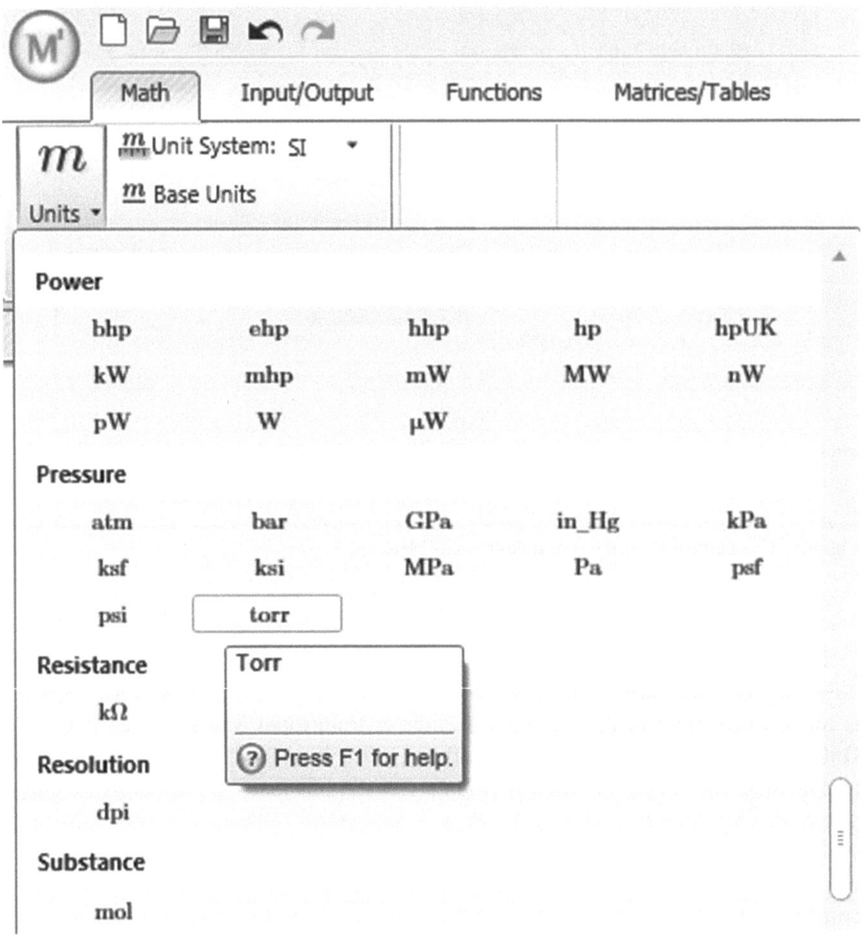

Fig. 6.6 Units in Mathcad Prime

There is another significant difference. The worksheet in Excel is divided into columns and rows and the data, numbers, test, formulas etc. are entered in the cells formed by these. The Mathcad worksheet is a "blank paper" where we can type anything anywhere. In addition, there is a possibility to insert information in table format (see Figs. 6.25 and 6.29).

The **Evaluate Numerically** operator = contains a large number of hidden defaults that can be overridden by clicking **Result** from **Format** menu or double-clicking the result itself. Figure 6.9 shows two inserts with dialog box for the formatting of numerical result.

The following is a list of numerical evaluation defaults, far from complete, which also evolved through the Mathcad versions.

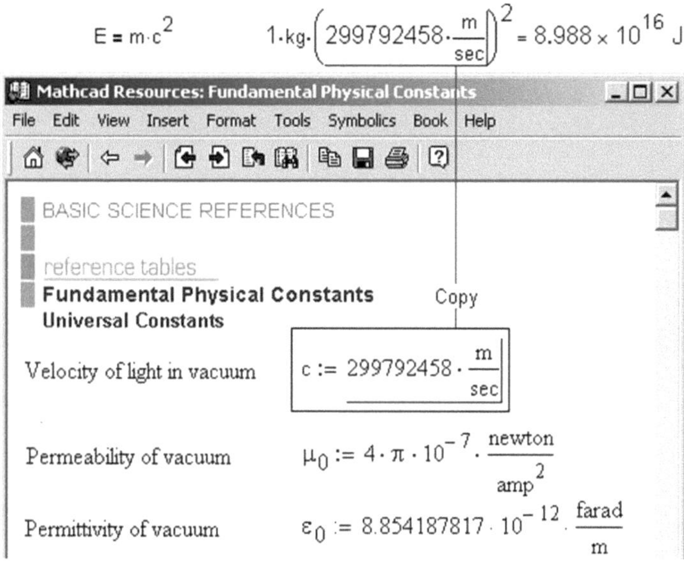

Fig. 6.7 Work with the Mathcad reference book

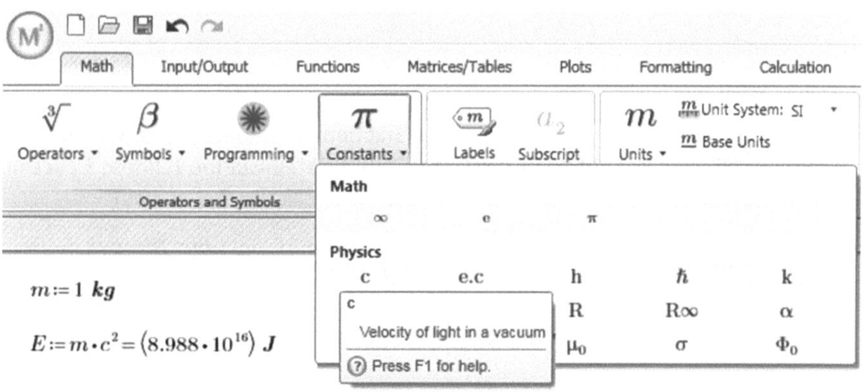

Fig. 6.8 Built-in constants in Mathcad Prime

- The numerical result is displayed as decimal but not as binary, octal, or hexadecimal.[6] We can change the setting via **Radix** from the **Display Options** tab shown in Fig. 6.9

---

[6]Windows' own calculator (Fig. 6.2) also tunes up the decimal system (Dec) by default but can work with hexadecimal (Hex), octal (Oct), and binary (Bin) systems. Entering numbers in Mathcad non-decimal system are shown by the suffix at the end of the number: *h* hexadecimal, *o* octal, *b* binary. The absence of a suffix is a sign that a number is decimal. Similar suffixes finish numbers shown by the operator =.

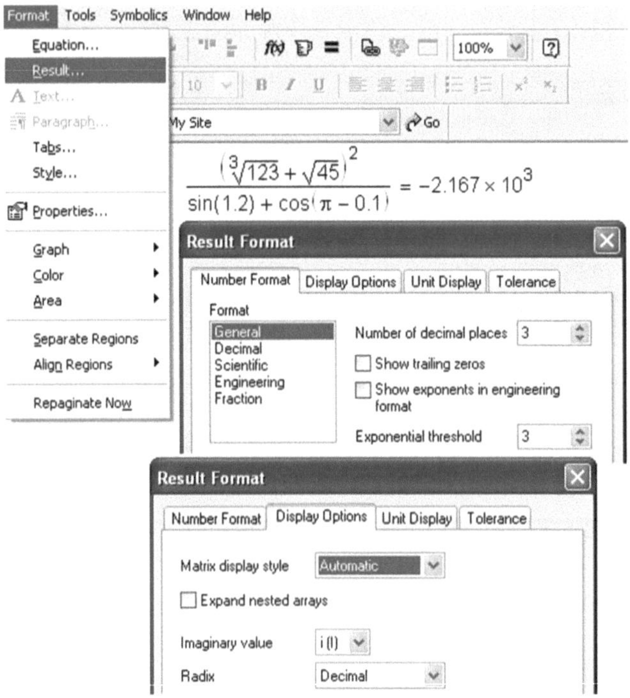

Fig. 6.9 Result format dialog box

- The numbers are displayed as decimal fractions (for example, 1.333) but not vulgar (4/3 or 11/3) although that is possible in Mathcad too (choose **Fraction** from the **Format** list in the **Number Format** tab as Fig. 6.9 shows)

Comment. *Inserting a number as vulgar fraction is possible through normal addition or division (for example, a:=1+1/3) or, if one prefers omitting the plus sign, by clicking*  *on the* ***Calculator*** *toolbar.*

- Only three numbers are displayed after point (**Number of decimal places** from the **Number Format** tab in Fig. 6.9)
- The number is displayed in exponential notation as it is less than 10^{-2} (**Exponential threshold** from the **Number Format** tab in Fig. 6.9)
- If the fractional part of a number ends in zeros they are not displayed (**Show trailing zeros** from the **Number Format** tab **in** Fig. 6.9)
- The number is approximated to the last digit in the fractional part
- The imaginary unit of a complex number is marked as *i* but not *j*
- The background is white and the numbers are black etc[7]

[7]One of the authors has a method of working with students on informatics seminars (see the program at http://twt.mpei.ac.ru/ochkov/Potoki.htm). The author asks students to list any

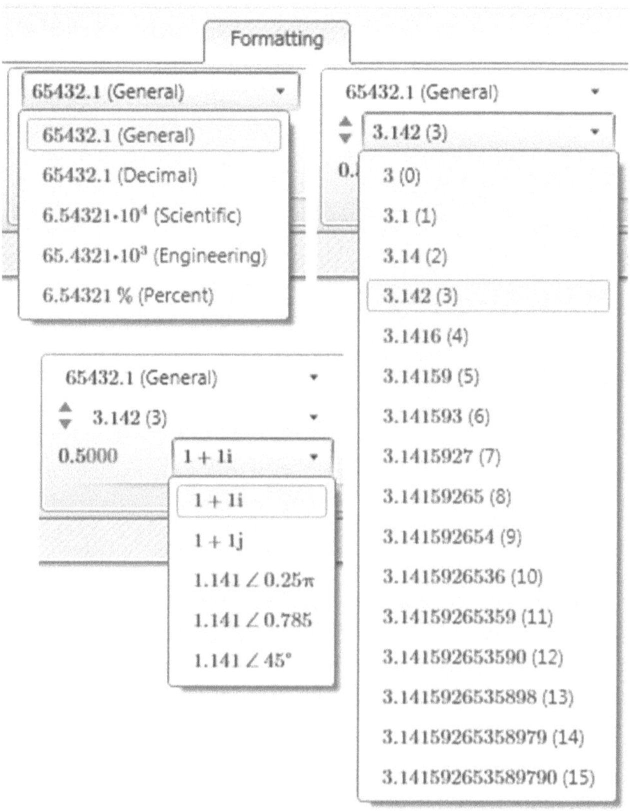

Fig. 6.10 Formatting results in Mathcad Prime

Mathcad users can both change these defaults by formatting the current result and change the worksheet default. After changing the settings, click **Set as Default** in the **Number Format** tab from the **Result Format** dialog box.

Mathcad Prime inherits all the capabilities of number formatting that were available in traditional Mathcad (Fig. 6.9). It has been appeared a convenient capability to display complex numbers in algebraic form (1 + 1i) and also in trigonometric form (1.141∠45°), see Fig. 6.10.

The second most important operator is the *assignment* (*definition*) designated as :=. To enter it just press <:> (colon). Mathcad conveniently adds the second sign (=) by shifting the cursor to the placeholder for entering a number or an expression. For example, Fig. 6.11 shows the insertion of three user variables, a, b, and c, two of which (a and b) are the input data, and the third is defined as the result of the calculation.

---

instruments, ideas, modes used in computer science; the same list of defaults in formatting the numerical result. The student who tells the last point gets a bonus.

Fig. 6.11 Entering numerical values into a variable

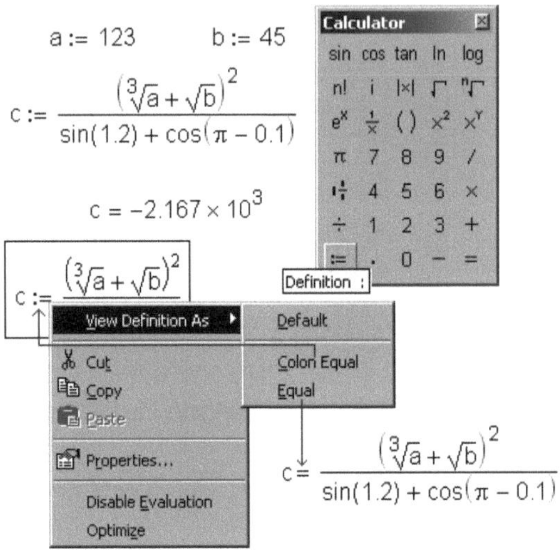

Comment. *Another way to enter the define sign in Mathcad is to click the corresponding button on the toolbar (for example, see left bottom corner on the* **Calculator** *toolbar in Fig. 6.1). The third way is to copy and paste the assignment operator entered before. In Mathcad any operation can usually be done at least three different ways.*

In Mathcad Prime definition operator is entered via ribbon **Math** and drop-down menu **Operators**, Fig. 6.12.

Starting with the seventh version users have the capability to change the definition sign from **:** = (style of the Pascal language) to = (BASIC[8]). The pop-up menu for changing the view of the definition operator is displayed by clicking on the operator with the right mouse button. Figure 1.7 shows that it has three positions: **Default**, **Colon Equal**, and **Equal**. Although, if the pointer is placed on the expression in which the assignment sign was changed from **:** = to = the previous default symbol **:** = is returned. The users seldom substitute the = sign for := since it results in the confusion: for example:

`a = 1` `a = 1` What is this?

The notation

`a:=1` `a = 1` is quite clear.

[8]A few may remember that in early BASIC the assignment operator should be written as Let a =. That was done to distinguish the operator from Boolean operator (in Mathcad as was noted before this operator has bold sign and is typed as < Ctrl>+<=>). Further it became possible to omit the codeword `Let` in BASIC programs but the interpreter routine adds it persistently after moving the cursor to the next line. Further it became possible to not use the word although the opportunity to write `Let a` = instead of `a` = exists even in Visual Basic.

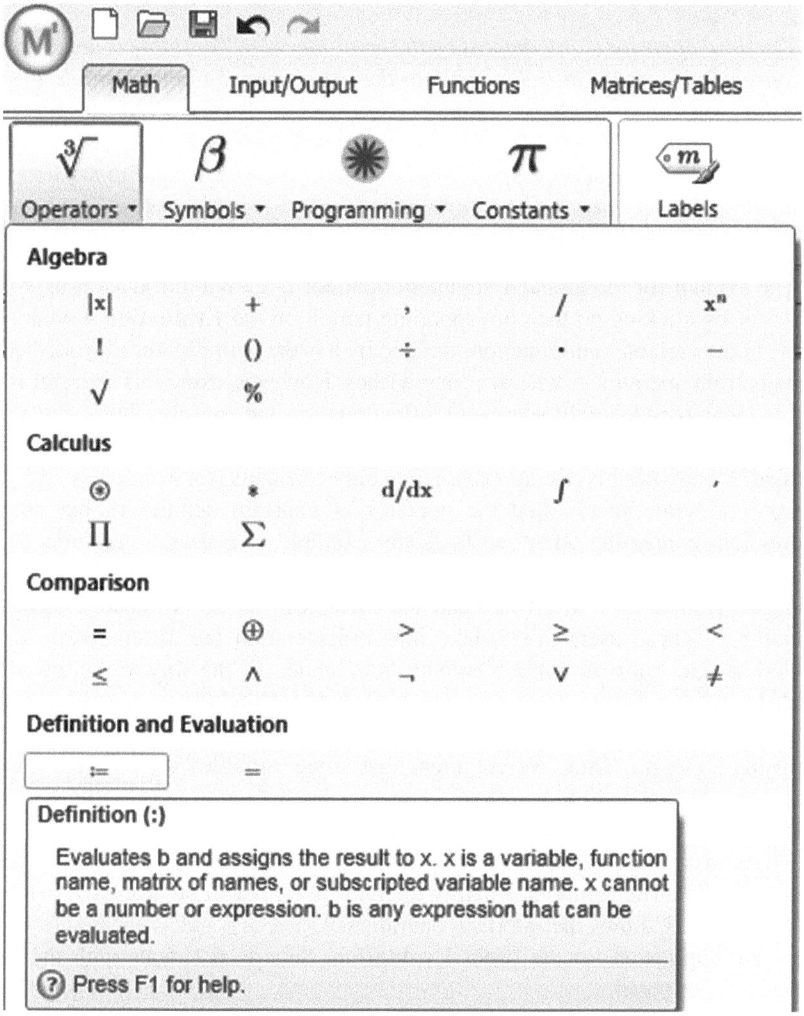

Fig. 6.12 Operator definition in Mathcad Prime

The operator := should be called more precisely *the semi-global assignment* because we can compute with the function defined by the operator anywhere to the right and below of the expression that defines it.

Comment. *The variables in the calculation are sure signs that distinguish programming from simple calculating – compare calculations on the Figs. 6.1 and 6.11. The variables in Mathcad are discussed in Sect. 6.2.*

Comment. *Here is a typical mistake made by a novice Mathcad user. The horizontal chain of operators is entered:*

`A := 1 b := 2 c := a + b`

The third operator is interrupted by the error message "Variable is not defined". An operator located a little higher than the others may be imperceptible to your eyes but will interrupt the calculation. To avoid such errors the recommended practice is to type the operators in columns pressing <Enter> after each expression. To make worksheets more compact we can use the second dimension in a finished worksheet, locating some operators in lines. It is also possible to have the third dimension in Mathcad (see Sect. 6.5).

The symbol for the global assignment operator is ≡. We introduce it by typing < ~> or by clicking on the corresponding button on the **Evaluation** toolbar. The scope of the variables and functions defined by it is the entire Mathcad worksheet; it spreads from operator ≡ wherever one wishes. However, using this operator is not highly recommended; it breaks cause–effect relation and confuses those who aim to study the document: a variable has a defined value but is not clear where it was defined. Moreover, this operator can define only constants (for example `a ≡ 3`, the operator is sometimes called the operator of constant definition), but not the expressions containing other constants (for example, `a ≡ 2b`). Some users define the variables that change the values in calculation using the operator := (for example, from `a:=1` to `a:=2`) and the variables that do not change using the operator ≡. These users fail to take into consideration the difference in scope. Global assignment operator can extend its influence all the way to the top of the worksheet, for example, distort the plots created by QuickPlot. Separating assignment operators ■ := *`constant`* from ■ ≡ *`constant`* can be useful before the keyword `Given.` Thus, we comment that some variables are constants, while others are the first approximation of the results. Such separation is useful to make a Mathcad worksheet well suited for Internet publication.

Mathcad can solve problems *numerically (approximately)* as well as *analytically (symbolically)*. The symbolic equality operator is →. It will be discussed below.

As Fig. 6.13 shows the interface operators =, `:=`, ≡, and → located (some of them are duplicated) in the panel **Evaluation**. Section 6.2 deals with the other buttons in this panel.

Mathcad also has a local assignment operator. Operator ← makes the variables defined only within a Mathcad-program.

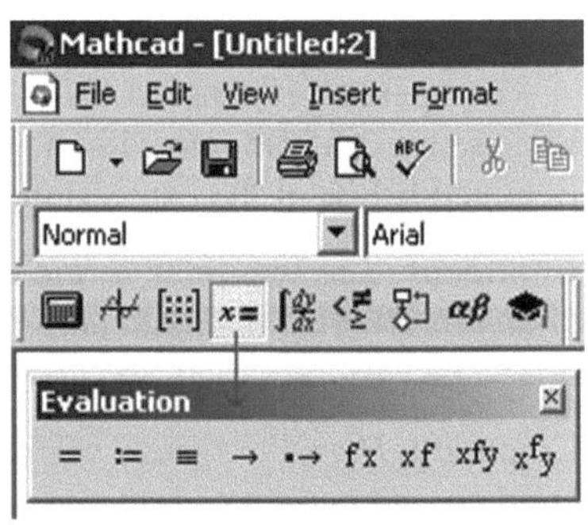

Fig. 6.13 Evaluation palette of Mathcad

Comment. *Both operators* ← and ≡ can be changed to the more customary =. Fig. 6.11 outlines how to do it.

Mathcad Prime keeps only two interface operators = and := (Figs. 6.2 and 6.11). The reason for it is that other interface operators (shown in Fig. 6.13) are rarely used and operator → (symbolic evaluation) will be introduced only in the next version of Mathcad Prime.

By means of the very convenient built-in feature, *Style of variables,* Mathcad allows distinct variables to have the same textual name. By default, the variable introduced into a worksheet is assigned the style Variables (Fig. 6.14).

A Mathcad user can change the style of a variable and insert a new variable with the same name into a worksheet. Figure 6.14 shows this feature of Mathcad: a certain moment of force, m, is calculated (the product of force (or, rather, weight – the product of a mass m and acceleration of gravity g) and arm length of the force application) using two variables of the same name. Speaking about the texts and comments, Mathcad offers a set of styles for dealing with text regions (*see* Sect. 6.3) allowing the user to format characters, words, paragraphs (title of the first level, title of the second level) practically the same as in MS Word.

In Mathcad Prime separation of like-named variables (m – mass, m – length unit, etc) is conducted via **Labels** menu of ribbon **Math**, see Fig. 6.15. At that, a variable or a function in Mathcad Prime can be marked as Variable, Unit (Fig. 6.15), Constant, Function, System, or Keyword.

Beginning from the eighth version, Mathcad allows to enter the evaluation equal sign typing not <:> but <=>. The point is: if the variable is *not defined* the operator = (displaying) is changed *automatically* into := (definition). Otherwise, if the assigning variable has already been defined its numerical value is represented. So, one can check that the name is *free*: neither a user nor Mathcad has defined it yet. Such technique allows us to omit a number of errors mentioned before. First, some *predetermined variables* could be changed: e:=5, m:=1, A:=2 etc. Variable e is the radix of the natural logarithm and variables m and A are units of length (meter) and current strength (Ampere) accordingly – see Fig. 6.14. Second, one-could forget that he has already defined the variable and assign it something new.

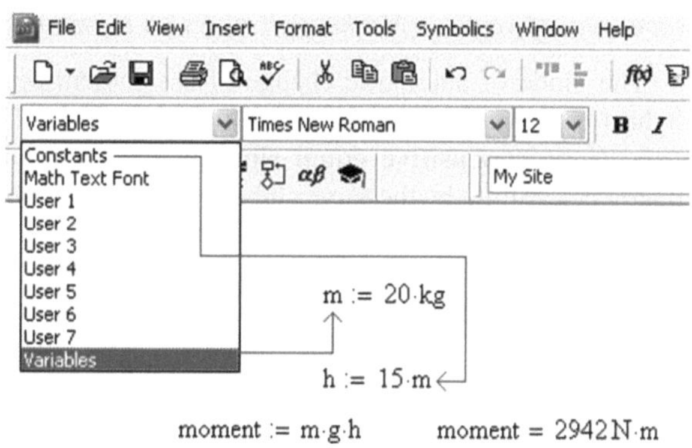

Fig. 6.14 Applying math styles to work with units

Fig. 6.15 Labels of variables and functions in Mathcad Prime

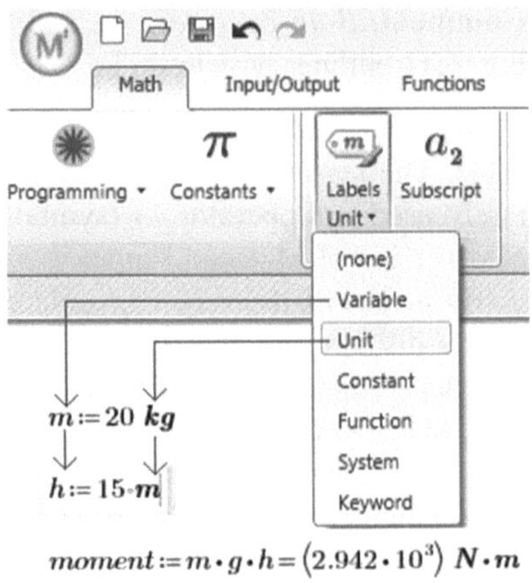

The situation when the same variable has one value in one part of Mathcad worksheet and another in another is not right.

The technique of variable redefinition was used when the memory was a limiting parameter in computing so that the spare variable was redefined right away to omit memory overflow.

Comment. *In Mathcad the variable keeping a massive (vector or matrix) can include text, scalar (real or complex number) and vice versa. A variable in Mathcad is not connected with one or another variable type that exists in most part of traditional programming languages.*

We use such technique now in locating the extensional *arrays* (vectors and matrixes, simple and compounded) in computer memory (including that working with Mathcad). Although, redefining the new smaller array or ever a scalar value does not change the volume of memory (the mechanism of static arrays but not that dynamic).

Comment. *Besides, this technique is not applicable to arrays. If one types* `v` *in a spare Mathcad worksheet* `[0` *and pushes the button <=> to enter zero element (constant) of vector v Mathcad displays an error message instead of equal sign.*

The mode of automatic change of assignment operator from = to := (such a hybrid of two operators called SmartOperator[9]) is disabled by throwing of a flag from **Context-sensitive equal sings** of the **General** tab from the **Preferences** dialog box calling by the same command from the **Tools** or **Math** menu in Mathcad 2001i and in earlier versions (see Fig. 6.16).

Figure 6.16 shows a very interesting and useful number 200 in the right bottom corner of the counter **Maximum number of undoable action per document**.

[9]It appeared in Mathcad 7.

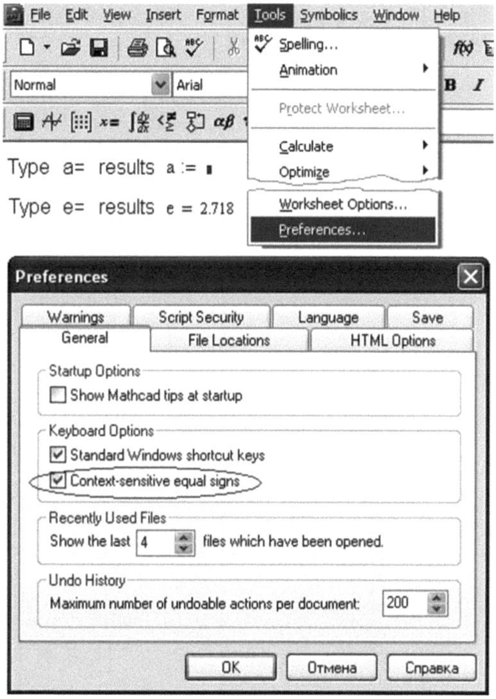

Fig. 6.16 General tab of the preferences dialog box

This is new feature appeared in the 11-th version. Here is a typical situation: one wakes up in the morning and thinks, "I made a stupid mistake yesterday, could I return "to yesterday" and "fix" it?" Alas (or fortunately), it is impossible in real life but is doable to some extent with computers. Just push the **Undo** button. Earlier versions of Mathcad allow us to do that only within the limits of a formula while the pointer is not taken away from it. Versions 11 and 12 of Mathcad allow the command **Undo** to affect the whole worksheet. In Fig. 6.16 the number indicated 200 is the number of such steps.

At a certain stage of Mathcad development it becomes possible to enter some interface operators in *tandem*, feeding one input/display operator into another. Thus, point 1 of Fig. 6.17 shows the use of the operators $\rightarrow$ and $=$ (symbol and numerical result) in solving the equation. The advantage of a symbolic result is absolute accuracy and numeric shows the distinct location on the number axis (or on a plane in the case of complex number). The tandem use of operators allows us to combine those advantages.

Comment. *Including one operator (one function) into another operator (function) on the place of operand (argument) is a traditional programming method. But it is quite exotic for input/display operators in Mathcad.*

Figure 6.17 at point 2 shows us the tandem operators at work. $:=$ and $\rightarrow$ (Definition and Evaluate Symbolically) allow us to enable symbol mathematics,

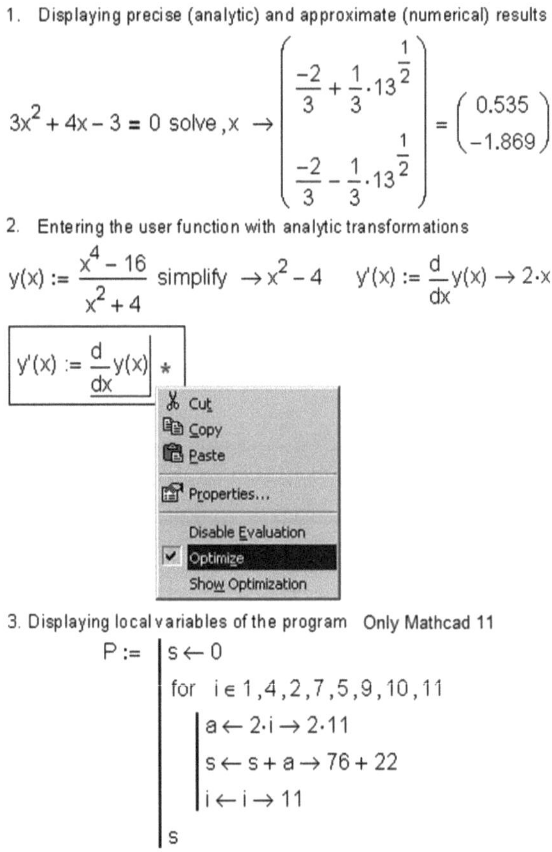

Fig. 6.17 Examples of Mathcad operators in tandems

for example, to create a user function or its derivative. Those two useful tandems (points 1 and 2 in Fig. 6.17) were in the category of undocumented techniques and later on became kind of half-documented: they are not described in the product documentation but are recommended to be used on different Mathcad sites including the developer's official site http://communities.ptc.com/community/mathcad. The tandem of operators $:=$ and $\rightarrow$ is the same as **Optimize** mode from the pop-up menu shown in point 2 in Fig. 6.17 which makes analytic transformation before calculation. If such transformation have been made, a red star appears to the right of transformed expression; but not the expression itself as is the case with the tandem $:=$ and $\rightarrow$.

Still undocumented is the tandem operators $\leftarrow$ and $\rightarrow$ (local definition and symbolic equal sign, see point 3 in Fig. 6.17) which enable us to look through the variables in programs and is very useful for verification.

The evaluation equal sign became a stumbling block for those who started Mathcad 10 or 15 years ago, hearing of its unusual abilities for calculating sophisticated formulae (see Fig. 6.1), making graphs (see Sect. 6.6), animations (see Sect. 6.7), solving equations and systems of equations. Through a habit formed by working with Fortran or BASIC users typed `a =` in Mathcad instead of `a:`[10] and... refused to work with this mathematical package further. An error, unintelligible at first glance, had emerged: Mathcad informed them that a variable is not defined. Users tried to define that error by operator of variable type being guided by experience of working with programming languages, but Mathcad has no such operator. By the seventh version, Mathsoft has "surrendered" and stopped demanding users to type `a:` instead of more convenient and customary `a=`. Now the pendulum has swung to the opposite side: now to assign the value to a variable it is recommended to type `a =` rather then `a:`. We could have recommended that Mathcad developers exclude the operator `:=` altogether, but it is required for changing the value of predetermined variables (for example, TOL:=10^{-7}, ORIGIN:=1) and for defining a user function or an element of a vector. Although, even in these cases one can omit operator `:=` coping it from the place where it was created by "smart" operator =.

It is also advisable to use the equal sign for evaluation in defining *a user function* in Mathcad. After typing the name of a new function it is better to press not < (> (a parenthesis opened a list of arguments) but <=>. The reason is the same: to protect a user from the possible errors connected with reassignment of the functions. For example, if a user wants to insert a function named F and types couple of symbols F= Mathcad may display the following:

- F = 1 F the unit mode is not disabled. In this case, Mathcad remind us that farad (unit of electrical capacity) equals to 1 farad and this variable name is used.
- F = *number* means that the user variable with name F already exists in the worksheet.
- F = *function* means the user function with name F already exists in the calculation.

Comment. *In Mathcad 12–15 the word "function" now appears in brackets: F = [function].*

- F := — means that the name is available for use as a function or a variable.

There is another variant: if the name of planning function coincides with that of the built-in function after pressing <=> it will be duplicated, for example line = line. The Mathcad user should decide what to do in this situation. In some cases, reassignment of a function solves certain difficulties of calculation.

[10]Of course, there were not any toolbars with button := (see Figs. 6.1 and 6.8) in DOS-versions (non-graphic) of Mathcad. Up-to-date Windows versions of Mathcad get as an atavism functional keys < @>, <&>, <$> and others from DOS-versions which introduce to Mathcad worksheet plots (@), integrals (&), sum ($) and others (see Fig. 6.47). These keys can be disabled by keystroke < Shift> + <Ctrl>+<k > that change the cursor color from blue into red (emergency state) and vice versa. When the cursor is red one can enter reserved characters noted before and some other into the variable and function name (see Sect. 6.2.2).

Comment. *The name will be duplicated in mode of "early" Mathcad calculation mode. In the* ***Higher Speed Calculation*** *mode Mathcad shows* `line = function`. *Besides, Mathcad 12–15 has only "quick" mathematics.*

Old habits die hard: it is impossible to break Mathcad users of the habit of using operator := to define variables or functions and force them to use = which, as we noted before, automatically chooses what a user want of him : definition or displaying. Because of it, the users continue to make the errors mentioned above in connection with reassignment of variables. Through it Mathcad 11.1 (Mathcad 11 was published and patch 11.1 was released soon after it) provided the user with the mechanism for checking the reassignment of variables and functions, built-in and user-defined, opened and hidden in closed regions (see Fig. 6.33) and in referenced documents (see Fig. 6.31). For this, the **Preferences** dialog box gets the new **Warnings** tab (see Fig. 6.18).

Developers added a green wavy line to the Mathcad armory; in other environments, for example, in Word it is used to mark incorrect punctuation. If such line appears in a Mathcad worksheet under the name of assigned variable or defined function the user should correct an error or just note that something is out of order and reassign the variable and/or the function name.

Starting with Mathcad 2001 the operator := assigns both the numerical values and text literals, for example, `c:=123` or `c:="text"`, and uses visual programming *standard controls*: **Check Box, Radio Button**, **Push Button**, **Text Box**, **List Box,** and **Slider** shown in Fig. 6.19.

Comment. *The command* ***Control*** *both inputs the data into the worksheet and shows the result (Fig. 6.22 shows such an example).*

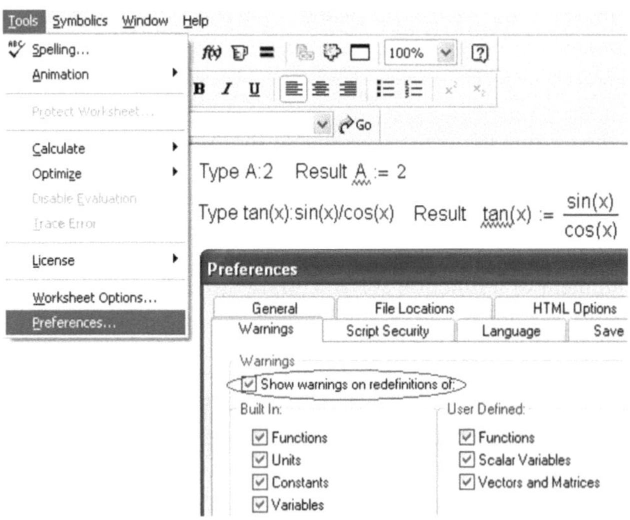

Fig. 6.18 Warnings tab of the preferences dialog box

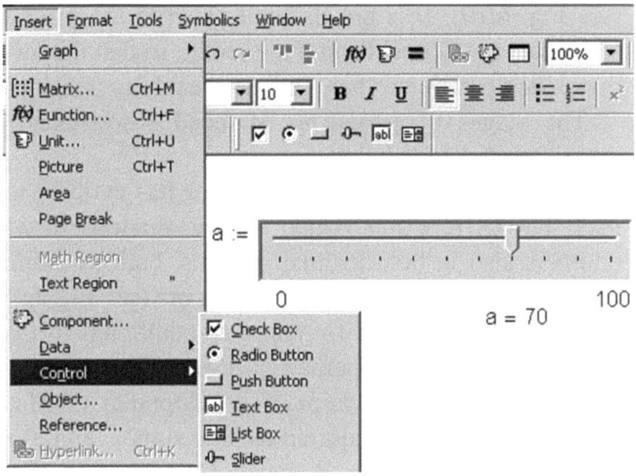

Fig. 6.19 Control components

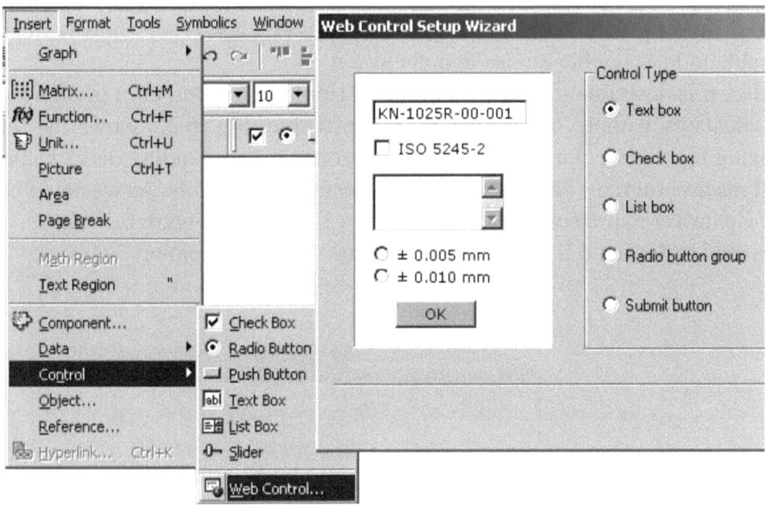

Fig. 6.20 Web controls

Controls appeared in Mathcad to simplify the input of new data. Figure 6.19 shows how the value of the variable a is changed by a simple mouse action without any typing. At that, Mathcad clearly shows the range of the variable (0–100), which cannot be overwritten by a user by mistake or knowingly and the current location of the variable within the range. The detailed description of these controls contains in *below*.

As the need arose to have worksheets accessible by the Internet Mathcad acquired the so-called Web Controls shown in Fig. 6.20. Unlike the standard Mathcad Controls, they do not require the user to write or edit the programs in one of the supported Active Scripting languages (JScript and VBScript) (see Fig. 6.19). It is not difficult to write the program but the point is that such scripts are not desirable on the Internet, so that they were left out. Besides, Controls make the documents "heavier" noticeably increasing the size of the file.

The stages of preparing Mathcad worksheets for Internet publication are described in details below.

The numerical evaluation operator has evolved too, or rather not the operator itself but its function. Thus, current versions of Mathcad allow users to show numbers also in non-decimal systems, both as common and decimal fraction, in scientific or in engineering notations, etc (see Fig. 6.9). However, operator = does not have any distinct differences or additional format options apart from "superimposing" it with the operator := (SmartOperator, see Fig. 6.16). Such changes are crucial from the standpoint of development of Mathcad Application Server. In particular, it is often required to hide sign = and everything to the left of it. This is possible in operator := using Controls and Web Controls (command **Hide Arguments** from pop-up menu shown in Fig. 6.21).

It would be so convenient if the numbers and texts would appear and disappear without showing their sources: `123` instead of `b = 123`. When showing text, it is desirable to hide double quotes that enclose it.

Also, it is desirable to change the format (font, size, color, etc.) of numeric and text constants without using macros. For example, Fig. 6.22 shows the way of changing highlight color of the variable (operator `b=`) depending on its value by a small macro program. The macros are used to format Controls. As we noted before, Web Controls require simple dialog boxes for formatting. Figure 6.21 shows one of these used to format a list; other examples are described below.

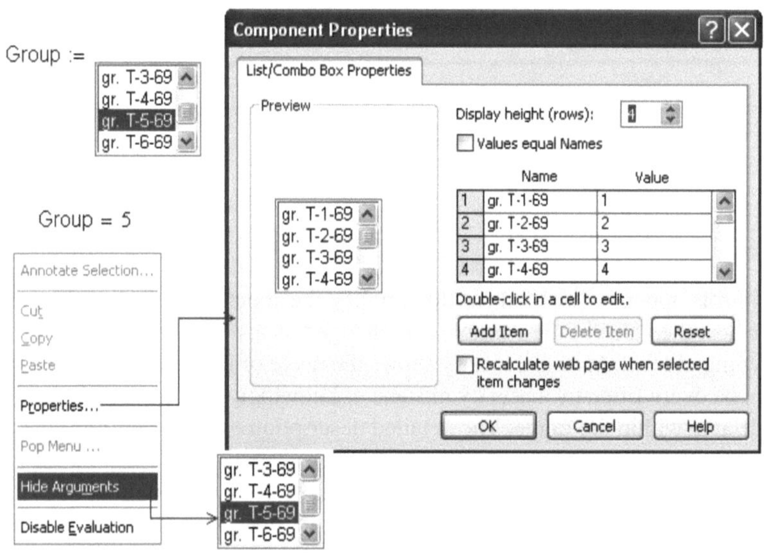

Fig. 6.21 Setting of a web control

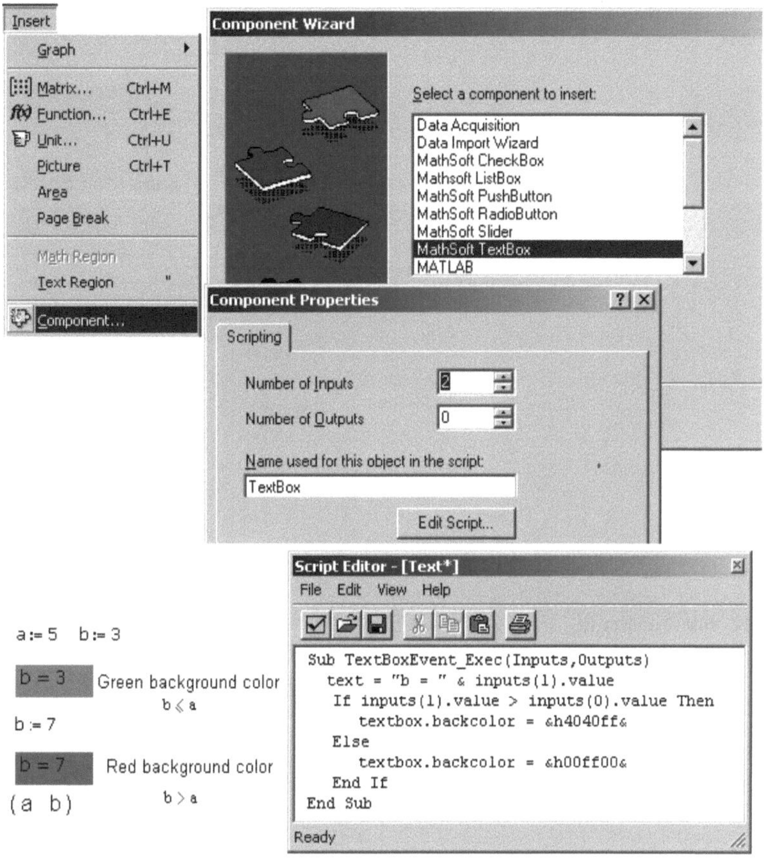

Fig. 6.22 Programming of a control

Especially we should take into consideration inserting the *arrays* (vectors and matrixes) into a worksheet, the data formatted in rows and columns. This storage method is widely used both in paper calculation and in computation.

The simplest method of inserting a matrix into a Mathcad worksheet is to choose **Matrix** from **Insert** menu, shown graphically in Fig. 6.23. However, it is not very comfortable to enter numbers with the **Matrix** command. First, we can insert at the most 100 (or 600 in Mathcad 13–15) array elements in such way. Because of this limit, many novice Mathcad users mistakenly think that 100 or 600 is the maximum number of elements in a vector or matrix although user documentation states that it can reach up to eight million.

Comment. *The vector (matrix with one column) up to 600 elements is inserted simply by command* `v:=stack(element1, element2, ...)`. The built-in function `stack` turns the list of its arguments (which can be both vectors and matrixes) into a vector.

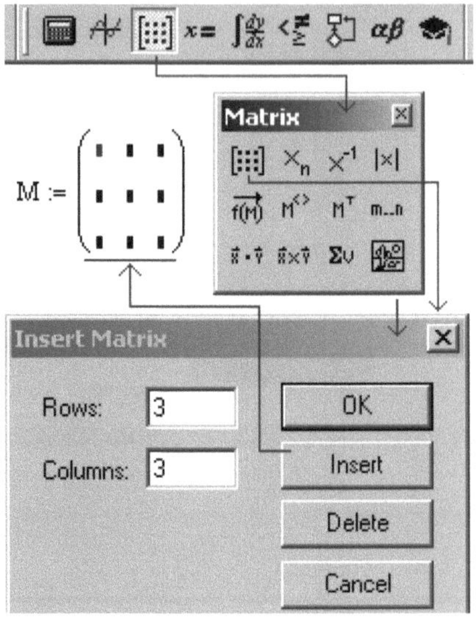

Fig. 6.23 Insert matrix dialog box

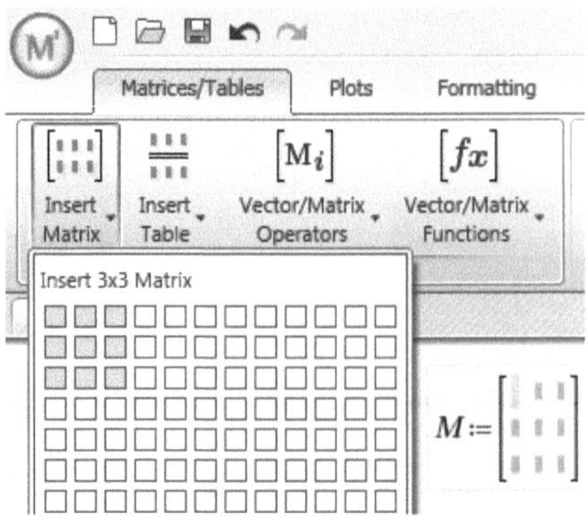

Fig. 6.24 Insert matrix in Mathcad Prime

Mathcad Prime enables us simpler inserting a vector or a matrix (array). To insert matrix choose **Insert Matrix** from **Matrices/Tables** ribbon and then drug-and-drop the table to specify a matrix dimensions, see Fig. 6.24.

It is more convenient to insert matrixes via **Insert | Data | Table** as shown in Fig. 6.25. This command inserts the assignment operator ■ := ■ into the worksheet. Its right operand is rather bulky table (an analogue of the Excel table); the user can type new information in the its top left region forming the picture on the screen via the **Component Properties** dialog box that also shown in Fig. 6.25.

Comment. *The earlier Mathcad versions have "table" in the list of components (see* ***Component*** *in Fig. 6.25). Later on this important component was located in a distinct menu item,* ***Data,*** *which contains two more commands discussed below.*

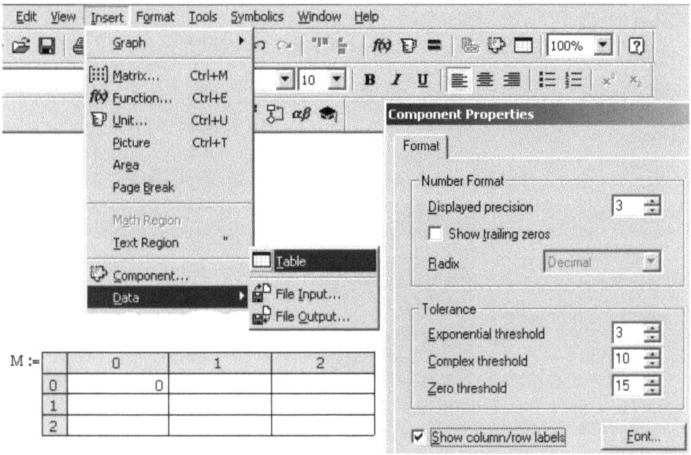

Fig. 6.25 Insert and formatting a table

Fig. 6.26 Insert a table in Mathcad Prime

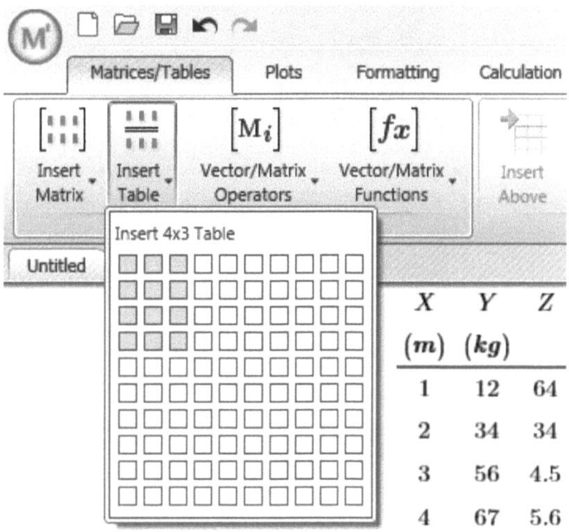

Mathcad Prime changed the procedure of inserting a table into a worksheet. After choosing **Insert Table** from **Matrices/Tables** ribbon a user sees not a table but a set of bounded vectors, for example X, Y, and Z as shown in Fig. 6.26.

Automated entry of large volumes of information is now possible if a user has installed both Excel and Mathcad software programs. Inserting of Excel tables into Mathcad worksheets choose **Insert | Component... | Microsoft Excel** and then in the **Excel Setup Wizard** dialog box choose **Create an empty Excel worksheet** or Create from file. After the insertion of a component, we can point the part of Excel table where the information from Mathcad is transferred (**Inputs**) or conversely the information is transferred to Mathcad variable from Excel table (**Outputs**).

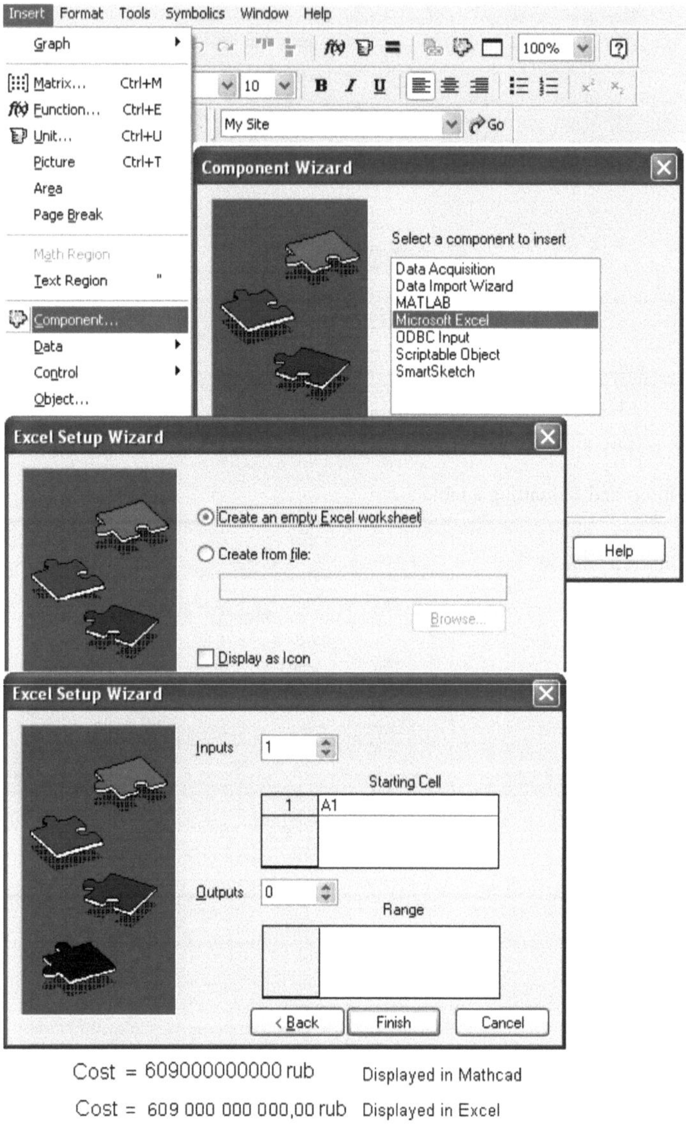

Fig. 6.27 Displaying a number in the excel format

Figure 6.27 shows that we "crushed a fly with a steam-roller" processed variable `Cost` in Excel using a special format of electronic worksheet which inserts thousands spacer, white space between three numbers, into numbers and thus, simplifying its reading. However, one can "crush" earnestly, for example, to access those Excel functions that Mathcad does not have (calendar functions) or to make Excel plot in Mathcad (pie chart). Besides, to enter data in Excel table is more convenient and quicker: one can use special Excel features – AutoFit and others.

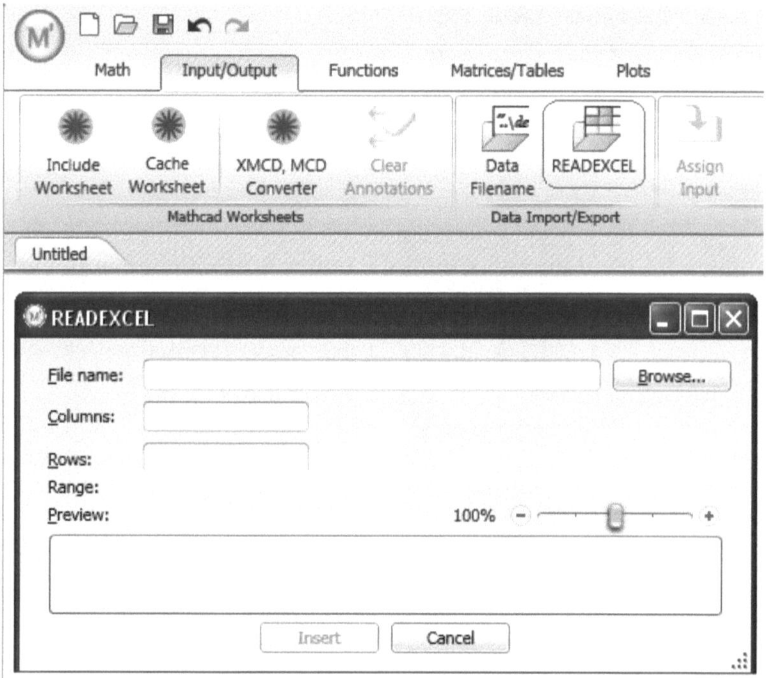

Fig. 6.28 Access to excel in Mathcad Prime

After entering the pertinent data, one may remove the Excel table later to make the Mathcad worksheet operable to those users who have not installed Excel to work with the worksheet.

Figure 6.28 shows how to read data from an Excel table in Mathcad Prime.

Besides Excel, one can insert other Windows applications components into Mathcad to expand its functionality and one can add Mathcad worksheet into other applications, Word, for example. Nevertheless, it is only worth inserting other programs components into Mathcad worksheet if they provide calculations and other features for which Mathcad is not sufficient. The point is that the Mathcad worksheets sent to another user who does not have the pertinent application, which used its components, the Mathcad worksheet will not fail. In addition, a Mathcad worksheet containing other applications components is difficult to publish on the Internet.

Comment. *To insert choose **Object** from **Insert** menu, common for all Windows applications.*

To save large tables to disk via text files connected to a Mathcad worksheet simply insert the operators of *writing/reading* a file to/from disk. Figure 6.29 shows how to create the operator of writing the data from file Tab_XV.dat and to send it to the variable `M`.

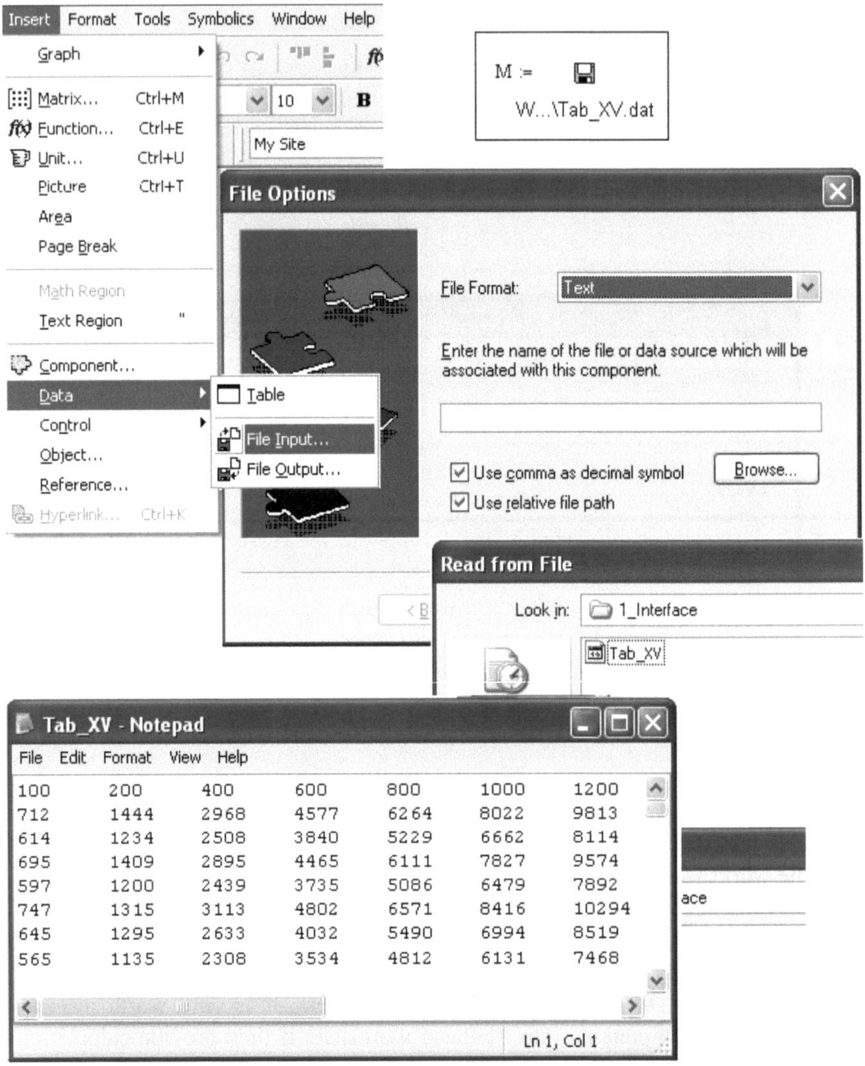

100	200	400	600	800	1000	1200
712	1444	2968	4577	6264	8022	9813
614	1234	2508	3840	5229	6662	8114
695	1409	2895	4465	6111	7827	9574
597	1200	2439	3735	5086	6479	7892
747	1315	3113	4802	6571	8416	10294
645	1295	2633	4032	5490	6994	8519
565	1135	2308	3534	4812	6131	7468

Fig. 6.29 Insert a matrix from a file

To assign data to the variable `M`, choose **Insert | Data | File Input**... which opens a dialog box, or rather file changing wizard (**File Options**), in which we can mark the file with required data in "file/disk/folder" Windows system clicking button **Browse** of the **File Options** dialog box. In addition, we can choose file format (in opening list **File Format**) and set the comma as separator instead of the tab character. The data itself (Fig. 6.29 shows it opening for review or editing in WordPad) can be entered in Mathcad and written to the disk via command **Insert | Data | File Output** or manually (for example, in Excel), or generated by another

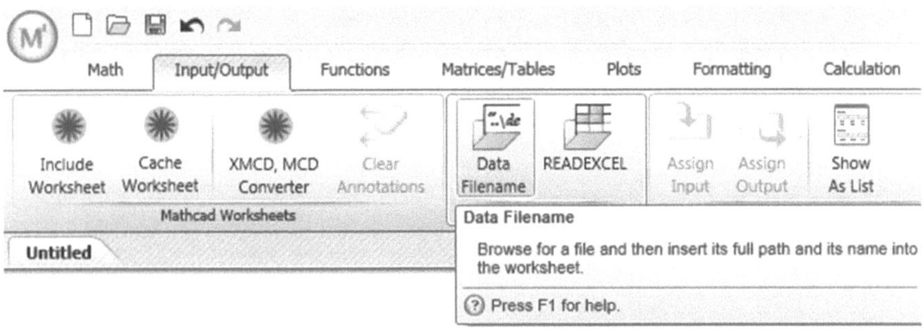

Fig. 6.30 Insert data from a file in Mathcad Prime

program, or received by e-mail et cetera. Besides the command for data changing shown in Fig. 6.29. Mathcad has a set of similar functions (readprn, writeprn etc) accessed by **File Access** from the **Insert Function** dialog tab. The length of numeric literal writing to disk depends on the value of built-in variable `PRNPRECISION`. The main disadvantage of saving data in a file, rather than in Mathcad is that one can send Mathcad worksheet and forget about the file.[11]

Figure 6.30 shows command **Data Filename** that enables a user to read data from the file stored in a user's own computer, local network, or internet.

To exchange information between Mathcad worksheets choose **Reference** from **Insert** menu.

It is not necessary to contain all the operators for a given calculation within the "current" Mathcad worksheet. Referenced operators saved in another worksheet in the user's disk or on any computer system network are accessible as referenced worksheets. To make them work for us, we must make a *reference-link* to the file containing the necessary data from the current worksheet. Figure 6.31 shows such situation. D disk contains Mathcad file sheet.mcd with address D:\Documents and Settings\user\My documents keeping a single operator `c:=a+b`; of course, the calculating is interrupted by an error message because the variables are not defined.

The other Mathcad worksheet which is also shown in Fig. 6.31 contains operators of source data (`a:=1 b:=2`) and the result (`c = 3`). The worksheet named sheet.mcd carries out the calculations contained in the referenced worksheet (which is connected to first worksheet by the command **Insert | Reference**). As a rule, the referenced worksheets contain constants and functions covering a *branch of applied sciences* or having some *general engineering significance*. Reference files allows, for instance, a dimensional set of user-defined functions in the form of DLL that have been turned into built-in, i.e. listed in the **Insert Function** dialog box (see site http://www.wsp.ru).

---

[11]That often happens in Mathcad forum when someone sends a "problem" worksheet but forgets to include the file with source data.

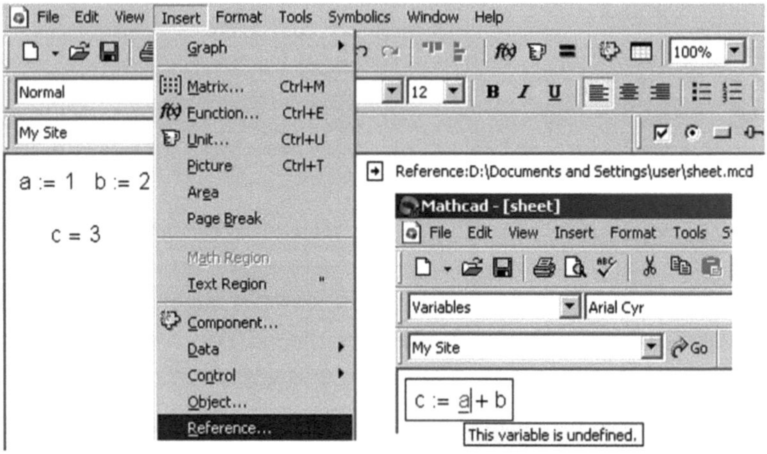

Fig. 6.31 Applying a reference

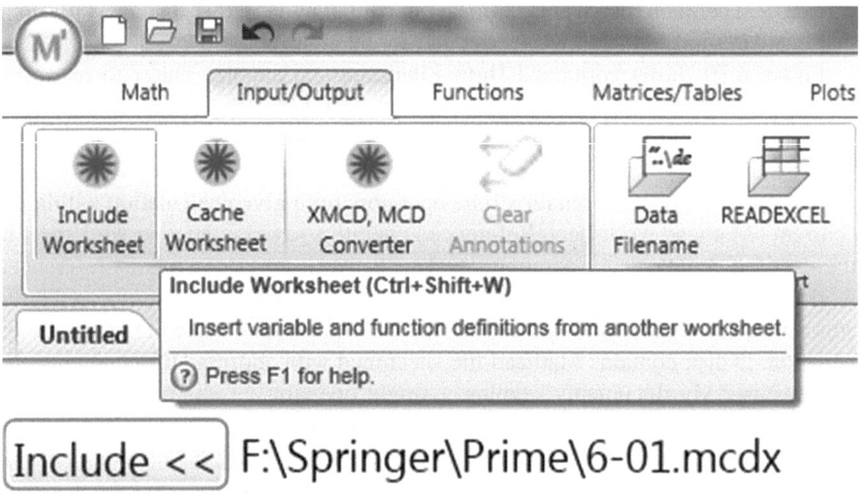

Fig. 6.32 Link to another sheet in Mathcad Prime

In Mathcad Prime, referring to another Mathcad-sheet has changed and now it calls **Include Worksheet**. Figure 6.32 shows this command in **Input/Output** ribbon.

One of Mathcad's interfaces features the ability to protect or to hide (a part of) the information content.

As a rule, Mathcad worksheet comprises of three areas: *source data area, calculation area,* and *area results*. We can delineate these areas in Mathcad worksheet clicking **Area** from the **Insert** menu. This command insert two horizontal

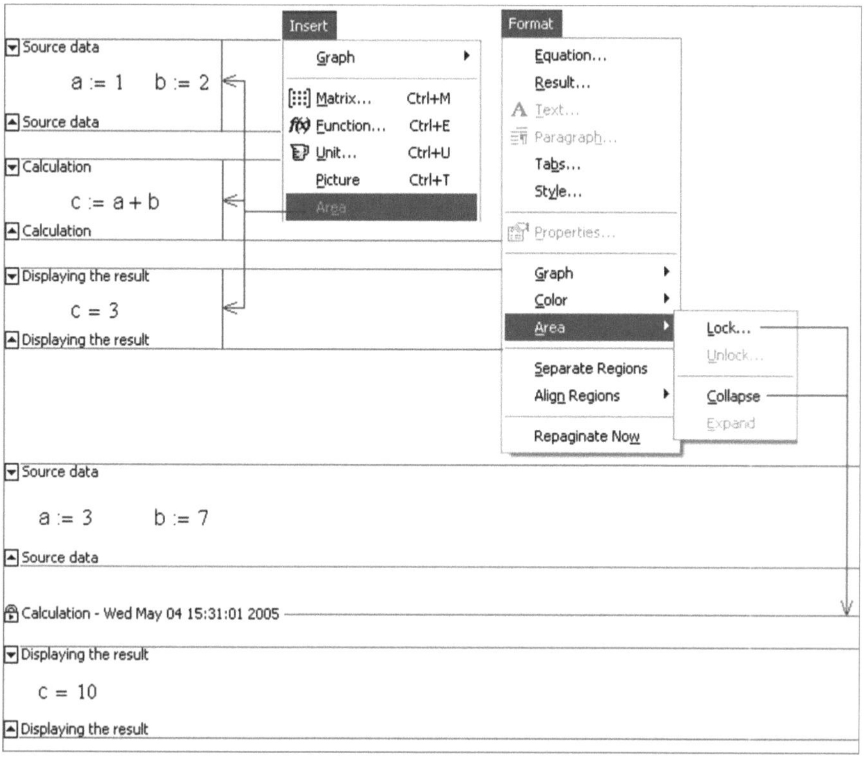

Fig. 6.33 Insert and formatting areas

lines near the mouse pointer, after which the user moves the top line to the start of an area they wish to create and the bottom line to the end of their area (see the top of Fig. 6.33). What can we do with it? First, we safeguard the information against unintended edits by command **Format | Area | Lock**. After such command, the region enclosed by the area allows the user to view it but not edit it.

The second command used separately or together with command **Lock** is **Collapse** and this command hides the chosen area from a user. The result of these two commands on the area named `Calculation` shown at the bottom of Fig. 6.33. The user types this name clicking **Properties** from the area pop-up menu. Otherwise, we can even hide a vestige of this area in Mathcad worksheet by this command. After such manipulations (inserting, locking, and collapsing an area) a user can change the values of a and b and see the result (variable c) but cannot see and edit the formulae. A Mathcad worksheet containing a collapsed area is like a piece of paper, the middle of which recedes into the background by several folds (Japan origami). The antonyms of the commands **Lock** and **Collapse,** – **Unlock**. . .and **Expand**, allow us to read their names. We can lock an area with a password, a row of symbols. The correct password is required to unlock and expand a **Locked** and **Collapsed** area.

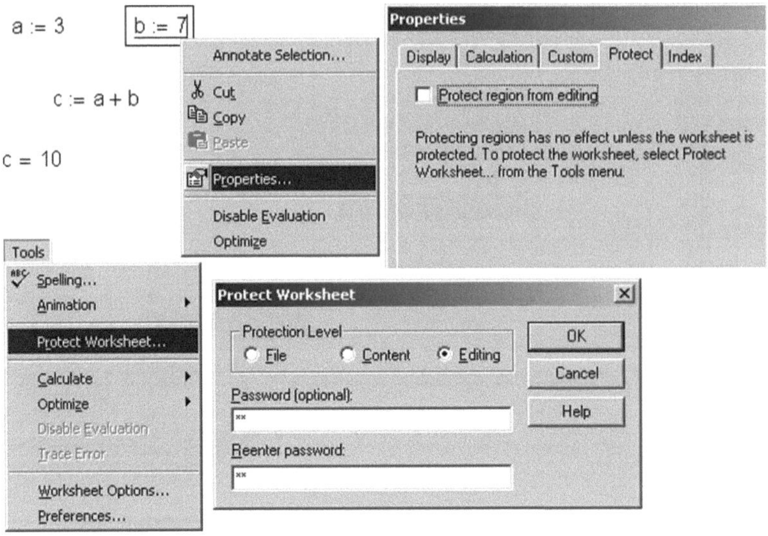

Fig. 6.34 Protection of a worksheet and its separate operators

We can safeguard Mathcad worksheet without inserting any areas, but safeguarding the whole calculation while keeping some required operators unlocked. Figure 6.34 shows the menu commands allowing us to protect all, or most of, the worksheet.

To protect the worksheet, choose **Protect Worksheet...** from the **Tools** menu, shown in the bottom part of Fig. 6.34. Worksheet protection comes in three levels presented by switches **Protection Level** in the **Protect Worksheet** dialog box. The first level, **File**, the lowest one, protect the worksheet from being saved in some formats, for example as earlier Mathcad versions. Figure 6.35 shows two lists of file formats for saving the current worksheet even without protection level **File**. In Mathcad 12, this protection level makes it impossible to save the worksheet as Mathcad 11, 2000i, or 2000.

That limitation works at two other levels also. The second level protects existing operators against change by the switch **Content** but new operators can be created. The third level, **Editing**, the highest one, does not allow the change to existing operators or to create new operators. Of course, we can make the "holes" beforehand, i. e. turn off protection from some operators (as usual, from those defining source information) by the commands shown at the top part of Fig. 6.34 with operator b:=7. In all levels we can protect the worksheet with a password or without it (options **Password (optional)** and **Reenter password**).

Some safeguarding mechanisms duplicate each other. We can protect a separate operator or several operators at once placing them into an area (Fig. 6.33) or by the technique shown in Fig. 6.34.[12] This duplication occurs because the protection

[12]We can protect twice Mathcad operators with two different passwords that two persons know. These two persons can open such banking lock, as each one of them knows only one password.

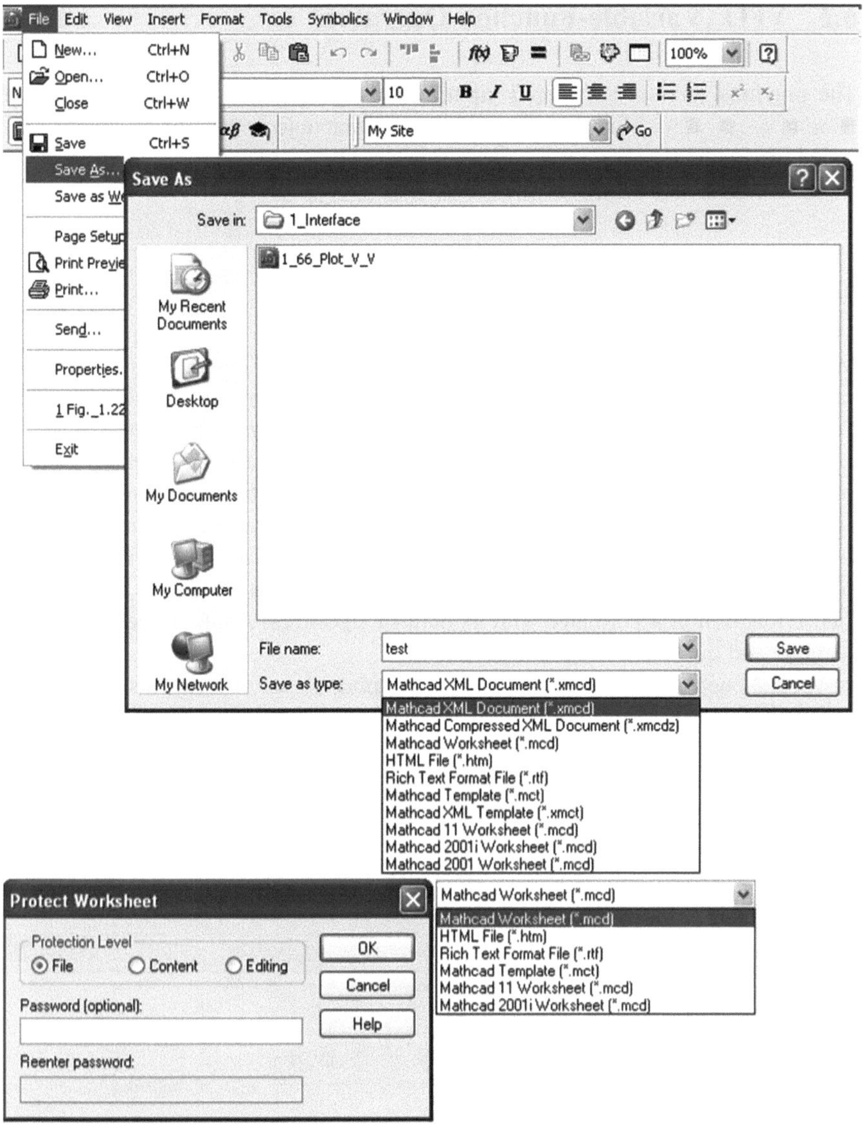

Fig. 6.35 Protection of a worksheet from saving it as an earlier Mathcad version

mechanisms were not introduced simultaneously but from earlier Mathcad versions to new. At the beginning, with Mathcad 2000, it became possible to insert areas into worksheets (Fig. 6.33) safeguarding it against editing only (**Lock/Unlock**). Then (Mathcad 2001) we could collapse these areas. Mathcad 2001i got the mechanisms to protect whole worksheet and some of its operators (Figs. 6.34 and 6.35).

6.2 VFO (Variable-Function-Operator)

The previous section considered input/displaying operators ■ = ■ ■, ■ := ■, ■ ≡, ■ ← ■, ■ → . This section will present what objects these black squares (■), called placeholders, may contain.

6.2.1 Function and Operator

Mathematics has a term *correlation*: for example, there are two enumerable sets and each element of the first *correlates* with a single element in the second one.[13] The particular case of such correlation is the *one-argument function* sin(x); for example, any value of an angle (x is the first set and argument of a function) correlates with value of sinus (y is the second set and the first function). A middle-aged reader will get this right away well knowing from Bradis tables such "sets" of angles, sines, logarithms, and other necessary values. Of course, Mathcad does not keep sets of angles,[14] sets of correspondent sines etc. but calculates this trigonometric function in accordance with its built-in algorithm. Another question is how accurate and quick these algorithms are.

We may note, for example, two sets of numbers, a set of functions and another set of variables, and set of definite integral values: each four-elements of the four-element sets correlate with an element of the fifth set. The case in point is that the definite integral operator with four operands that is built in Mathcad (Fig. 6.36)

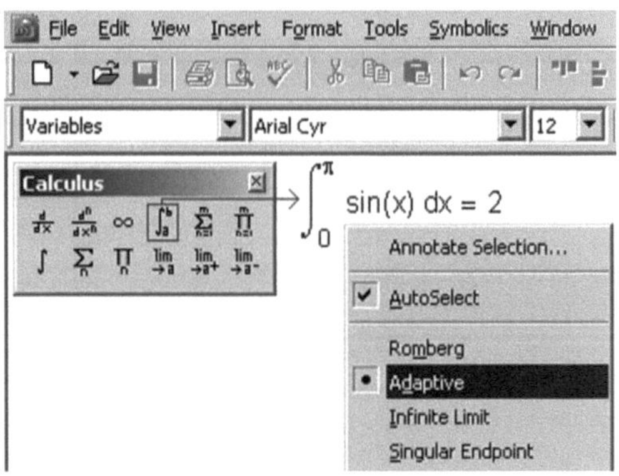

Fig. 6.36 Dropdown menu for a built-in operator setting

[13] One may say that any number of elements of the first set (any number of sets from the first group) correlates with any number of the second (any number from the second group).

[14] However, Mathcad could keep these sets if they do not contain too many elements; treat them in obtaining continuous functions.

About 30 years ago, a discrepancy appeared between mathematicians and programmers in terms "function" and "operator".

A mathematician reading this book may justly conclude that the authors do not fully understand what an operator is and what a function is. Moreover, there is no unity of treatment of these terms in programming. "Operator" in Mathcad has another meaning in BASIC, for example, and vice versa. Thus, BASIC has a convenient operator `Swap(a, b)` that changes values of the variables `a` and `b`: `c = a : a = : b = c`, but without enabling variable `c`.

Comment. *This operator would not be superfluous in Mathcad, either. However, it would upset a stable system functions and operators as it does not return a value but executes a certain procedure.*

As this operator does not return a value, it cannot be called operator from the point of view of a Mathcad user. On the other hand, Mathcad operators and functions themselves may not return values either (for example the equality `a = sin(x)`) but be a peculiar comment (see Sect. 6.3) and expect to be treated.

Comment. *Besides, function `sin` of the operator `a:=sin(x)`, for example, does not return a value either. It will return sine of `x` if the operator `a =` appears after the assignment operator.*

Let us agree that the terms "operator" and "function" are application dependent and will discuss not their essence (let the theorists dispute about it) but their differences in Mathcad.

If terms "operator" and "function" are considered with respect to the mathematics that we use in calculations but not with respect to Mathcad features (see instances below) we can mark out some aspects dividing Mathcad mathematical mechanisms into operators and functions.

- A definite function is distinguished from others by its name `sin(x)`, `cos(x)`, `log(x)`. Operators differ from each other by symbols `n!`, $\neg x$, $|x|$, (see Figs. 6.11 and 6.12) etc. Mathcad has three operators with invisible (absent) symbol: x^y (power), X_n (element of array or text index), and 2 K (multiplication). Section 6.2.3 describes visible operators.
- Some Mathcad operators have ambivalent contents. For example, x^2 what is it? Is it a power operator with two operands, with the second operand equals to two, or squaring operator with one operand? The second example: `2 K;` is it a multiplication operator or just a variable named `2 K` (a name of this type is possible in Mathcad, see Sect. 6.2.2) etc. Mathcad also has mathematical operations executed both as operators and as functions. For example, the exponent functions call e^x operator and exp(x) a function. It would be convenient if all Mathcad operators have such twins. First, it would give the users additional freedom in choosing a suitable expression and second, enable them to introduce formulas in the text format noted *below*.
- All Mathcad functions are equal. However, operators follow the standard hierarchy of operation (given a priority status of execution). Thus, a compound

operator `2 + 2·2` returns 6 but not 8 as multiplication operator has a priority over addition. Parentheses are utilized the alter the basic hierarchy

- The attribute, the fundamental characteristic of a function is of parentheses framing a list of arguments: `sin(x)`, `min(1, -7, 5, 4)`, `Find(a, b)` etc.

Comment. *It should be noted that the parentheses themselves are a kind of operator in Mathcad and in other computing applications as well as in mathematics as a whole changing the order of operator executing:* $(2 + 2) \cdot 2 = 8$ *but not 6 as without them.*

The parentheses may frame an operand in an operator, not as attribute but as a new operator to combine their operands: `(5)!` (Five factorial – the parentheses are superfluous here but do not result in error), `(1 + 4)!` (Factorial of the sum – here will be an error without parentheses)

Comment. *It should be noted in the previous comment that parentheses are not so much an operator for changing the order of operator execution but rather an operator for processing (functional) block and they can be inserted in each other. Such insertion in Mathcad may result in the appearance parentheses changing to square brackets that make it easier to understand them. Besides, that noted above, in Excel the parentheses change color when editing an expression (see Fig. 6.3).*

- Operators always have a fixed number of operands

Comment. *Number of operators ranges from 1 to whatever. Speaking of common mathematical operators the maximum number is 4 (definite integral (see Fig. 6.47), sum and others). However, if we consider that a plot in Mathcad (see Fig. 6.11) is represented by an operator the maximum number of operands is questionable. In case of a plot, we can tell about variable number of operands too.*

Some built-in Mathcad functions have variable number of arguments. Thus, the function `root` (it returns the zero of an analytic function) can have two or four arguments. Actually, there are two functions of the same name using different algorithms. The function `log` usually has one argument but if we write the second, it changes the logarithm radix from 10 (by default) to another determined by the user. Function `Find` returns a solution to a set of equations and inequalities and can have from 1 to 50 arguments.

Comment. *This number 50 is not the limitation of function* `Find` *but of all Mathcad functions with variable argument number.*

Mathcad has a function documented in version 12, the argument number of which, one could say, equals zero. It is function `time` and it returns time (in seconds) passing since some date.[15] The date itself is insignificant because users do

[15]This is a calendar date of 1970, which the reader can calculate himself. Currently, when this text was typed into the computer the function returned the value 1093238417.077.

not generally work with it, but with the period between two calls to it, for example we can use the difference in calculation testing. It should be noted that the function `time` has a formal argument, which does not affect its result.

Comment. *We can call this function without parentheses and a formal argument but as a variable* `time`. *The argument of the user functions* `y(x):=sin(x)` *or* `y (t):=sin(t)` *is called formal too: we can use any other name for the formal variable not only* `x` *or* `t`.

We can consider Mathcad built-in mathematical constants π and `e` as functions having no arguments and that returns constant values.

- Mathcad has some particular functions and operators. We can understand how they work only when we size up their mathematical meaning and the techniques of their execution. These functions return the value depending on values of arguments, operands,[16] which is situated near them, and on additional specific adjustments. Thus, the function `Find` returns different values for the same arguments depending on the first approximation in finding the roots of analytical system of equations (that is the function `Find` is appropriated). The second example, Fig. 6.36 shows what is contained in the pop-up menu appearing after right clicking the mouse for the definite integral operator. Here the additional operands of the operator are listed – algorithm adjustments (options) for numerical solution of this problem. One may say that there is another method of penetrating the inside, or interior of the function bypassing formal entrance, the list of the arguments. Side entrances of this kind are often made by a user who forms a new function in the following way: `a:=2 y(x):=x^2-a` instead of more correct notation `y(x, a):=x^2-a`. Advantage of the first form is that where there are many arguments we can single out one or two and list them as formal arguments of the function. The rest of the arguments we can consider by convention as constant. The disadvantages of such non-closed (unlocked) functions are difficult to transfer: we may forget about some external constants, and lose them. A function of that kind is similar to Mathcad worksheet having some data in external files (see Fig. 6.29)

A method of entering the functions into the worksheets is by clicking the correspondent keys: <s>, <i>, <n>, < (> for sine). However, it is better to use the **Insert Function** dialog box shown in Fig. 6.37 and 6.38.

Figure 6.37 shows the dialog box in Excel for comparison. One of the fundamental distinctions between these mechanisms in Mathcad and Excel is that Excel list contains both built-in and user-defined functions. User functions in Mathcad can appear in the **Insert Function** dialog box only if transformed into built-in functions in the form of a Dynamic Link Library (DLL).

Figure 6.38 shows inserting built-in functions in Mathcad Prime.

[16] See the beginning of the section with description of correlation between the elements of the sets.

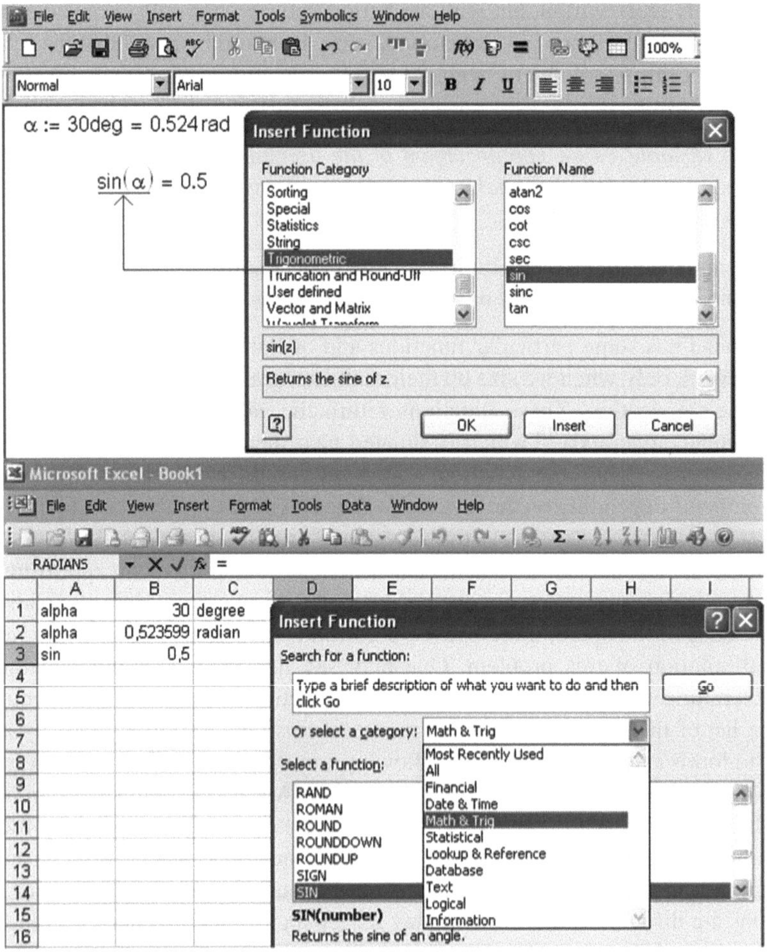

Fig. 6.37 Insert function in Mathcad and excel

As to built-in operators, we can insert them clicking the buttons with their pictures on the appropriate toolbar (see Figs. 6.39 and 6.40).

Comment. *The fact that almost all buttons for inserting mathematical operators have their doublers as corresponding key combination (<Shift> + <2> for plots) practically have been forgotten now.*

As we saw in Sect. 6.1, the input/output operators are the special operators. The mechanisms for working with symbol mathematics programming instruments are called operators too (buttons → and ▪→ on the toolbar **Evaluation** in Fig. 6.39).

Some menu commands are instruments used to solve the problems. Thus, we can solve an analytical equation or inequality by the command **Symbolic | Variable | Solve** if we previously type the expression in a worksheet and mark the variable to

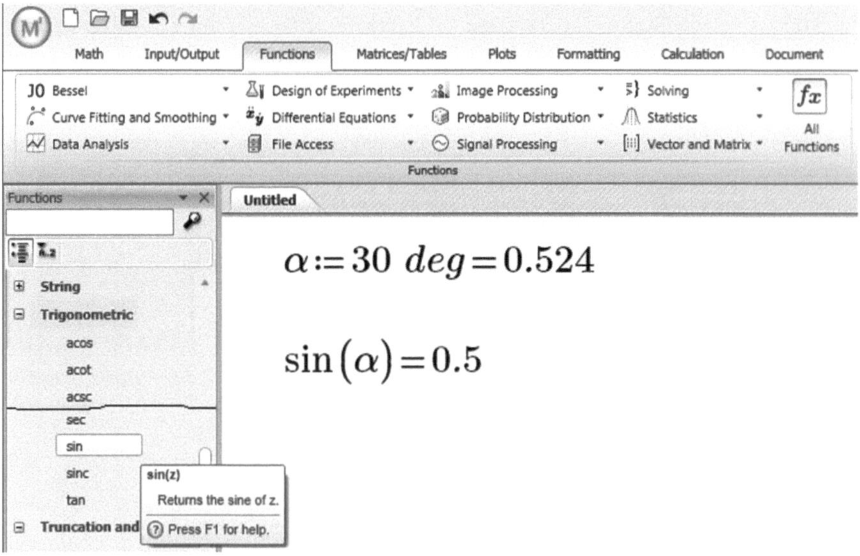

Fig. 6.38 Insert function in Mathcad Prime

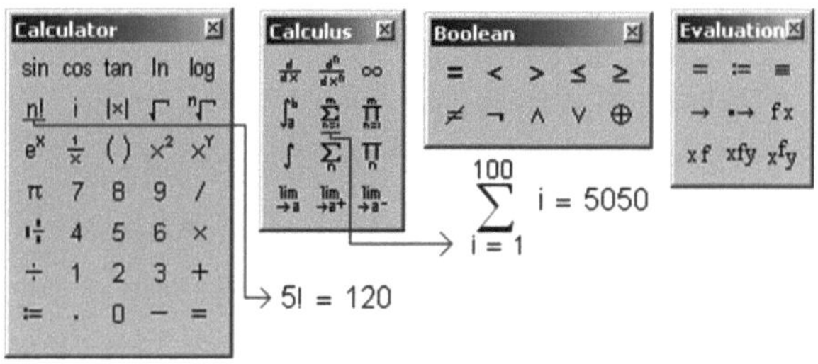

Fig. 6.39 Math toolbars of built-in operators

solve for with the cursor. Mathcad displays the result below (by default), to the right, or in place of source expression. Menu commands are seldom used to solve mathematical problems as operators have almost substituted them. Figure 6.40 shows built-in operators of Mathcad Prime.

The list of built-in operators has been modified a little. For example, there is no operator x^2 in Mathcad now is x^n. Derivation operator enables us to work with derivates of any order: the first, the second, etc. Indefinite and definite integration are united into one operator – if it is required to make indefinite integration (antiderivative) a user should keep placeholders for upper and lower limits empty. Mathcad Prime has some new operators, for example call of a matrix string but not only a column as it was in traditional Mathcad.

Beside the common method of calling functions, we can call the one or two arguments functions (user and built-in) by clicking buttons **fx**, **xf**, **xfy** и **x$^{\mathbf{f}}$y** in the

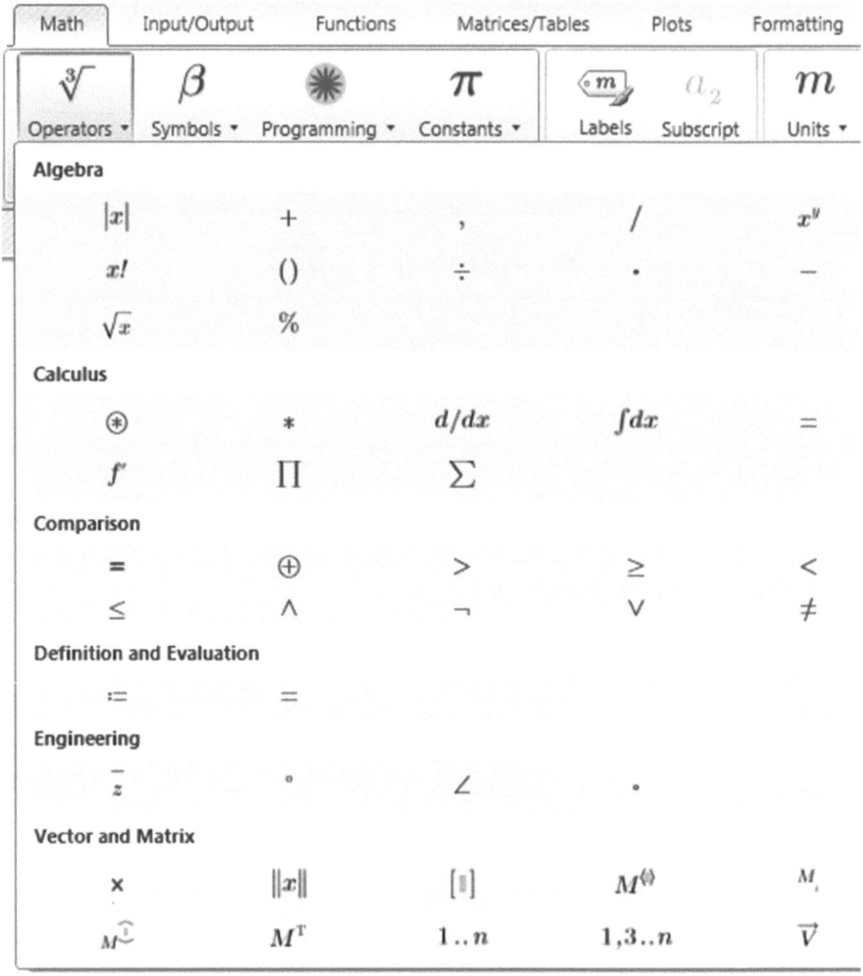

Fig. 6.40 Built-in operators Mathcad Prime

Evaluation toolbar. At that, Mathcad displays the placeholders for postfix(**xf**) and prefix (**fx**) operators with one operand, infix (**xfy**) and tree ($\mathbf{x}^{\mathbf{f}}\mathbf{y}$) operators with two operands. That enables us to insert the user operators into a Mathcad worksheet.

Figures 6.41–6.44 show the use of these operators to solve some particular problems.

The function `mean` returns arithmetic mean of arrays (vector, matrix or list). The first call of this function in Fig. 6.41 is made in common form: as a function. Two parentheses appear (it is like saying salt is salty) that may confuse a novice user when the `mean` function is invoked. The user will try to delete excess parentheses not understanding why it is impossible to do. The solution is to call the "matrix" function (the function whose

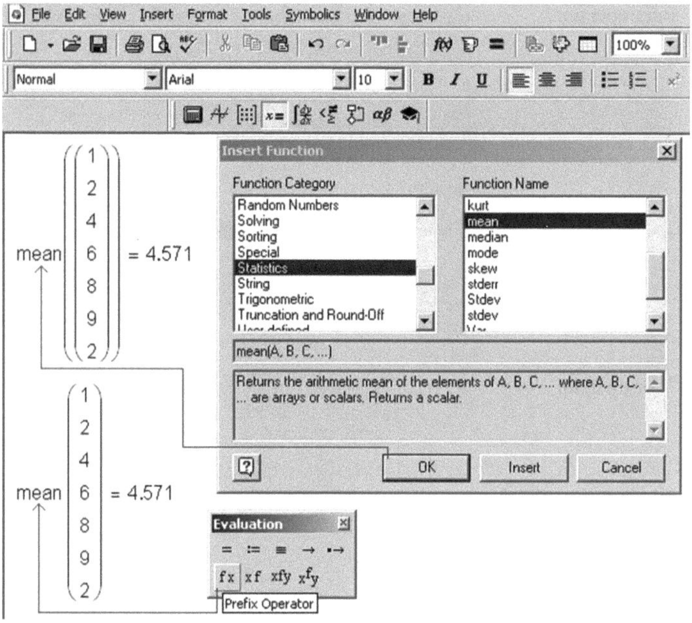

Fig. 6.41 Prefix operator

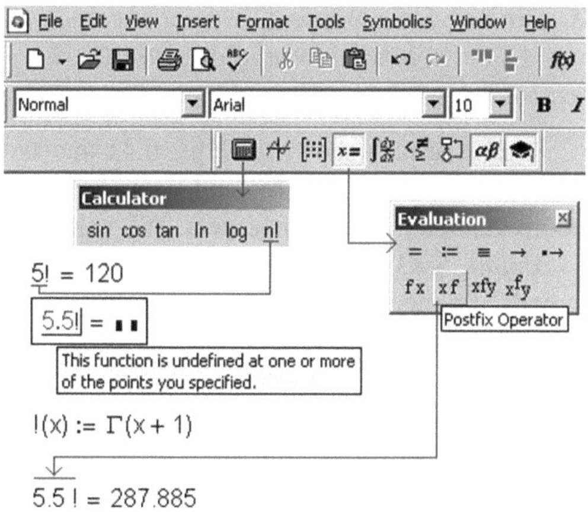

Fig. 6.42 Postfix operator

argument is a matrix) not as a function but as prefix operator, which allows operand without parentheses (see the second operator in Fig. 6.41).

Figure 6.42 shows how to redefine built-in factorial operator to make it work with fractional operands. For that we insert the function `!(x)` into worksheet being equal to built-in gamma function which argument is shifted to one. We can call this new function in traditional form `!(5.01)` = but also to call it as shown in Fig. 6.42, as postfix operator, is better.

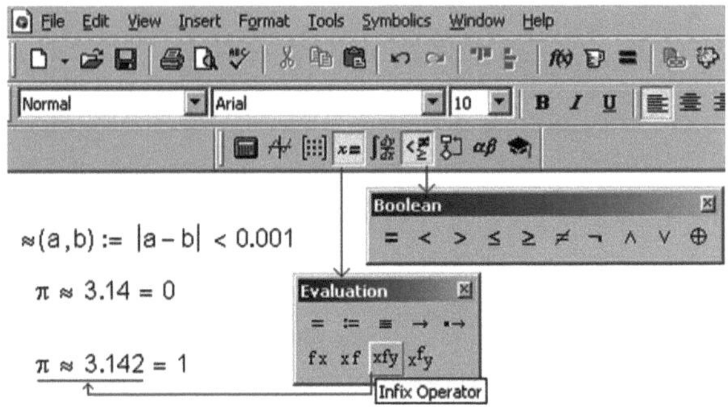

Fig. 6.43 Infix operator

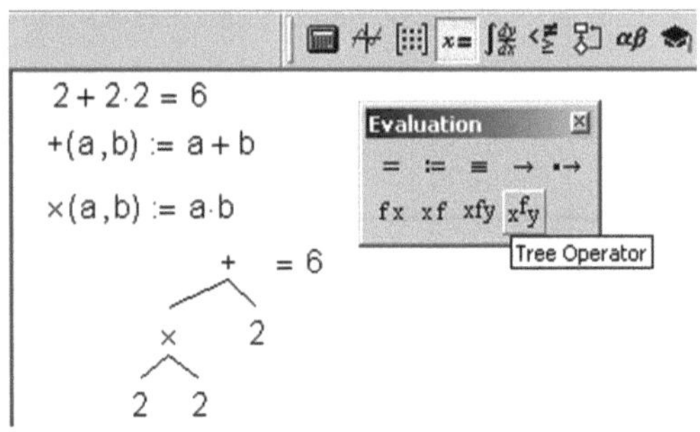

Fig. 6.44 Tree operator

Comment. *How to insert the symbol ! and other reserved characters into Mathcad worksheet not in the form of operator (in this case, factorial sign) but as a function name is described in Sect. 6.2.2.*

The examples shown in Figs. 6.41 and 6.42 are simple and have no practical value. Still, using postfix and prefix operators to work with relative scales of temperatures is very convenient. We will discuss it in Sect. 6.4.

Figure 6.43 shows how to insert additional Boolean operator "approximately equals" by means of infix operator into a worksheet. It is very useful in performing iterations where loops stop execution by "approximately equals" instead of "not exactly equals to". Figure 6.43 shows that π approximately equals to 3.142 but the value 3.14 (which we remember, as a rule) is not "approximately equals" to π.

Figure 6.44 shows that after we have redefined two-operand built-in operators addition and multiplication the hierarchy of the expression 2 + 2·2 is opened (we discussed it above) by means of the tree operator.

One of the reasons why Mathcad has become popular is that a user can insert operators or functions into a worksheets depending which he may have got accustomed to when learning mathematics in school. This makes the Mathcad worksheet looks like a technical paper with calculations made by hand or in a text processor environment (Scientific Word, ChiWriter, etc.).

Nevertheless, advantages in some instances can lead to disadvantages in other instances. Live Mathcad equations using many-storied operators, instead of text functions, cause difficulties in Mathcad Application Server technology.

6.2.2 Variable Name

While the names (symbols) of built-in variables, functions, and operators are fixed, we may give any name to a user object. The limitations here are connected, first, with certain traditions (discussed in Sect. 6.3), and then, with the features of Mathcad itself.

Figure 6.45 shows symbols, Greek letters and special characters from the **Mathcad Resources**, which we may use in addition to the keyboard characters to name variables, functions, and operators.

Comment. *The Greek toolbar also contains two mathematical instruments – constant* π *and gamma function* Γ.

Figure 6.46 shows built-in special symbols in Mathcad Prime.

Combination keystroke < Shift>+<Ctrl> + <k> allows us to insert variable names first, the symbols prohibited from using in traditional programming (blank, dash, comma, etc.), and, secondly, symbols fixed for some operators in Mathcad ($, @, etc.). After pressing this combination, the color of pointer changes from blue to

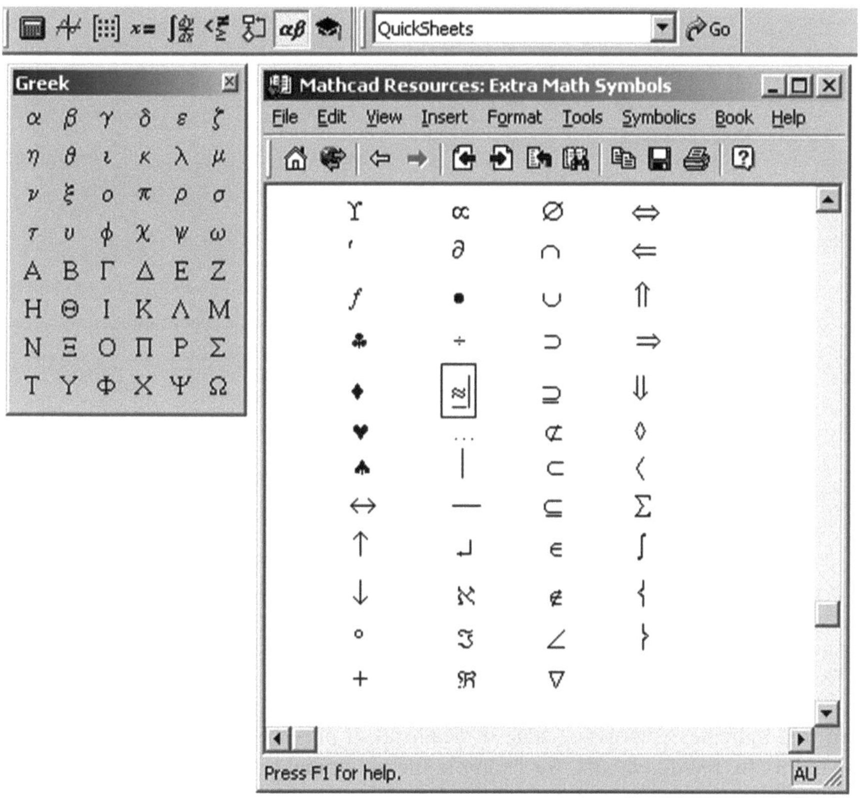

Fig. 6.45 Greek letters and special symbols in Mathcad

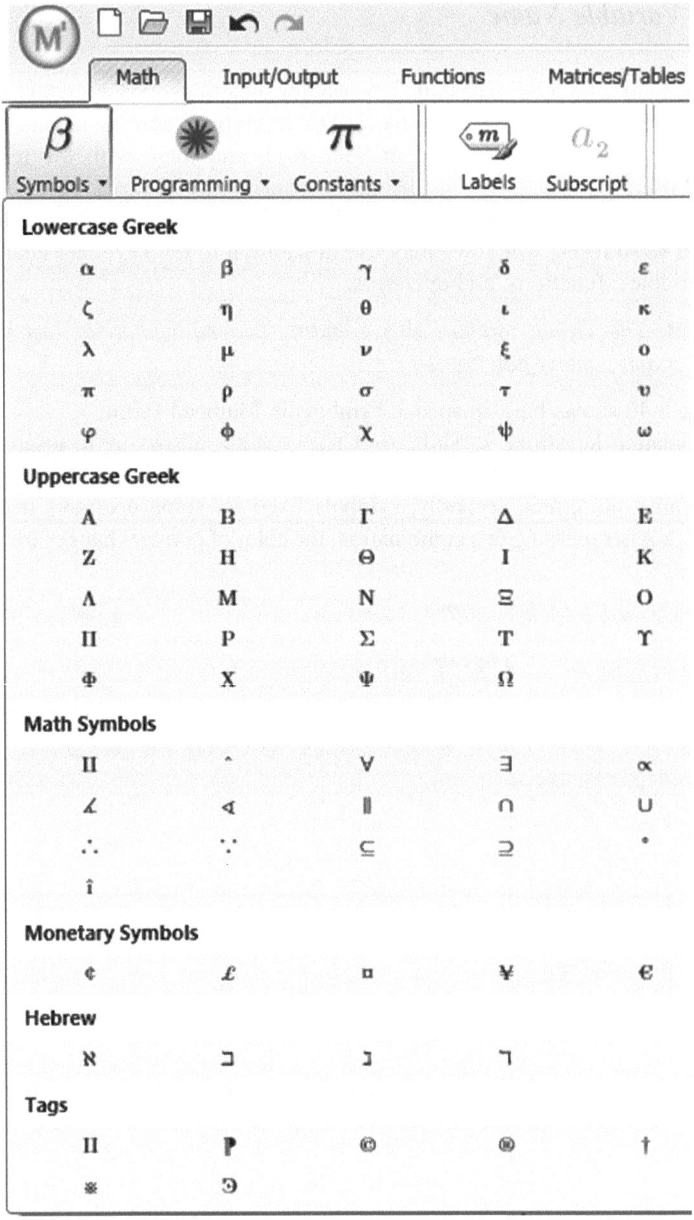

Fig. 6.46 Greek letters and special symbols in Mathcad Prime

red that indicates emergency state of Mathcad. It prevent us from inserting certain operators by fixed symbols, for example the assignment equal sigh by typing <:> (see Sect. 6.1). The symbol just will be added to the variable name which has been finished already as we type symbol <:>. To change the pointer color back to blue we should type the symbols <Shift>+<Ctrl>+<k> again. Figure 6.47 shows this combination allows us to insert nonstandard but colloquial variable names: `US$`, etc.

Only one character, which we cannot insert into the variable name (or rather, we can type but cannot see it then), is a period. It divides the name into two parts – the name itself and its subscript, for example, typing `t.in` we obtain t_{in}. Nevertheless, we can do that creating a variable with a literal (text) subscript, which has an invisible blank as a name before subscript and periods as subscript.

A reader can find a blank (space) at the beginning of same variable names. This blank is not fortuitous: some characters cannot stand at the beginning of variable names, e.g., digits 0 through 9. If the variable name consist of one character, which is a digit, that follows to such curious thing `3:=7 7:=3` – the variable named 3 assigned the value equals to 7, and the variable 7 equals to 3, etc. Sometimes (in certain Mathcad versions in combination with certain Windows versions) some letters of Cyrillic alphabet cannot stand at the beginning of a name (that restriction does not apply to a blank). Therefore, it is desirable to start a questionable name with a blank. Using a blank as a variable name makes it invisible (see Sect. 6.2.3).

In Mathcad Prime, a user can enter special symbols into a variable name in such a way. At first, enter a variable, for example, amount of substance, mole, as shown in Fig. 6.48. Then insert the variable into a **Text Box** where it will be noted as **@LABLE Unit mole**. This notation requires correction: entering/L in the end of the note. After that, insert the changed notation into the worksheet as the name of the variable **mole/L**.

Another method of entering a sophisticated variable name into Mathcad worksheet is by pressing < Shift> + <Ctrl>+<j> (Fig. 6.49).

Figure 6.49 shows how to enter the variable with a rather complex name such as $H_2PO_4^-$(one-valent ion of orthophosphoric acid) that practically consists of three variables: the variable H_2 (`H.2`) multiplied by variable H_2 (`H.2`), which, in turn, raised to power minus `--` (<Shift> + <Ctrl> + <k>+<−>). There is a limitation: such complex names are enclosed in square brackets.

Mathcad Prime enables a user to introduce a part of a variable name as a subscript by **Subscript** from **Math** ribbon. Figure 6.50 shows inserting variable ρH_2O (density of water).

Mathcad 12–15 has the third key combination < Shift> + <Ctrl>+<n> that enters a system index into the variable name. This index contains one of four key words: `mc`, `unit`, `user`, and `doc` enclosed in square brackets.

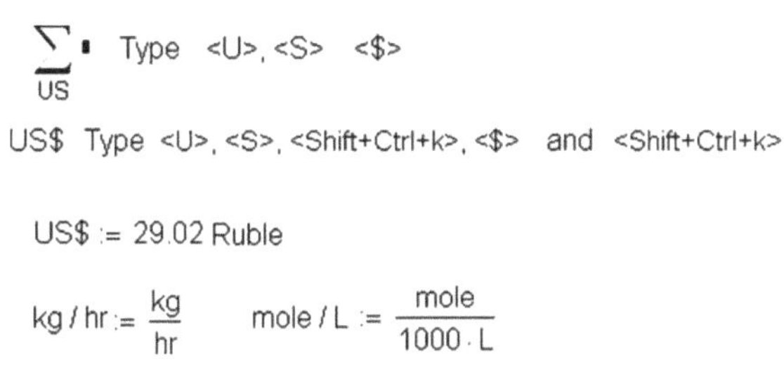

Fig. 6.47 Repressing the special keys

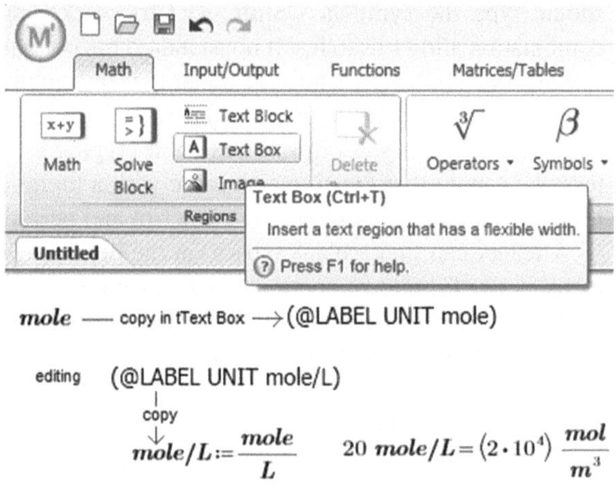

Fig. 6.48 Special variables name in Mathcad Prime

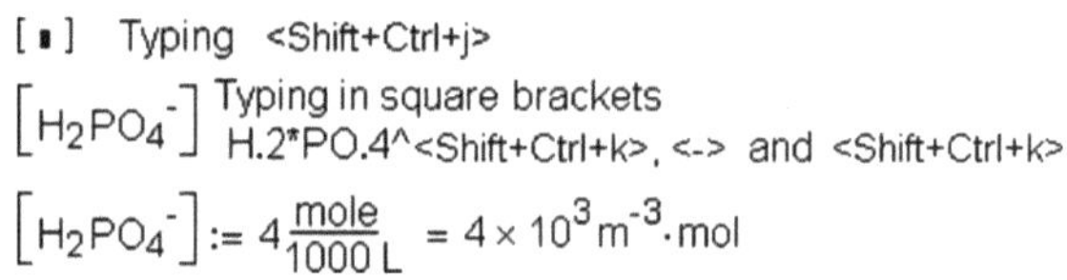

Fig. 6.49 Inserting compound names

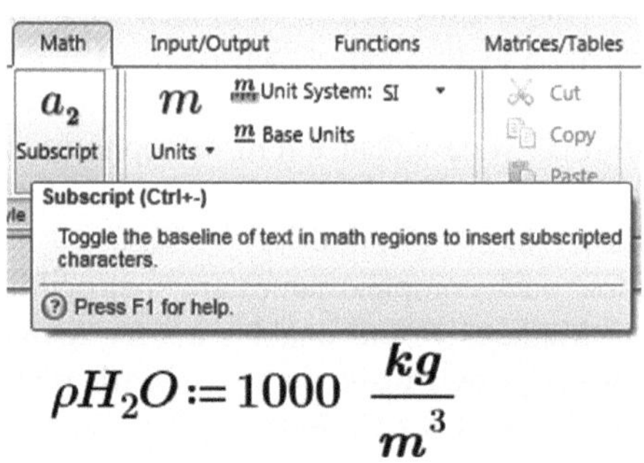

Fig. 6.50 Mathcad Prime variable name with subscript

Comment. *System index is the third type of that in the variables. The previous indexes are text (*`t.i`*) and digital (*`V[i`*).*

Figure 6.51 shows defining (redefining) `sin(x)` ≡ `sin[mc](x·deg)` by the new system index. Mathcad users sometimes need to have the function sine work with degrees but not with radians. In this case, the native Mathcad function is used to omit memory overflow through recursive function calls.

There is one more reason to introduce the system index into Mathcad. As we noted earlier, the worksheet may contain different variables of the same name. This is a typical example beside of those shown above – the conventional mathematical notation `f:=f(x)`: the variable `f` is assigned the value of function `f` with the argument value saved as variable `x`. To avoid errors these two objects must be

Fig. 6.51 Namespace operator

$\sin(x) \equiv \sin(x \cdot deg)$

It is better to do in this way:

$\sin(x) \equiv \sin_{[mc]}(x \cdot deg)$ Typing [Shift+Ctrl+n]

$\sin(30) = 0.5$

Model "Hares and wolves"

Reproduction of hares in wolves' absence $a := 4$ Hare decrease $b := -2.5$

Reproduction of wolves in hares' absence $c := -2$ Wolf increase $d := 1$

$Hare := 0$ $Wolf := 1$

$$F(t, Animal) := \begin{pmatrix} a \cdot Animal_{Hare} + b \cdot Animal_{Hare} \cdot Animal_{Wolf} \\ c \cdot Animal_{Wolf} + d \cdot Animal_{Hare} \cdot Animal_{Wolf} \end{pmatrix}$$

$X := rkfixed\left[\begin{pmatrix} 3 \\ 1 \end{pmatrix}, 0, 8, 400, F\right]$  $:= X^{\langle 1 \rangle}$  $:= X^{\langle 2 \rangle}$

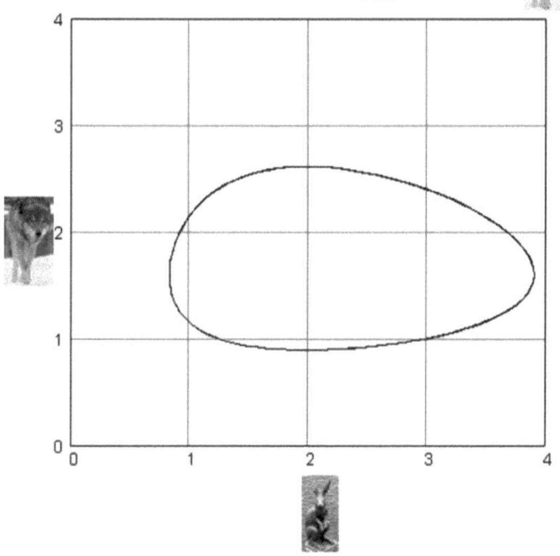

Fig. 6.52 Pictures as variable names

divided by styles. Although, such Mathcad worksheet is impossible to enter "at sight" in which we cannot see a style of a variable or a function. That is why the notation `f:=f[mc](x)` is better. Although, that may be worse as the excess information makes a worksheet hard to read and study.

Comment. *Using styles is a kind of coding, encipherment of a worksheet: everything is computed right but is impossible to renew a worksheet.*

As will be discussed in Sect. 6.5, Mathcad worksheet is three-dimensional. That allows us to overlap the variable name by a picture, its graphic pseudonym, and remove the restrictions on the variable names, for example, insert a period of two indexes without across shift (see Fig. 6.49). Figure 6.52 shows the problem on population growth of wolfs and hares; the pictures of the animals substitute the variable names.

It is impossible to use variable names with "pseudonyms" on Mathcad worksheets assigned for further modifying. Those are appropriate in Mathcad worksheets opened on the Web.

6.2.3 Invisible Variable

This section considers an unusual problem: is it possible and expedient to have invisible symbols on the screen? The answer: it is possible and expedient. Moreover, this technique does not hide something from a user but makes a worksheet easier to read.

Most people know the history of invisible man from the novel by Herbert Wells and by numerous movies. Here is a story of an invisible variable (constant, function, operator). Its life is possible and in a number of cases is expedient not only in Mathcad but in other applications also.

As was noted above, Mathcad allows us to change the color of the variable font. White color is a mixture of seven rainbow colors in existence but a color equal in rights and appropriate to paint variables in Mathcad. If a white variable is located on a white background, it becomes invisible.

In Mathcad worksheets, by default, the user types in black or blue text on a white background: mathematical expressions are black and comments are blue (see Sect. 6.3). Besides, by default these two objects classes have different fonts: mathematical expressions have it san-serif, text – that allows us to make them out in black-and-white hard copies, for example in prints.

Comment. *Default choices for font and its color refers to a template, an empty worksheet that we see on the screen when we first open Mathcad or click the button* ***New*** *on the Standard toolbar. The name of the template is normal. When Mathcad executes command* ***New*** *from the* ***File*** *menu it displays the dialog box containing the list of built-in and users templates that differ from the standard (normal) in arrangement and filling. We can create a user template (another name is "Blank Worksheet") choosing command Save As. The file has extension mct and contains in Templates folder.*

The background of Mathcad worksheet is white[17] (we type in black and blue on white). A user may change it for green, for example.

Comment. *Green color is good for vision contrast (green lampshades, spectacles with green glasses etc.). The Herculean displays typing in green on black were widespread 10 or 15 years ago.*

[17]We used to type white text on black background in DOS versions of Mathcad to save fluorescent layer of a screen. Now people replace their screens as soon as they became obsolete. This is the reason why white screens substitute for the black.

Moreover, a user may change the background color of some expressions to make them more conspicuous for those who will study a worksheet. Otherwise, one may hide the distinct expressions changing its background from white to black (the invisible expressions: we write in black on black[18]).

As we noted before a Mathcad worksheet may contain different objects of the same name through the different styles.

`A:=3  A:=4  A:=A+A  A = 3  A = 4  A = 7`

This example shows (also see Fig. 6.14) not one but three variables A which save their values equal to 3, 4, and 7. Our example is rather artificial, but real Mathcad worksheets quite often contains two variables A one of which is a user (such variable name is very popular) the second is built-in (A is a unit of current strength).

Comment. *Mathcad is not just a mathematical environment but also physical environment. It allows us to assign the variables not abstract numerical values (as in traditional programming) but the unit values of the physical quantities (mass, time, length, energy).*

To apply both ampere and operator A := we must assign these variables different styles. To omit the confusions we may change some font characteristics of the variable style: size or color. The color may be white as well. That is the invisible variable, the hero of our discussion. In the Herbert Wells's novel, the invisible man became visible when got dressed. We can make visible such variable in whole Mathcad worksheet highlighting some operators or changing the background color of the worksheet.

Let us consider the examples that justify using invisible variables and show the benefit of them.

Example 1: Invisible addition Mathcad allows us to change the multiplication sign. A user may select the one from the following:

`2·a  2•a  2 × a  2 a  2a`

The multiplication sign is invisible in last two examples that conform to the tradition existing in mathematics do not place a sign between factors, if the first factor is a constant and the second factor is a variable.

Comment. *Therefore, a variable name cannot start with digits as Mathcad will assume that a multiplication sign exists between the numerical factors in front of a literal factor.*

However, blank space between the two values may mean an addition or a multiplication. For example, 2 h, 30 min, 1 km, 200 m etc. Here the invisible addition sign stands between the same quantities (time and length), and multiplication sign – between the constants and the units. Figure 6.53 shows how to solve it in Mathcad.

The first two operators in Fig. 6.53 insert the user function named + into a worksheet duplicating the built-in addition operator. We cannot change the style,

[18]Such black rectangles overlap police officers' faces in the newspaper photos or in TV shots.

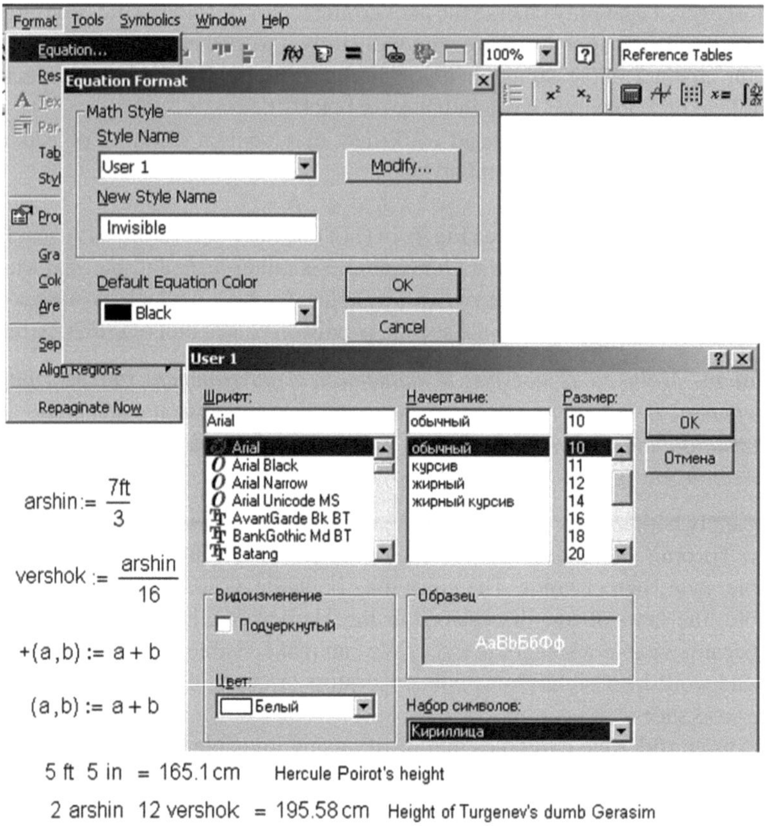

Fig. 6.53 Invisible addition

therefore the color, of built-in addition operator (that is not advisable: we need the "visible" addition), but to change a user-defined function is allowable via the second operator. Mathcad allow us to call a function with two arguments as an infix operator adding up invisibly 5 ft and 12 in., "the size of an averaged Englishman".[19] Figure 6.53 also shows how to change the name of variable style from `User 1` to invisible and the color of the variables to white (see **New Style Name** in dialog box **Equation Format**).

Mathcad Prime enables us to change the colour of variables. Figure 6.54 shows that color of the variable with label System changed to white.

Besides, the built-in operator of invisible addition for a vulgar fraction appeared in Mathcad since version 2001 (Fig. 6.55). We can use it by pressing the particular

[19]In the former Soviet Union (now Russia), a person's height was measured in arshins and vershoks. Telling the height the first component was omitted because that of an average person amounted from two up to three arshins. Gerasim from Turgenev's story "Mumu" was twelve vershoks in height. In addition, both the sign and one of components can be invisible.

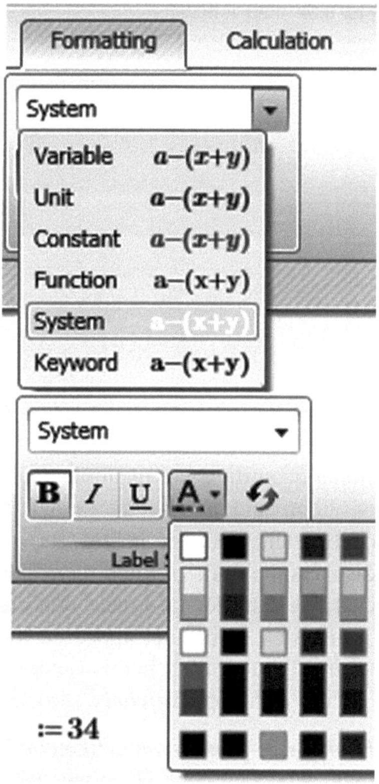

Fig. 6.54 Color of variables names in Mathcad Prime

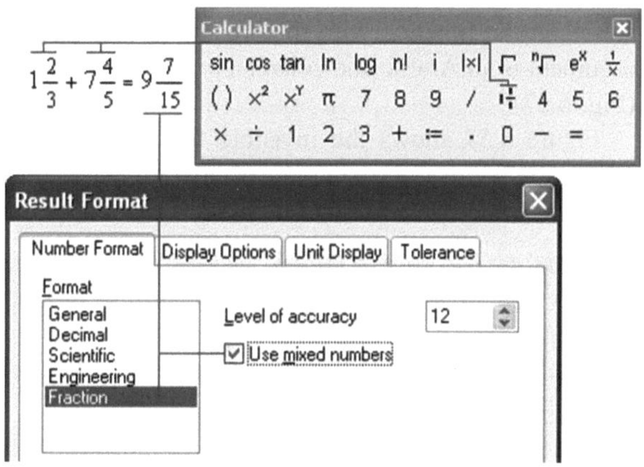

Fig. 6.55 Built-in invisible addition

button on the toolbar **Calculator** before fraction introducing. It is also possible to use the invisible addition in the result inserted by the evaluation equal sign = between the integer part and the fraction after the corresponding formatting (format **Fraction**, also see Fig. 6.9).

Fig. 6.56 Invisible unit

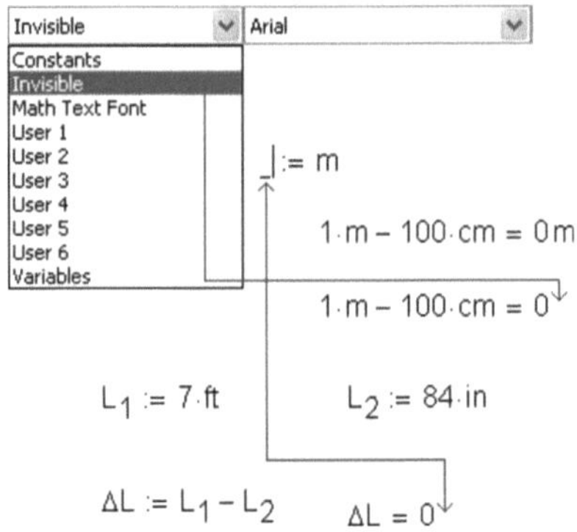

Example 2: Zero dimension quantity Sometimes Mathcad is too pedantic in dimensional quantities. For example, one says that some equipment is located at a height of 20 m and another at zero and not specify the units of that zero (meters, centimeters, feet, or inches etc) reference level. Nevertheless, Mathcad always displays the units of the dimensional values even when it is not necessary. In that case, we can hide an excess unit converting it to invisible (Fig. 6.56).

Invisible units first appeared in Mathcad 12. Now the operator `1 m–100 cm` returns `0` but not `0 m` in this version. Nevertheless, that technique remains valid. We can use it in approximately zero solution.

Example 3: The roman arithmetic Mathcad works with decimal, binary, hexadecimal, octal numbers. However, we may need to make Mathcad work with forms more exotic, for example, with Roman numbers. For Roman numbers operations, we insert the function with the invisible name that returns a Roman number if its argument is an Arabic and conversely the Arabic number if the argument is Roman (Fig. 6.57).

Figure 6.57 shows the invisible function called as postfix or prefix operator which arguments are not in brackets that gives the illusion of Roman arithmetic (see also Fig. 6.41). Only quotation marks weigh down the Roman numbers.

Example 4: The sparse matrix Mathcad has powerful instruments to work with vectors and matrixes (arrays). There is one limitation: these arrays should be completely filled. In practice, we sometimes deal with nonrectangular matrixes, for example, with triangular. The matrices may be even more complex. Figure 6.58 shows how to imitate working with sparse matrix.

The empty elements of the matrix in Fig. 6.58 keep the number that cannot be a matrix element. We do not see it; it was assigned invisible style. Usually, before working with such matrices they are turned into vectors eliminating empty elements by the small program that shows Fig. 6.58.

Example 5: Displaying a dimensional value in several units Often we display the result of a computation in different units `P = 760 mm Hg`, `P = 1 atm`,

```
Invisible function name→(n) :=  | AV ← (1000 900 500 400 100 90 50 40 10 9 5 4 1)^T
                                | RV ← ("M" "CM" "D" "CD" "C" "XC" "L" "XL" "X" "IX" "V" "IV" "I")^T
                                | if IsString(n)
                                |    | (A ← 0  i ← 0)
                                |    | while strlen(n) > 0
                                |    |    | while RV_i = substr(n, 0, strlen(RV_i))
                                |    |    |    (A ← A + AV_i  n ← substr(n, strlen(RV_i), strlen(n) − strlen(RV_i)))
                                |    |    | i ← i + 1
                                |    | return A
                                | otherwise
                                |    | (R ← ""  i ← 0)
                                |    | while n > 0
                                |    |    | while AV_i ≤ n
                                |    |    |    (R ← concat(R, RV_i)  n ← n − AV_i)
                                |    |    | i ← i + 1
                                |    | return R

1948 = "MCMXLVIII"          "MMIV" = 2004

2^"VII" = 128               ∫_"II"^"III" sin(x) dx = 0.574
```

Fig. 6.57 Invisible function-operator

Fig. 6.58 Sparse matrix

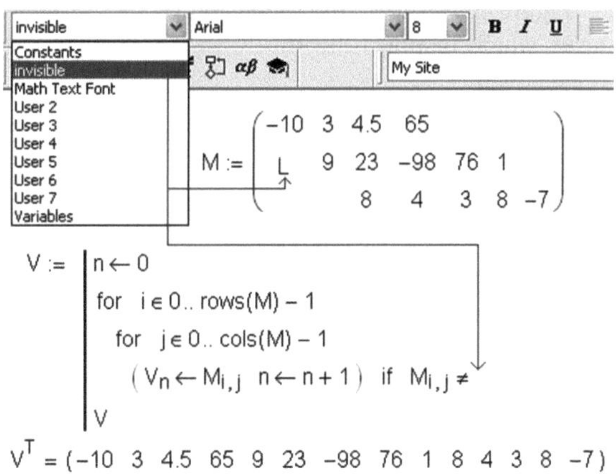

`P = 101.32 kPa` etc. It is better to display here only the first variable and put away the rest.

Comment. *That is a good general principle for all documents including Mathcad. If we can put something away, we should do it.*

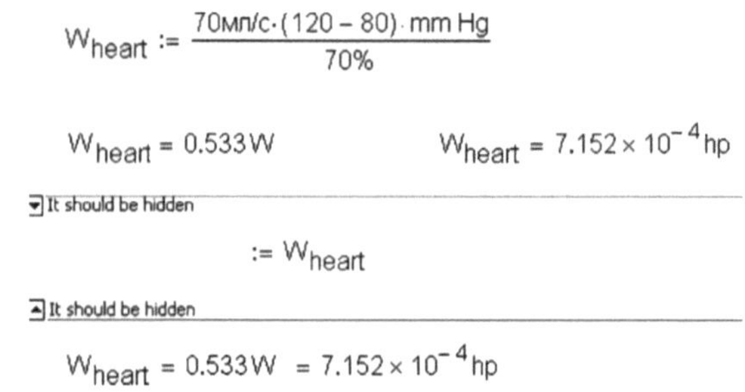

Fig. 6.59 Duplicating of a unit

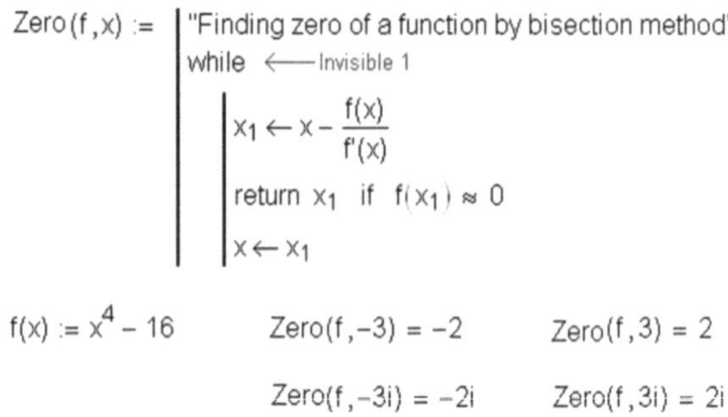

Fig. 6.60 Invisible symbols on a program

Figure 6.59 shows the extension of the problem about capacity of the human heart which displays the file `W`$_{\texttt{heart}}$= = in two capacity units, in watts and in horsepower.

Example 6: An endless loop Mathcad provides programming operators for creating `for` loops and `while` loops. To create `until` loop or interrupt a loop during execution, we should create an endless loop and insert statements `break`, `continue`, or `return`. To create an endless loop we use "infinite" operand in a `while` statement. The infinity symbol (it may be any non-zero number) is typed in white on white (Fig. 6.60).

We can also insert other invisible symbols into the program, for example, to insert an empty string or to shift an operator to the right for fixation of the cycle nest.

Example 7: A Mathcad user's dream (invisible operator explicit) It is known that introducing a variable value by the operator $:=$ we can display a numerical value by the operator $\rightarrow$. This calculation technique was shown in Fig. 6.17. At that, if all the variables of an expression have their numerical values operator $\rightarrow$ displays a result and not the expression. Still, (this is a Mathcad user's dream) it is desirable to see not only the resulting value of a variable (that is easy to do by the operator =) but the values of all variables forming the result (values of variables `Re`, `Pr`, and ξ from the example in Fig. 6.61). Such displaying is useful especially if the

1-st way:

$$Nu := \frac{\frac{\xi}{8}\cdot Re\cdot Pr}{1+\frac{900}{Re}+12.7\cdot\sqrt{\frac{\xi}{8}}\cdot\left(Pr^{\frac{2}{3}}-1\right)} = 942.99$$

where $Re = 4\times 10^3$ $Pr = 7$ $\xi = 11$

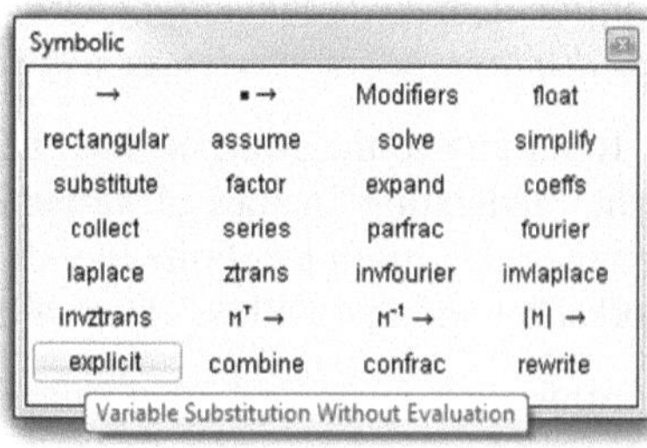

2-nd way:

$$Nu := \frac{\frac{\xi}{8}\cdot Re\cdot Pr}{1+\frac{900}{Re}+12.7\cdot\sqrt{\frac{\xi}{8}}\cdot\left(Pr^{\frac{2}{3}}-1\right)} \text{ explicit, ALL } \rightarrow \frac{\frac{11}{8}\cdot 4000\cdot 7}{1+\frac{900}{4000}+12.7\cdot\sqrt{\frac{11}{8}}\cdot\left(7^{\frac{2}{3}}-1\right)} = 942.99$$

Hide the keywort explicit and chahge → to =

$$Nu := \frac{\frac{\xi}{8}\cdot Re\cdot Pr}{1+\frac{900}{Re}+12.7\cdot\sqrt{\frac{\xi}{8}}\cdot\left(Pr^{\frac{2}{3}}-1\right)} = \frac{\frac{11}{8}\cdot 4000\cdot 7}{1+\frac{900}{4000}+12.7\cdot\sqrt{\frac{11}{8}}\cdot\left(7^{\frac{2}{3}}-1\right)} = 942.99$$

Fig. 6.61 Formula explicit

operators forming variable values `Re`, `Pr`, and ξ (speaking of Fig. 6.61) are far from the operator forming the variable `Nu` that is of interest to us.

Mathcad 14 and 15 enable us to develop an expression and substitute numerical values of the variables, see Fig. 6.61.

6.3 Comments in Mathcad Worksheets

The previous Mathcad worksheets shown as pictures are without any comments, texts or pictures. Comments do not affect to computation results but help to understand its essence. Figure 6.7 being an exception, shows Mathcad worksheet with fundamental physical constants as an assignment operator (for example, `c:=299792458 m s`$^{-1}$) and with text comments to the left `Velocity of light in vacuum`. Besides, in the left top corner of Fig. 6.7 there is a name of the worksheet and small graphic "adornment".

Many users of Mathcad do not insert the comments to the worksheets thinking that they are created for personal use and elucidative fragments could be inserted later. Often this "later" never occurs: all "non-comments" (mathematical operators) are inserted into the worksheet, it works and gives precise result; there is no time to insert comments, we should go further to develop this worksheet or to create the new one. Still, if the worksheet is intended for personal use some comments in it will not be superfluous. We may tangle even in our own worksheet opening it after a time if it has no comment, for example a name.

Comment. *We can name the operator non-comment if the computation interrupts and displays an error message when it being withdrawn from the worksheet.*

Returning to the point of Sect. 6.2 we can contend that the best comments are right, "indicating" names of variables and functions were fixed long ago on the definite quantity in a definite branch of science. It will be enough to name such a worksheet and that will be clear without comments.

Comment. *A name and other data concerning a worksheet (the time of creation etc) can be placed in the heading and the "basement" of a worksheet by the command **Header and Footer**... from the **View** menu. Mathcad 12–15 provide advanced features to save metadata (information about information).*

We can transform the pure mathematical operators to the comments clicking on it with the right mouse button and selecting **Enable/Disable Calculation** from pop-up menu (Fig. 6.62).

The indicator that a mathematical operator is turned off is when the computation results are black rectangles upward and to the right of it. We can disable operators to transform it to a comment and, for example, to select the formulas for calculation from the list available or to speed up computation. For example, a three-dimensional plot can be disabled in checkout and be enabled again in ready worksheet. If one function is defined twice its first definition can be considered the comment under the certain conditions. There are Mathcad worksheets containing whole pages that are the comments in fact. The developer of such worksheet suggests users to study the calculation, after that insert their data, and make computation. In principle, here the operator ■ ≡ ■ should work which definitions applies above it (global definition) but it inaccessible in Controls (see Fig. 6.19) and in Web Controls (see Fig. 6.20) being used more often to design an interface. Therefore, we ought to type again (duplicate) calculation operators after inserting the source data and hide this area as shown in Fig. 6.33, for example. Another way to transform the assignment operator to comment is to

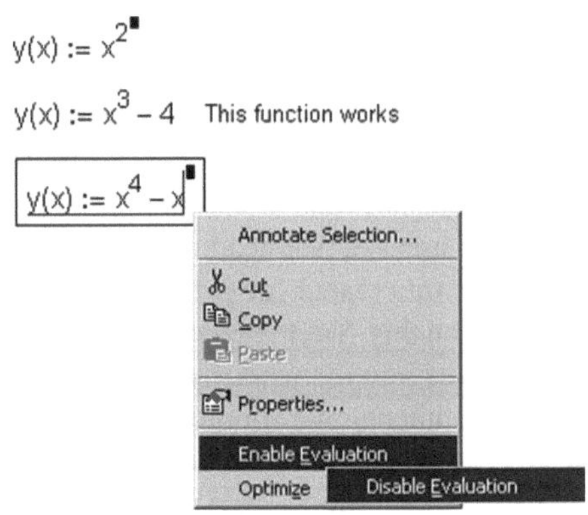

Fig. 6.62 Transformation of an operator to comment

substitute = for := (Boolean equals, but not the equal sign for evaluation): c = a + b^2 for c:=a+b^2.

This is a general method of inserting comments. We type several symbols that is a name of a variable or a function by default but it are transformed into a text after pressing a blank. Thus the fact is marked that blanks can be in comments only but not in the names of variables or functions.

Comment. *Pressing the button < "> (quotation marks) before inserting a comment or choosing* ***Text*** *from* ***Insert*** *menu we create a text region immediately. If the new comment is practically similar to a previous, we should copy the old one and edit it.*

We know from Sect. 6.2 that the names of variables can contain blanks and other reserved characters inserted by keystroke < Shift> + <Ctrl> + <K>. In this case, we can include any symbol allowed in comments except for the period, which is known to turn to invisible and indicate the beginning of literal subscript.

On the one hand, the comment consisting on variable names is a typical mistake of a novice user who does not know how to insert it right.[20] Still, there is another extreme of this phenomenon (the comment consisting of variable names).The most experienced users make all the comments or part of them as names of variables preparing their worksheets to publication in the Internet. The point is that such comments are transferred in the Internet as graphics without distortions while texts are displayed with wrong character coding. Through that, many of Mathcad worksheets opened in the Internet have such comment in the title (as picture, not as the text): "If the texts are distorted though wrong coding change it with the correspondent browser command."

The pictures are very informative in Mathcad worksheets. We create a picture elucidating the calculation in graphic applications (in Paint included into Windows) or scan it from a book and insert it as part of one object of Windows application to another.

Comment. *Often Mathcad includes the SmartSketch application that works with vector graphics but not with bitmapped one. In addition, Mathcad can exchange data with applications such as AutoCAD.*

If we double-click this picture in Mathcad, causing in-place activation of the originating application we can edit it, for example, in Paint and return to Mathcad (Fig. 6.63).

We can insert parts of Mathcad worksheet itself into the picture "freezing" (button < PrintScreen>) it and transferring the "freezing" to graphic application. This way, for instance, we can transfer names of some variables from Mathcad worksheet without changing their fonts and other attributes.

[20]The authors often encounter student's works with comments formed with several variable names.

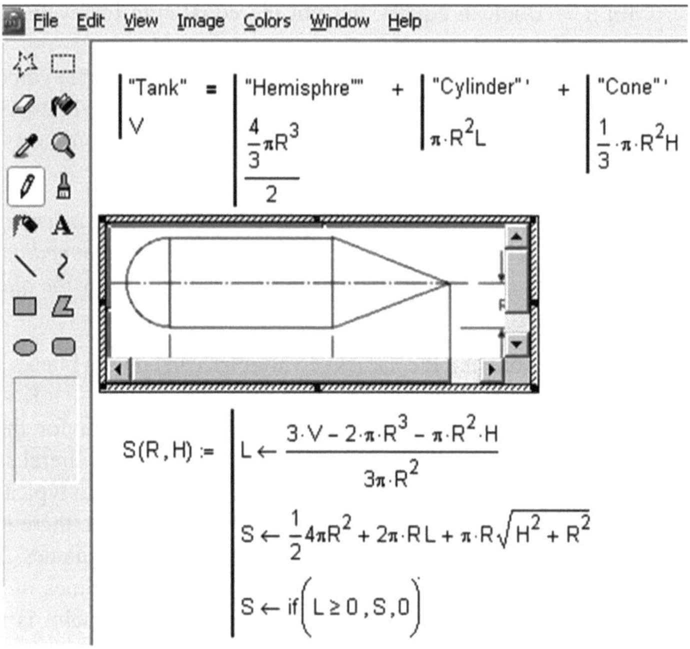

Fig. 6.63 Editing a paint graphics image within Mathcad

We can insert the picture both manually and automatically selecting **Picture** from **Matrix** toolbar – see Fig. 6.64.

Figure 6.64 shows the Mathcad worksheet where variable `F` is assigned the value 1 or 2 (not 1) which change the value of variable `Flow` from `Direct_Flow` to `Counter_Flow` following the scheme displaying of cocurrent or counterflow heat exchange. These two pictures made in advance in the graphic application (Paint) were saved as Direct_Flow.gif and Counter_Flow.gif. Figure 6.64 shows "edge" of screen displaying two image files in File Manager "FAR" (the analogue of Norton Commander).

To change the pictures in Mathcad worksheet according to the way of calculation is a very useful technique (one may say change of decor). That allows us, for example, to change from Russian into English, modify a set of visible formulas of calculation, etc.

"Change of decor" may imply change of the languages. Thus, Fig. 6.65 shows the content of the hidden operators that display the correspondent text according to the chosen language (choose defined switch). The text regions are saved as separate image files. Internet version of the file is located at http://twt.mpei.ac.ru/MAS/Worksheets/Therm/V_balloon_add_ER.mcd.

Figure 6.65 shows part of a screen with Explorer containing these four image files.

Although we will return to text regions discussion later, this brief description is the lion's share of comments.

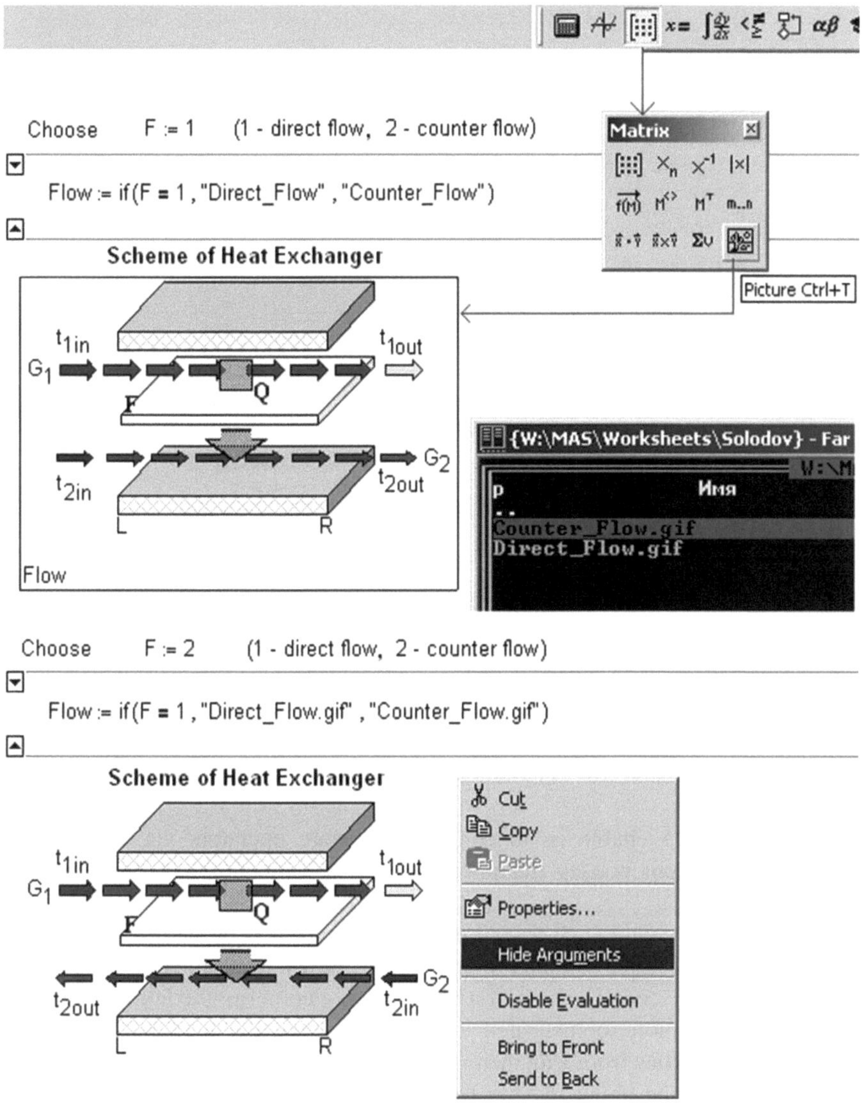

Fig. 6.64 Automatic changing of picture-comment

Mathcad has the spell-checking but for English texts only.

For that reason to create text regions in Mathcad, it is better to type that in Word or insert Word itself into Mathcad worksheet as shown in Fig. 6.66.

Incidentally, finally Mathcad 12–15 allow us to choose language in menu commands, references and other comment environments of this mathematical program.

Figure 6.66 also shows new **Language** tab available in Mathcad 12–15. We can change not only spell check language via the **Language** tab (clicking **Spell Check**

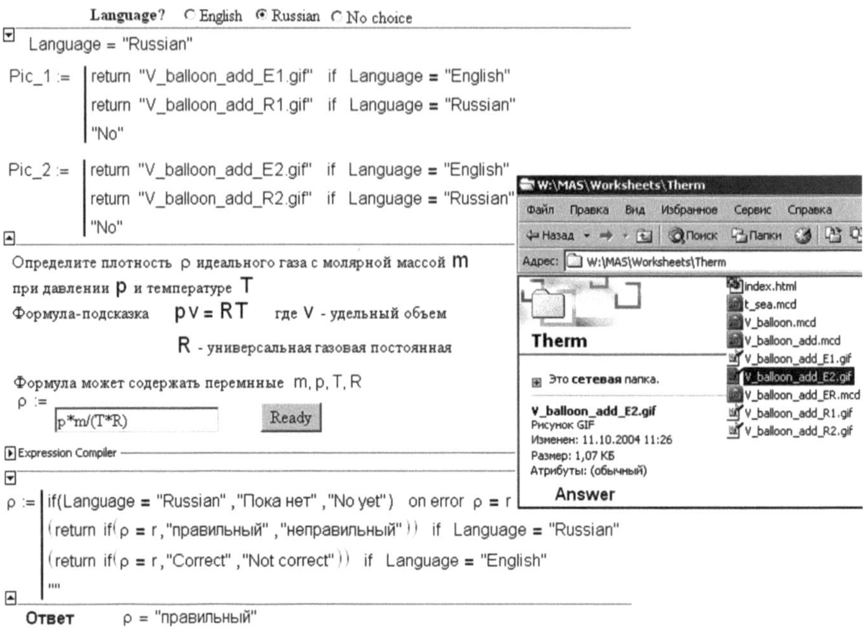

Fig. 6.65 Changing of languages in Mathcad

Options) but language in menus and dialog boxes (top scrolling list **Menus and dialogs**) and also some mathematical expressions (top scrolling list **Math language**). Earlier versions (8–11) Mathcad have only British and American dialects in spell checking.

Mathcad 12–15 enable us to comment separate operators via **View | Edit Annotation**... from pop-up menu clicking on it with the right mouse button (Fig. 6.67).

The operator having such a comment is distinguished by additional brackets appeared when we move a pointer on it. Besides, we can recall Excel allowing us to comment the table cell. The sign of a comment is a corner in the left top of a cell.

We can attach more sophisticated "comment" information in Mathcad 12–15 via command **Properties** from **File** menu (Fig. 6.68).

Such information about the file (metadata) can be necessary to distribute in Internet with technology Mathcad Application Server. Such information helps browsers to find this file (cite) on the Internet.

6.4 Calculation with Physical Quantities: Problems and Solutions

In the beginning of this chapter, we noted that Mathcad is not just for mathematical applications only but for physical applications also (see Figs. 6.5 and 6.8). Frequently, the real "physical" calculations, i.e. where almost all values have units

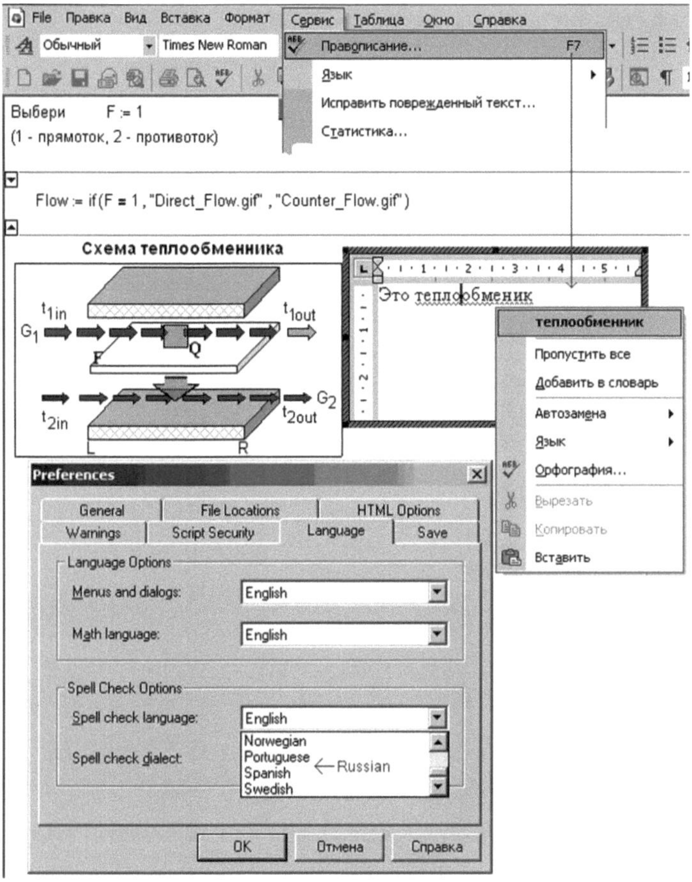

Fig. 6.66 Checking of Russian texts in MS Word within Mathcad

(mass, length, force, etc.), have units not as a multiplier, simplifying and allowing us to avoid some units-conversion errors, for instance: `Enter pressure value in atm P:=120` instead of more convenient `Enter pressure value P:=120 atm`.

Comment. *We tell below how to simplify and automate the unit choice from the offered list by interface elements Controls and Web Controls changing the operators of the type variable = quantity·unit* to more convenient.

What are the reasons for imperfect use of Mathcad capabilities? The first reason is, of course, that some users do not know about such useful Mathcad tool as built-in constants of physical quantities. The users apply techniques in Mathcad, in physical and mathematical programs, based spreadsheet experience or even computer programming (such as C++, Fortran, etc.). Such "non-physical" languages where the

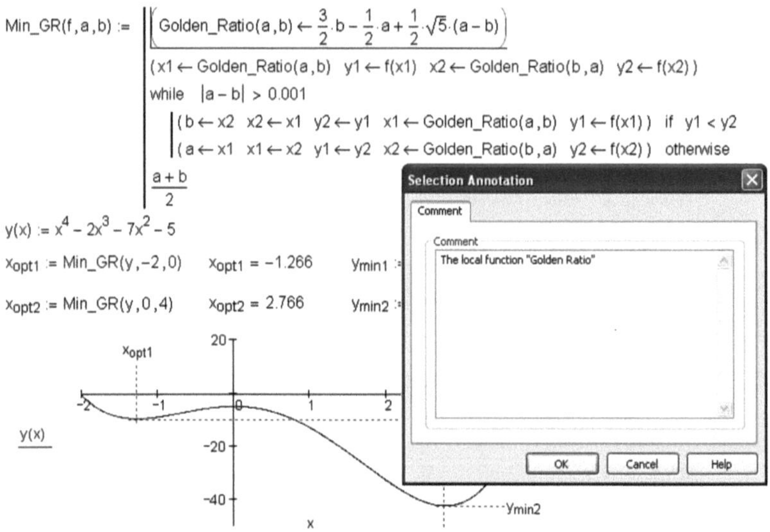

Fig. 6.67 Comment of an operator

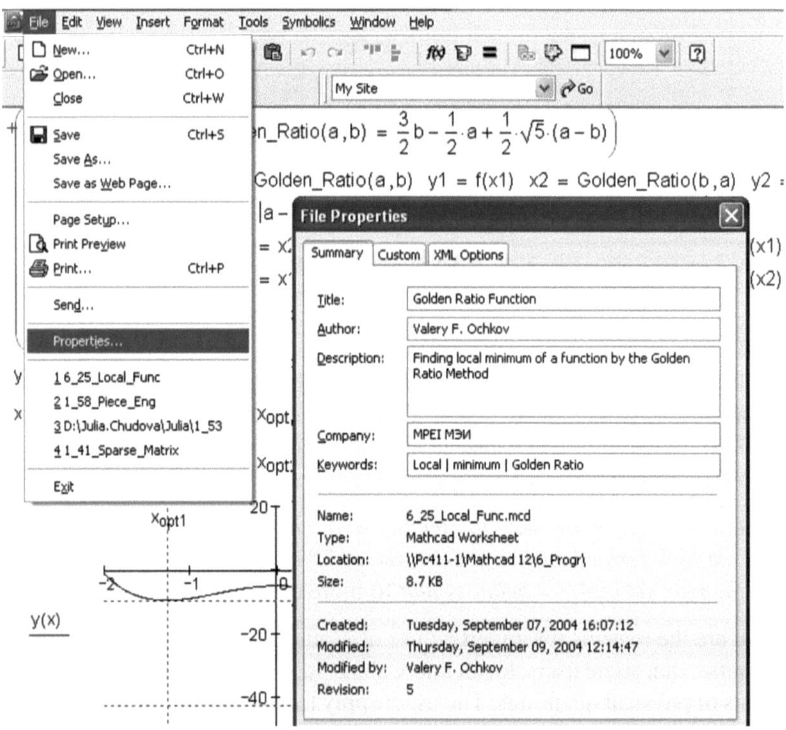

Fig. 6.68 Information about information. Summary of a file

variables keep only numerical values while units are not associated with the numerical values of the variables (left in comment regions or cells, if ever).

The second reason for imperfect use of Mathcad capabilities is from a group of Mathcad users who do not use units justifying such decision with the idea that their quantities expressed in the basic units of one unit system (for example, in SI) and they have no difficulties with unit conversion. This reinforcing this notion is that without units Mathcad worksheets are simpler to compile to environment of the programming languages (such as C++, Fortran, etc), for conversion of customized functions into built-in DLLs (Dynamic-linked Libraries). The third reason is of major importance. It is connected with some peculiarities and limitations of unit tool that cause a skilled user to extract units from a worksheet, almost finished, and turn them into comments: instead of

`Enter pressure value P:=120 atm`

there is

`Enter pressure value in atm P:=120`

Physical values in a calculation are appropriate comments in a Mathcad worksheet (see Sect. 6.3) without going into their "physical" root. The operator `P:=120` is silent but that `P:=120 atm` does not require additional comments. One more feature, connected, if it is possible to say, with "psychology" of Mathcad worksheet creating, brings together units and comments – they postpone the writing, draw away the core. Insertion of units tend to be deferred for future implementation just as comments. Afterward, calculation gives acceptable result.

The tool of units in Mathcad is rather simple. We just type a variable consisting of a number followed by a unit name, user or built-in. We type the unit or insert it from **Insert Unit** dialog box shown in Fig. 6.69.

Comment. *If we type the unit name, Mathcad writes multiplication sigh, visible or invisible, between value and unit. Changing the form of this sign (point, cross) is easy to implement.*

Figure 6.69 shows **Unit System** tab from **Worksheet Options** dialog box with SI unit system set by default. Mathcad checks calculations for dimensional consistency: we cannot add meters and kilograms. That allows us to avoid errors and misprints in formulas, for example, wrong power or addition sign instead of multiplication.

Comment. *Mathcad does not perform dimensional checking in plots but Mathcad Prime does.*

Figure 6.70 shows choosing system of units in Mathcad Prime. Only three of them are now available: international system of units (SI), the system of units customary in the US (USCS) and centimeter–gram–second system (CGS), conventional for a number of physical calculations. It is impossible to disable units in Mathcad Prime (see slider **None** in Fig. 6.69)

Displaying dimensional value by operator ■ = ■ ■ we can insert another unit into the third operand (into the third rectangle appeared when we hold the pointer over operator) to obtain required result or, rather, to change a unit by default to another. We can also duplicate dimensional value with different units to give a

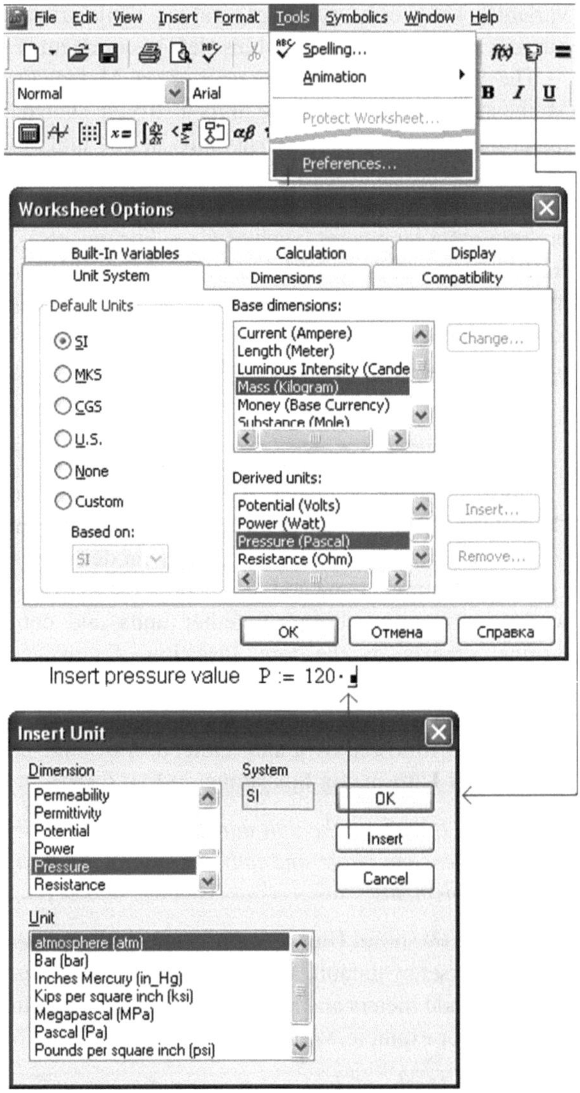

Fig. 6.69 Choosing unit system and inserting units

choice for a reader: joule and calorie, atmosphere and megapascal, etc. (for example, capacity of human heart expressed in watt and in horsepower – see Fig. 6.59).

What cause the skilled Mathcad users to keep of units or seize that from almost finished worksheets?

There are some pitfalls:

1. Some Mathcad tools are not applicable for working with dimensional values. The interruption error message "Can't have anything with units and dimensions here" (finance functions or functions working with splines) will appear or return incorrect answers (function `line`). In that case, we should remove the units

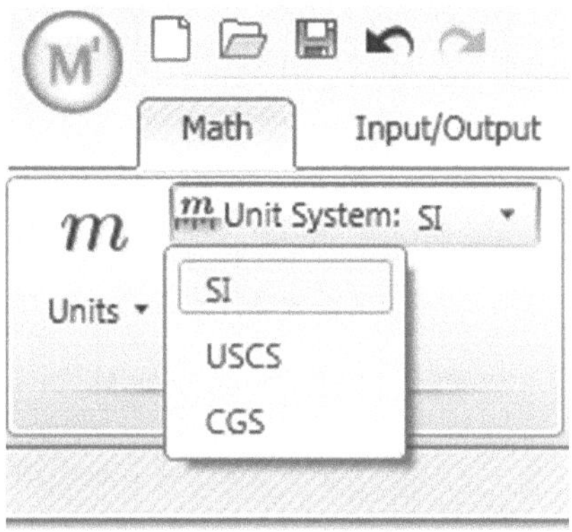

Fig. 6.70 Choosing unit system in Mathcad Prime

dividing the variables with the correspondent basic units and then return required units multiplying it by the value. If the work with dimensional values is not planned, unit tool should be enabled via command **Tools | Worksheet Options**. . . and switch **None** in **Unit System** tab. Otherwise, unit tool may result in errors. For example, a user forgot to define a variable that has the similar name with a unit; nevertheless, the calculation is not interrupted. Here is incomplete list of such "crafty" variables: A, K, T, S, R, m, L, s , etc. Another example. A user defines a function and wants to represent it as QuickPlot, i.e. without defining a range variable previously, for example m: ¼ □3, □2.9.. 5. As a result, a plot is not displayed or, rather only one invisible point is represented because variable m keeps the length unit – meter.

2. Mathcad symbol mathematics works with units as with variables neglecting that 1 m amounts to 100 cm or 60 min are in an hour, etc. Symbol mathematics is foreign element in Mathcad; it was taken from Maple where unit appeared in the eighth version. Although, symbol mathematics is auxiliary tool in Mathcad integrated seldom into a calculation but operates auxiliary functions. If it is necessary to have units worked in symbol mathematics, we should make efforts: enter to Mathcad via operator `substitute` that 1 m = 100 cm, 1 h = 60 min, etc.
3. Mathcad arrays contain only dimensionless quantities or the quantities having the same unit, i.e. the same quantities, time, force, mass, etc. Only one exception from this rule is known and that is block-solver unction Find returns a vector of values with different units if its arguments have different units. If it is necessary the function created by program to return several values with different units in one vector or matrix we may take them off and later return them back. Only Mathcad Prime can have in both, vectors and matrixes.
4. Sometimes we have to use so-called empirical formulas[21] connecting not just quantities but quantities in the appointed units. In that case, we should also take

[21]Empiricism is experimentation. Empirical formulas are obtained not with theoretical analysis of a phenomenon but with statistical manipulation of the data received from experiments.

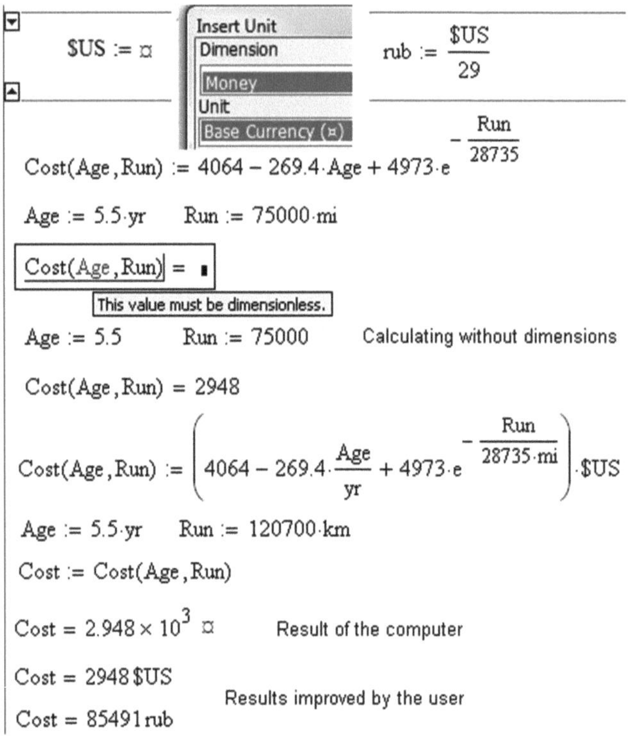

Fig. 6.71 Working with empiric formulae

steps similar to that described in point 1 to avoid fails in unit use. As example, Fig. 6.71 shows the calculation of a secondhand car cost depending on its age and run.

The example below shows how the similar formulas can be enlarged to make unit tool of Mathcad work in them.

Introducing the empirical formulas, we always clearly define the units for source data and dimension of the result. In that case, car age should be expressed in years and run – in miles. The formula returns cost in US dollars.[22] These units should be added to the formula and we must disable unit tool temporarily (see point 1) in it. We did it when we redefine the function Cost(Age, Run) := … dividing the arguments to their units and the function itself was multiplied by required unit as shown in Fig. 6.71.

1. Mathcad list of units includes not all physical and other quantities that we have to deal with. For example, there is no units of information (bit, byte), units of

[22]Here we calculated price of one car brand. This formula was changes in Fig. 6.89 to return not a cost but the percent of a new cost. In that case, we can apply the formula to all used cars.

Fig. 6.72 Working with pseudo-empiric formulae

1. Calculation without units

Power plant efficiency, % $\eta := 32$

Equivalent fuel consumption, g/(kW*hr), where η in percent

$$b_{ef} := \frac{12300}{\eta}$$

$b_{ef} = 384.4$ g/(kW*hr)

2. Calculation with units

Power plant efficiency $\eta := 32\%$ or $\eta := 0.32$

Constant

Combustion value of equivalent fuel $Q_{ef} := 7000 \cdot \frac{kcal}{kg}$

Constant

Equivalent fuel consumption $b_{ef} := \frac{1}{Q_{y\tau} \cdot \eta}$

$b_{ef} = 1.066 \times 10^{-7} sec^2 m^{-2}$ Result of the computer

$b_{ef} = 383.9 \frac{g}{kW \cdot hr}$ Result after inserting necessary units

cost (in old version of Mathcad –dollar, ruble, euro, etc.) may be required in calculations. Mathcad works only with "The magnificent seven" of SI (length, time, mass, current, temperature, luminosity, and substance) and with their derived-units (force, power, energy, etc.). What we have to do in old versions of Mathcad to make technological and economic calculations dealing with dollar, ruble, euro, etc? There is a number of available techniques. Dollar (`$US`) is assigned a unit which is not used in this calculation, and generally rarely used in calculations, for example, *candela*[23] `$US:=cd`. Then we can attach other currencies to dollar, for example `rubl:=$US/29`. After that, the cost will be displayed in candela which we should replace with the required unit (pseudounit) of the cost. Later versions of Mathcad can work correctly with currency units (see Fig. 6.71).

2. A user sometimes meets formulas purely physical in essence and empiric in their form. Figure 6.72 shows the example of working out such formula in the following problem: Given an efficiency of a power station (η) and it is required to find consumption of equivalent fuel (b_{ef}) for power output. All thermodynamic courses give the formula for such calculation as $b_{ef} = 12300/\eta$ where (η) should be expressed in percents and result (b_{ef}) will be in gram per kilowatt-hour. An example: $12300/32 = 384.4$ – power station having efficiency 32% burns 384.4 g of equivalent fuel to produce 1 kW-h of energy.

[23] In earlier versions, we could use steradian (sr) but it becomes dimensionless quantity in Mathcad 12–15.

Listing the appropriate units in the comments allows this formula to work correctly by the method described in point 1. Although there is another method – such pseudo empiric formulas were derived to free us from additional calculations connected with units. That is, if the coefficients obtained are easy to memorize, as in our case – one, two, three, and two zeros, such simplified formula settles down in the thermodynamic courses. Source formula for calculations of equivalent fuel consumption is $b_{ef} = 1/(\eta \cdot Q_{ef})$, where Q_{ef} is heat of combustion of equivalent fuel, the accepted value of it amounts 7000 kkal kg^{-1} (easy to remember, also). If we insert this formula, but not simplified, work of Mathcad unit tool will not be disturbed (see Fig. 6.72 at the bottom).

3. Mathcad uses only *absolute scale* for units. Particularly, that means if the quantity equals zero we need not have to assign a unit. Nevertheless, we should do it in Mathcad, for example, `l:=0 m`, for proper work of dimensional checking.[24] Although, there are *relative scales* also. For example, we measure temperature in centigrade degrees (relative scale) but not in Kelvin (absolute scale). Particularly, that mean the expression `t:=25°C` contains not the multiplication sign between numerical constant `25` and unit `°C` but something else masked by blank.

Figure 6.73 shows one of solutions of relative scales based on a simple problem. It is given the input temperature t_1 of a certain heater and difference of temperatures in input and output Δt. We should find the output temperature t_2. The problem is clear to be not arithmetical but metrological. To solve it, we insert three objects named `C`: the function named `C` and two constants of the same name, the first (`°C:=1`) has style (`°C:=1`), the second (`°C:=K`) has style (`°C:=K`) and we also insert function with invisible name `(t):=(T/K – 273.15)` (see Sect. 6.2.3).

Comment. *There is not "degree" sign in the keyboard but the keystrokes < Alt> + 0176 generates it or by coping from Mathcad set of mathematical symbols.*

All objects are of the same name `C` but they are different having different styles (see Sect. 6.2).

Working with temperature expressed on the relative scale three situations occur. The functions and constants described above help to solve these problems.

- **Situation 1.** We need to insert temperature value in degrees centigrade. The function °C with style Variables is called (the first object named °C) as the postfix operator in the right part of assignment operator := – t1:=120°C. That is, we assign temperature value on the absolute scale to the variable t_1.
- **Situation 2.** We need to insert temperature difference. Here we can use common Mathcad rule and multiply the value by the unit `K` or `°C` equal to `K` (the second object named `°C`).

[24]We can make variable `m` invisible (see Sect. 6.2.3).

Fig. 6.73 Temperature scales

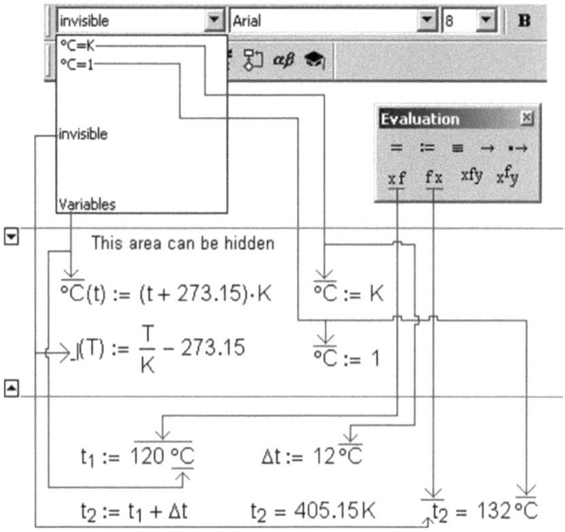

- **Situation 3.** We need to display temperature value in centigrade degrees. It is recommended to turn the variable into operand of the prefix-operator type with invisible name in the left part of operator :=. That is, if we multiply displayed value by the constant °C equal to 1 (the third object named °C) we will get illusion that temperature is represented on relative scale $t_2 = 132°C$.

Three techniques and three objects of the same name described above allow us to work with temperatures: insert temperature value on any scale, display temperature value, insert and display temperature difference. That is, we should apply a simple but important rule: inserting and displaying temperatures in any scales use only absolute scale, Kelvin, in computing. Mathcad worksheet with temperature conversion on different scales is accessible at http://twt.mpei.ac.ru/mas/worksheets/Temperature_Recalc.mcd.

Figure 6.73 shows work with temperature scales when Mathcad had not built-in Celsius and Fahrenheit temperature scales. The last versions of traditional Mathcad and Mathcad Prime enables us to work with built-in relative temperature scales that shown in Fig. 6.74.

There is another unit that may cause difficulties in Mathcad – decibel (dB). Bel is the decimal logarithm of the ratio of two quantities with the same units[25] and decibel corresponds to a one tenth of bel (dB := 0.1 Bel). Measuring something in decibels[26]

[25]Natural logarithm of the ratio of two quantities with same name is neper. This dimensionless unit is called after John Napier (1550–1617), Scottish mathematician, inventor of logarithm.

[26]Usually decibels measure sound pressure levels (SPL). Sound intensity is compared with the least I_0 – usually $I_0 = 0.01\ W\,m^{-2}$ which human ear can hear. The ratio of measuring sound intensity to the least is possible to calculate but the range is too wide that cause inconvenience. This is removed by introducing of logarithmic scale with decimal multiplier $0.1\ lg(I/I_0)$.

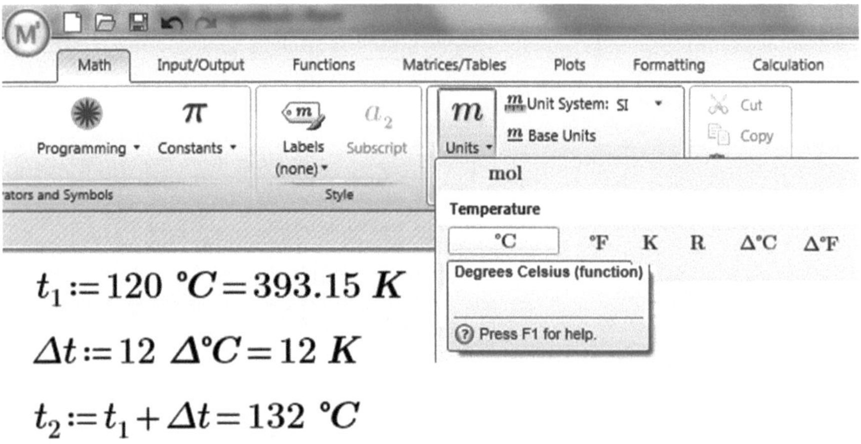

Fig. 6.74 Temperature scales in Mathcad Prime

we create a scale (logarithmic) of the quantity values. That is, we should choose the base from which to begin subsequent calculations. In Fig. 6.75 we created such scale for capacity. The accepted base was capacity of human heart (see Fig. 6.5).

Mathcad requires the introduction of two functions to work with decibels. Their names are the same, dB; name of one function is invisible written in white on white.[27] These objects are different because their styles are different. The invisible function is used to display the value in decibels and that visible to insert. As with temperature, it is called a postfix operator: not `p:=dB(0)` but `p:=0 dB` that imitates dimension.

Working decibels (bels, nepers) the base (source value) is indicated sometimes too; thus, `p:=100 dB (re 0.533 W)` where `re` is the initial letters of *reference*. Such way of working with decibels requires not prefix but infix operator shown in Fig. 6.76.

Comment. *Changes made in Mathcad version 12 to the* `on error` *operator do not allow this technique to work.*

We can note some difficulties in using so-called dimensionless physical values: plane and solid angles, mass, mole, and inclusion volume fractions, etc. These problems occur in Mathcad, also: in version 12 steradian has become dimensionless (`sr = 1`), earlier it had dimension (`sr = 1 sr`). On the one hand, the developers restore logic by that: radian (the ratio of two lengths) was dimensionless earlier

---

The very loud sound of pneumatic drill has level 80 dB, talks inside a room – 60 db, and hardly heard sound of leaves rustling – 10 dB.

[27]Highlighting the operator defining such function makes the function name visible. Calling the function as prefix operator we do not change background color and a user see not `dB p = 5 dB`, but `p = 5 dB` (constant `dB` is typed by a user; it equals to one and has no effect on the calculation but just imitates the unit).

$$\text{base} := \frac{70\,\frac{\text{mL}}{\text{sec}}(120\,\text{torr} - 80\,\text{torr})}{70\%}$$ Power of a human heart $\text{base} = 0.533\,\text{W}$

Two functions and one constant

$\text{dB}(x) := 10\cdot\log\left(\frac{x}{\text{base}}\right)$ Style of this function is Invisible (it is typed in white on white)

$\text{dB} := 1$ Style of this constant is Constants

$\text{dB}(x) := \text{base}\cdot 10^{\frac{x}{10}}$ Style of this function is Variables

Two functions and one constant

Displaying of power in dB (invisible function dB + prefix operator + dB = 1)+ dB=1)

$p := 1\,\text{W}$ $\text{dB}\ p = 2.73$ $p = 2.73$ $p = 2.73\,\text{dB}$

$p := 1\,\text{hp}$ $p = 31.456\,\text{dB}$ Horse power

$p := 1\,\text{kW}$ $p = 32.73\,\text{dB}$

$p := 6 \times 800000\,\text{kW}$ $p = 99.543\,\text{dB}$ The biggest power plant Surgutskaya GRES-2

Insert of power in dB (“visible” function dB + postfix operator)

$p := -\infty\,\text{dB}$ $p = 0\,\text{W}$ $p = 0$

$p := 0\,\text{dB}$ $p = 0.533\,\text{W}$ $p := 2\,\text{dB}$ $p = 8.452 \times 10^{-4}\,\text{kW}$

Power addition

$p := 0\,\text{dB} + 0\,\text{dB}$ $p = 1.067\,\text{W}$ $p = 3.01\,\text{dB}$ Power of two hearts

$p := 12 \times 200000\,\text{kW} + 6 \times 800000\,\text{kW}$ Surgutskaya GRES-1 + Surgutskaya GRES-2

$p = 7200000\,\text{kW}$ $p = 101.304\,\text{dB}$

Fig. 6.75 Using decibels

one function and four constants

$$\text{dB}(x,\text{base}) \equiv \text{base} \times 10^{\frac{x}{10}} \text{ on error } 10 \times \log\left(\frac{x}{\text{base}}\right)$$

$\text{re} \equiv 1$ $\text{dB} \equiv 1$ $\text{B} \equiv 10 \times \text{dB}$ $\text{Np} \equiv \frac{2 \times \text{B}}{\ln(10)}$

one function and four constants

Inserting power in terms of power

$P_1 := 100\,\text{kW}$

Conversion power value into logarithmic units

$P_1 := P_1\ \text{dB}\ (\text{re}\,0.533\,\text{W})$

$P_1 = 52.733\,\text{dB}$ $P_1 = 5.273\,\text{B}$ $P_1 = 6.071\,\text{Np}$

Inserting power in logarithmic units

$P_2 := 50\,\text{dB}$ $P_2 = 5\,\text{B}$ $P_2 = 5.756\,\text{Np}$

Conversion pressure into units of pressure

$P_2 := P_2\ \text{dB}\ (\text{re}\,0.533\,\text{W})$ $P_2 = 53.3\,\text{kW}$

Fig. 6.76 Using decibels and reference to the base

while steradian (the ration of two areas) had dimensions. On the other hand, that clears the way for some errors of the kind: radians and steradians addition, sine of steradian, etc.

We can face the problem of lack of dimension checking while working with other "dimensionless" quantities that have different dimensions. While we cannot add mass and moles, addition of mole and mass fractions will not result in an error message.

The dimensionless parameters are often encountered in calculations. As a rule, it is inserted into the worksheet as it = 1 and then work with it in the known way: `Number of tubes in a heat exchanger N:=2,200 it`, for example. That is, the variable N remains dimensionless. Although we may recall that Mathcad keeps a built-in unit for items (base units) – that is mole. A Mole divided by Avogadro constant is an item that could be used in subsequent calculations. Figure 6.77 shows finding velocity of water in a certain heat exchanger; the tubes number (n_p), interior diameter (d_p), and volume consumption of water (Q) are known.

Dimensional items and other quantities with "dimensionless units" seem to be just curious objects. Thus, in a calculation shown in Fig. 6.77 cross-section area of one tube and total area of all tubes have, strange as it may seem, different dimensions: mm^2/it and mm^2 and they cannot be added. That is correct – such operation can be done only by error. The example shown in Fig. 6.77 is strained, of course. Nevertheless, the authors know of actual examples when the errors in sophisticated technical and economical calculations were discovered only after the heading operators `ruble:=cd` and `it:=mole/`N_A were substituted for `ruble:=1` and `it:=1`. It turns out, that in an incorrect calculation two added quantities had dimensions `ruble` and `it`.

▼ Units

$$N_A := 6.02214199 \cdot 10^{23} \cdot \frac{1}{mole} \qquad it := \frac{1}{N_A}$$

▲ Units

Source data:

Number of pipes $n_p := 2200\,it$

Pipe diameter $d_p := 15\,mm$

Flow in heat exchanger $Q := 1200\,\frac{m^3}{hr}$

Calculation of flow rate in pipes

Sectional area of one pipe $f_p := \pi \cdot \frac{d_p^2}{4} \cdot \frac{1}{it} = 176.7\,\frac{mm^2}{it}$

Sectional area of all pipes $f := f_p \cdot n_p = 0.389\,m^2$

Flow rate $v := \frac{Q}{f} = 0.857\,\frac{m}{s}$

Fig. 6.77 Dimensionless quantity – item

Assigning relative temperatures with lower case and absolute with upper case, temperature variables facilitate their distinction. That is, for example, the temperature in thermodynamic calculations where variable `t` assigns centigrade degrees (a relative-units variable) and `T` – Kelvin (an absolute-units variable). Mathcad units tool allow us to avoid this bisection.

When the authors see a student's calculation of a "physical" problem made in Mathcad without units it is analogous to situations when one writes calculations in Word: types formulas there and calculates manually, on Windows calculator (see Fig. 6.2) or often on the single one lying not far from computer keyboard.

6.5 Three Dimensions of Mathcad Worksheets

As was noted in Sect. 6.1 Mathcad has one special feature standing out from other applications. Mathcad worksheets are "three-dimensional": beside the coordinate axes above/below (let us call it x-axis) and right/left (y-axis) there is the third one: close/farther (z-axis). In Mathcad, it is possible to overlap two objects and select which will be in front of another. Figure 6.78 shows the situation where the texts `Hyperbola` and `Parabola` are overlapped on the plot and indicate correspondent lines.

The features indicating the third dimension of Mathcad worksheet are two additional commands, **Bring to Front** and **Send to Back**, from pop-up menu. Choosing a command from the menu, we can move an object.

Introduced in version 2000 of Mathcad, the third dimension became widely used immediately.

Comment. *Mathcad adopted this feature from Word and Excel that enable objects overlapping, too.*

The example shown in Fig. 6.78 (inscriptions on the plots) is the most widespread.

Comment. *The inscriptions on the plots can be made transparent if needed, not in Mathcad but in Paint, for example, if we insert it there to redesign (see Sect. 6.6).*

Here are two examples less known to Mathcad users. Figure 6.79 shows Mathcad function created by a program `Min_GR` that returns coordinate of the point `x`, the minimum of another (analyzed) function dividing the segment in golden ratio.

Mathcad programming tool have no means of making comments in program-blocks and it does not enable us to have a local function inside program-blocks (henceforth, programs) until version twelve (12). Nevertheless, Fig. 6.79 shows us comments: texts with different fonts, Mathcad operators (`a/b = b/(a + b)`), and also a picture that is an exotic object for comments in programs. Besides, the program contains subroutine `Golden_Ratio` that has all features of a local function although the original worksheet is in Mathcad 11 version in which local functions are impossible. What is the issue? The issue is that all these objects (texts,

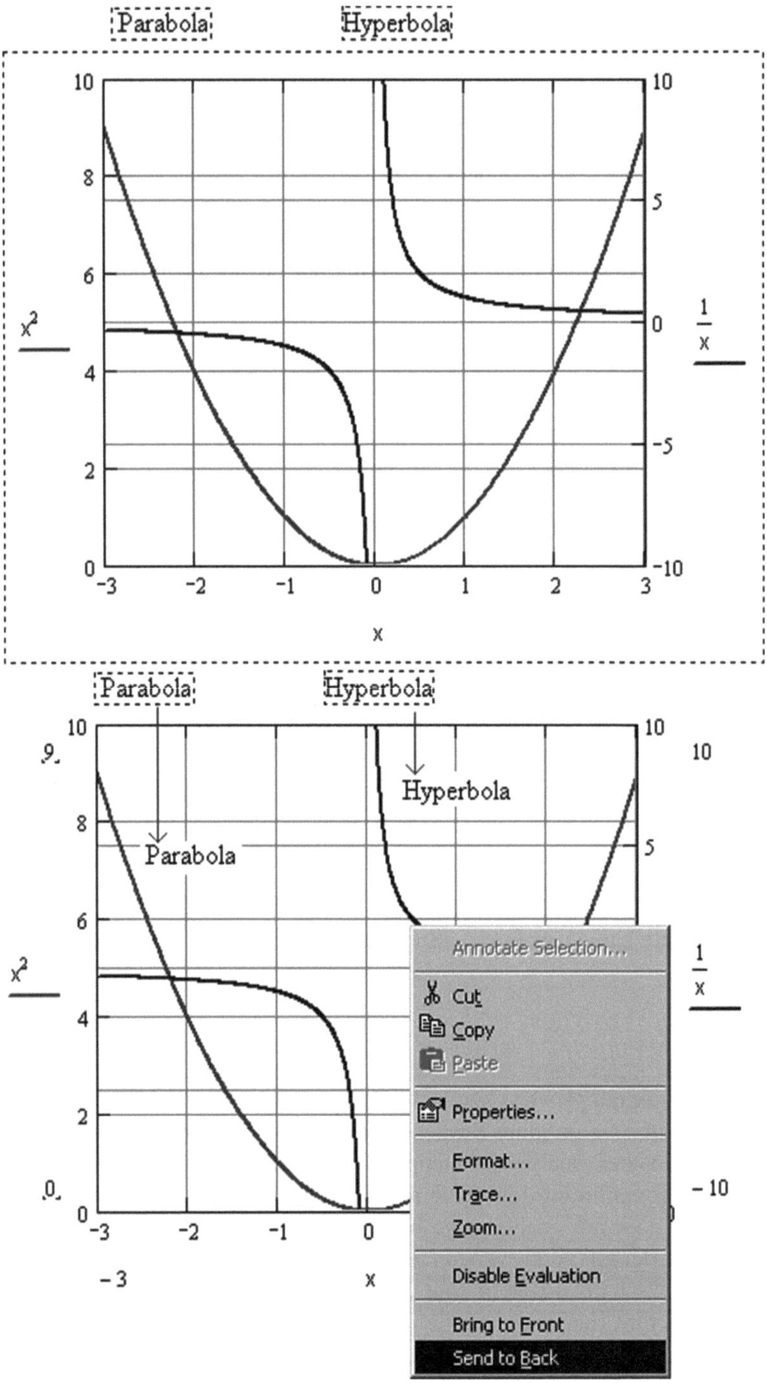

Fig. 6.78 Title over the plot

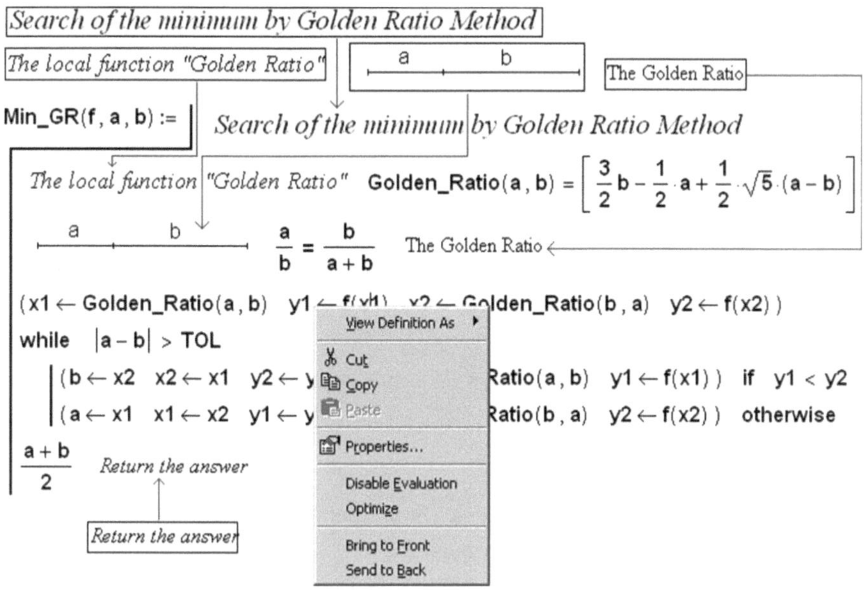

Fig. 6.79 Overlay comments in a program

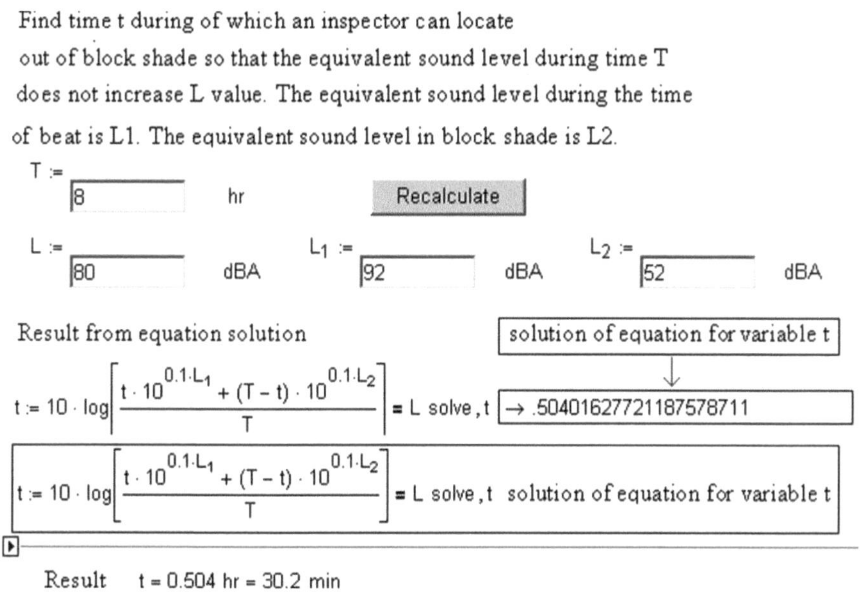

Fig. 6.80 A fig leaf in Mathcad worksheet

pictures, operators and even working function `Golden_Ratio`) placed on function `Min_GR` that gives an illusion of a single well-commented program.

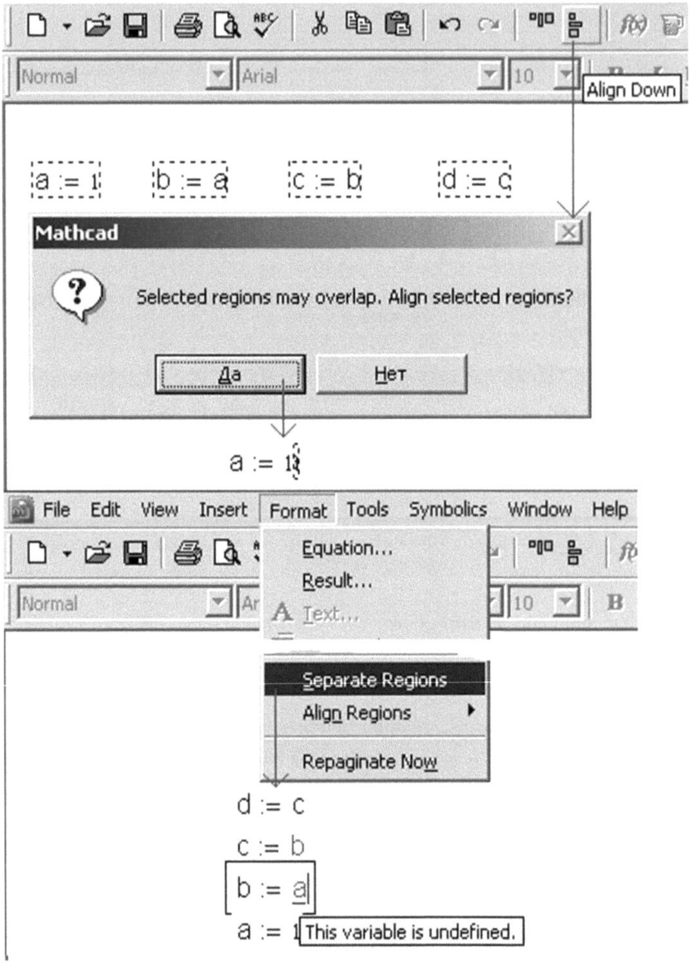

Fig. 6.81 Assembling and separating operators in z-direction

Comment. *To make the function* `Golden_Ratio` *visible within function* `Min_GR` *we should use operator* ≡ *but not* := *which sigh is replaced by* =.

The third dimension is useful in hiding some information in Mathcad worksheet. There is a well-known literature rule, and the author[28] of this book try to follow it, to omit a word, a sentence, a paragraph, a chapter if it is possible to do. In informational technologies this principle (its motto is known:"Brevity is the soul

[28]The authors thought much about this principle writing this book. If all the reader needs is detailed description of Mathcad tools, then a chapter containing the thoughts of the authors and their works may be (should be) omitted.. Then it will be not a book but manual.

of wit") results in demand to omit unnecessary, overhead information that prevent to read and understand the document probably more than absence of the comments. Such unnecessary information is hidden in close areas of the Mathcad worksheets or moved over to the right border of a screen. If it is impossible to do, we can overlap this information by additional or neutral comment as shown in Fig. 6.80.

Figure 6.80 shows the problem, from safety engineering at power stations, coming to the equation solved symbolically by Mathcad operator ■ `solve`, ■ →. This operator "shoots" both to the left (assigns the result to variable `t`) and to the right (displays answer, turned out to be unnecessary, which is overlapped in Fig. 6.80 by the comment).

Figure 6.81 shows the way in which four operators aligned horizontally can be realigned in z-direction, axis perpendicular the display plane.

6.6 Mathcad Plots

One of the most effective methods to view and represent source, temporary, and obtained data is by plotting. To continue comparing Excel and Mathcad started in this chapter, we may say that the first application possesses business and the second scientific plotting.

In Mathcad, we can create plot via command **Graph** from **Insert** menu or via the toolbar of the same name shown in Fig. 6.82. Users can classify Mathcad plots into the following groups:

- The plots displaying dependences (functions and arrays) of one variable: X-Y Plot, Polar Plot, and 3D Scatter Plot
- The plots displaying dependences (functions and arrays) of two variables: Surface Plot, Counter Plot, 3D Bar Plot, 3D Scatter Plot, and Vector Field Plot
- Flat plots (2D): X-Y Plot, Polar Plot, Counter Plot, and Vector Field Plot
- Three-dimensional plots (3D): Surface Plot, 3D Bar Plot, and 3D Scatter Plot

It is the opinion of the authors that such classification is better than given in documentation and references:

- 2-D plots: X-Y Plot, Polar Plot
- 3-D plots: Surface Plot, 3D Bar Plot, 3D Scatter Plot, Counter Plot, and Vector Field Plot

There are more plot types than noted above where we listed practically the buttons on the plotting toolbar (see Fig. 6.82) rather than types of plots. Independent group plots (e.g., 2-D Bar plot of graph of errors), are displayed by correspondent formatting of the others, **X-Y Plot**.

The authors believe that 3-D plots are rather promotional thing than a method to represent results of large calculations, having practical aims not presentation or advertising, especially, if we should print such plot where we cannot rotate it to see from another point.

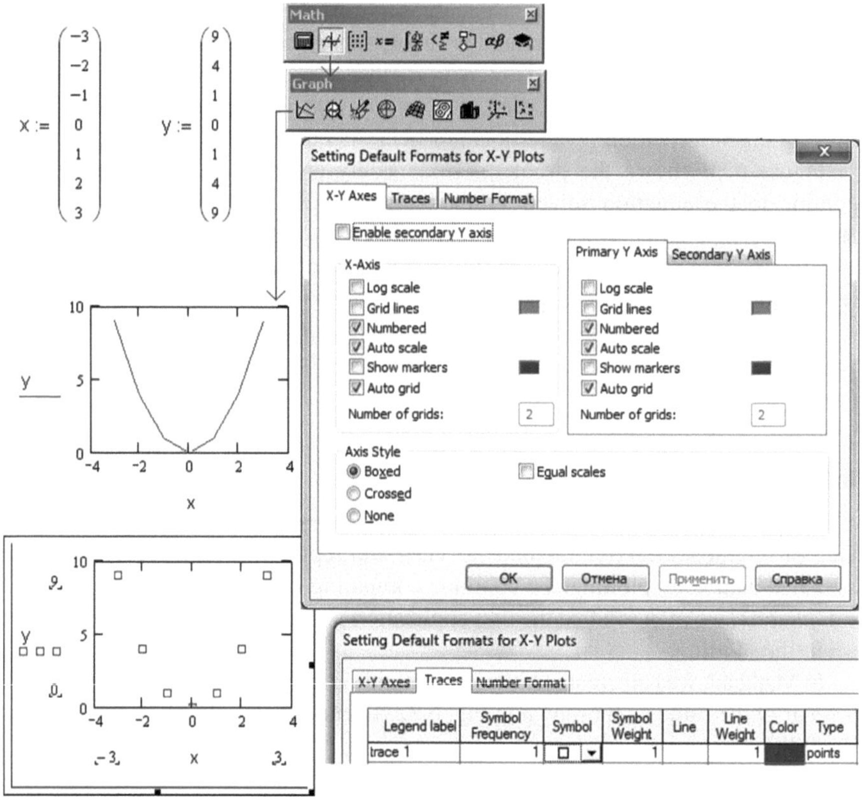

Fig. 6.82 Plotting two vectors

Comment. *We can rotate 3-D plots by specifying coordinates, changing values of three numbers in the dialog box or with the mouse. If the mouse has a center wheel, we can rotate the wheel to zoom in or out of a three-dimensional plot.*

Therefore, considering screen and paper of printer to be flat, we should prefer simple 2-D dimensional plots creating Mathcad worksheets. Real sophisticated problems may base on set of functions with three, four, and five arguments which plots we must represent correspondently in three-, four-, and five-dimensional space. We should remember that complex plots often hide rather trivial calculation that masks itself with perspective, lightning, fog and other options. Complex calculations, having practical value, as a rule, illustrate the results with simple, plane plots, which represent the basic law of calculations. Such calculations include set of curves, for instance, instead of surfaces looking agreeably but hard in work.

The most commonly used plot in Mathcad worksheets is **X-Y Plot**. They represent locations of a couple of elements (components) of two, three and more vectors in the plane, tabulation of functional dependencies of one argument. Figure 6.82 shows that two vectors of equal size (length), x and y, are inserted, then a pointer is moved to blank space where the graph appears, the button **X-Y Plot** was pushed (or < Shift>+<2>, or <@> was typed). Variables x and y were placed in the appeared blank X-Y Plot (we can also insert there vectors themselves). After that simple manipulations (vectors x and y can be read from the disk or obtained in a calculation, not just inserted manually) Mathcad represent a plot having a set of defaults. In particular, the points of the plot representing the location of the couple of vector elements are connected with a broken line starting from the first point (the first elements of vectors) to the last one (the last elements of vectors). We can change the default settings via the dialog box **Formatting Currently Selected X-Y Plot**. It is invoked, as the other dialog boxes for formatting Mathcad objects, with mouse double-click the graph.

Mathcad Prime 1.0 enables us to work only with 2-D Plots according to the author's gradation described above. Mathcad Prime allows only three types of plots: **XY Plot**, **Polar Plot**, and **Counter Plot**. Figure 6.83 shows the simplest point-by-point plot in Mathcad Prime using the points analogous those shown in Fig. 6.82 (traditional Mathcad). Figure 6.83 also shows some instruments for plot formatting, choosing **Line Style** and **Symbols**.

Inserting the points of plots not through two vectors as shown in Fig. 6.82 but using tables appears in Fig. 6.26.

The plot shown in Fig. 6.82 was reformatted in the following way: the lines were put away (the word "lines" from the column **Type** was replaced for "points") and the boxes were substituted for periods, almost invisible, "none" from the column

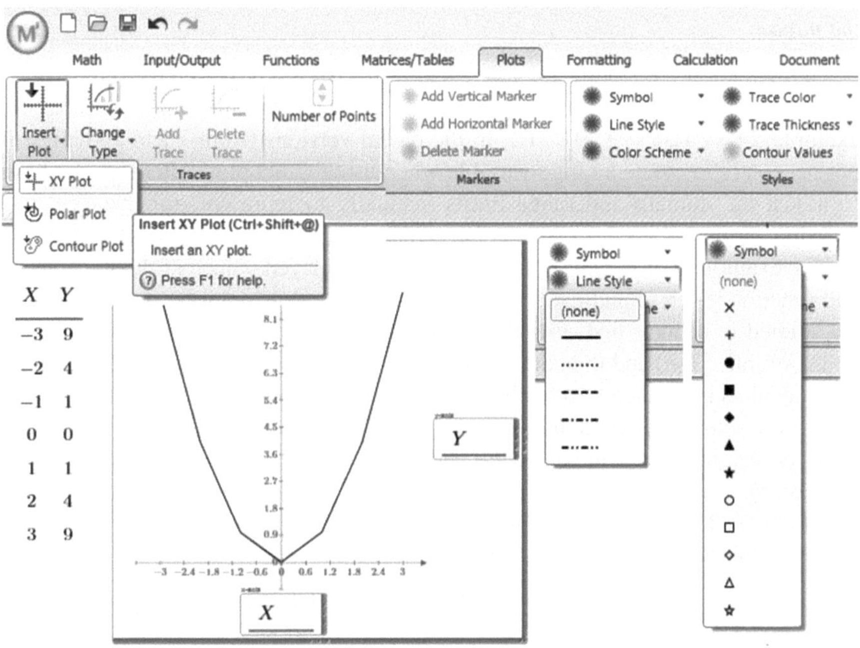

Fig. 6.83 Plotting a table in Mathcad Prime

Symbol. It is needless to list in this book all accessible and possible changes in plots since the documentations are sufficient. Mathcad references and user's guide describe these possible changes in details in. We shall note only new features that appeared in version 12:

- Two y-axes in an X-Y plot (see Figs. 6.87 and 6.88)
- Change of the marker color, four dotted lines (two horizontal and two vertical) by which we can indicate special points in a plot, minimums, maximums etc.
- Adding legends in plots, in one of four spare corners of a plot (and not just below as in earlier versions) that makes a plot more compact.

X-Y plots can represent both couple of vectors and functional dependence themselves. For that, it is enough to press **X-Y Plot** and fill in the placeholders of a blank X-Y plot with the function defined above `y(x)` and its argument `x`, as shown in Fig. 6.82 but not the names of vectors. After moving a pointer, the plot is produced over a domain from −10 to 10 if the correspondent function returns real numbers in this range. For example, square root of x will be represented within the range from 0 to 10 because it returns imaginary numbers in the left part of x-axis. That is, the problem comes to plotting of two vectors again: Mathcad divides the segment of plot to points, the number of which depends on plot size on the screen (on average it is 50) and on monitor definition, and computes the values of an argument and a function in these points, i.e. forms two vectors to plot them.

Comment. *A user can resize a plot holding the mouse button down and dragging the mouse.*

Up to version 7 of Mathcad, a user had to format a plot before creating: range-variables of the argument, number of points on the plot, to define range variable, for example `x:=−10, -9.9... 10`. In the recent versions (from 7 and above), it is not necessary to create a range-variable when using Quickplot. Although, we have to adjust the domain and range limits manually to represent, for instance, two functions in the different ranges of the variable, as shown in Fig. 6.84.

The example to the right of Fig. 6.84 shown displayed values of the variable x_1 and function $y_1(x_1)$ indicates that Mathcad draws plots point by point but not as we studied in school: first, find the critical points (zeros, maximums, minimums, points of inflection) and then draw a qualitative plot. We should remember about it creating plots in Mathcad. Now, the surfaces discussed below are plots of a function of two variables given in tabulated form. Then the correspondent grid of a matrix is risen at height proportional to the values of function `f(x,y)` in the point (Fig. 6.85).

After that, Mathcad formats this node in the following way: fills its cells with the different colors over the various schemes, removes the grid, etc.

Comment. *Number of defaults of 3-D plots (additional tools of formatting) is a dozen higher than that of 2-D plots.*

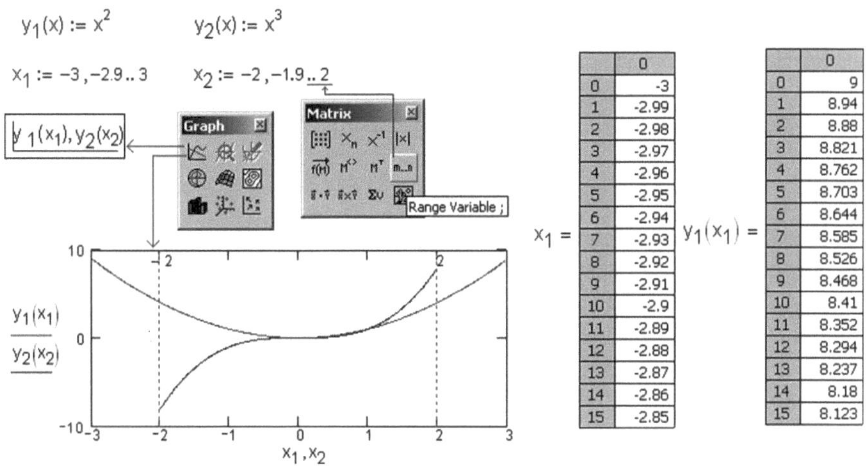

Fig. 6.84 Creating a 2-D Plot using a range variable

Figure 6.86 shows **Counter plot** in Mathcad Prime. Color scale with numerical values pointing "height" of the contours on the plot itself represents "magnitude" of the variables.

Figures 6.87 and 6.88 show the "graph" novelty of Mathcad 12 and next versions (13, 14, and 15): the second axis y of 2D plot in the example of finding the volume (V) and square of the liquid surface (S) contacting with air in the tank, contained from a cylinder and two half-spheres, by maximum depth of a liquid (h).

The second y-axis allows the user estimate the volume of the rest of a liquid both in m^3 (the left axis) and in barrels (the right axis). Then the question arises concerning the second x-axis to represent the height of the liquid layer h (the argument of function V) in an alternative unit, not in meters (the first x-axis) but in feet, for example.

Figure 6.88 shows the plot in which different y-axes represent different physical quantities: the volume of the liquid layer and the square of its surface. We may say that the essence of the second y-axis is the second *grid* of y-axis (see Figs. 6.87 and 6.88). Earlier versions allowed the representation of two and more functions with single y-axis (up to 16) also (see Fig. 6.84).

Besides, we should remember of such a case. Here is a typical situation in mathematical analysis: two plots are created in one coordinate system, a function and its derivative. That could be considered an error, which Mathcad is to interrupt with such error message when meters are added with kilograms.[29] In addition, we can recall a problem from the book for novice: find the values of x of the given function in which this function is higher than its derivative. For the reason that we cannot compare a function and its derivative because these values are different

[29]Suggestion that is still more cardinal is to forbid us to add the value, even dimensionless, to its square, for example.

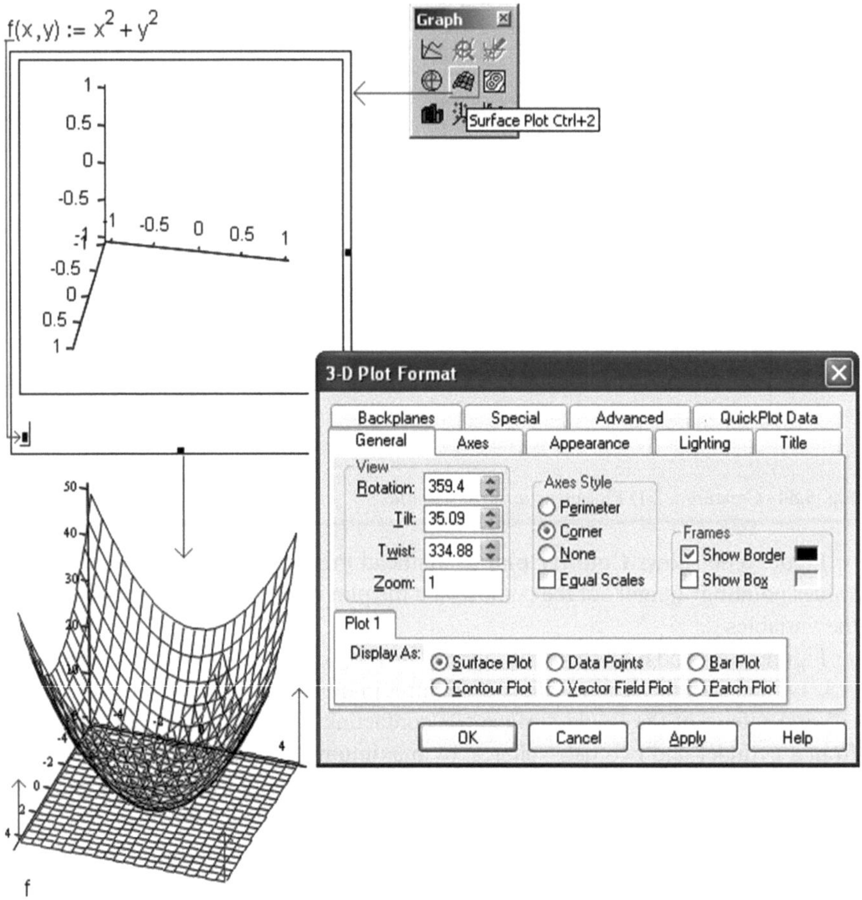

Fig. 6.85 Plotting a surface in Mathcad

(the same meters and kilograms) the authors say that "pure" mathematics does not connect a function and its derivative with any physical quantities, for example with distance and speed (due to units discrepancies, meters versus meters/seconds). They also do not agree to change the form of the problem but to keep the essence (derivation and solving the inequality): to find values of x in which the derivative is greater (or less) than zero. From this concept, it follows that we can draw two or more curves in a single plot only if they are of the same dimension (length, time, force, etc.). Otherwise, a reader will apply himself to the meaning of their cross-points while they have no meaning.

Often we need to represent only definite combinations of arguments and functions. The following example illustrates this thesis. The function returns the cost of a used car depending on its age and run[30] considering the mean speed (i.e. run

[30]We have discussed this function considering its units (see Fig. 6.71)

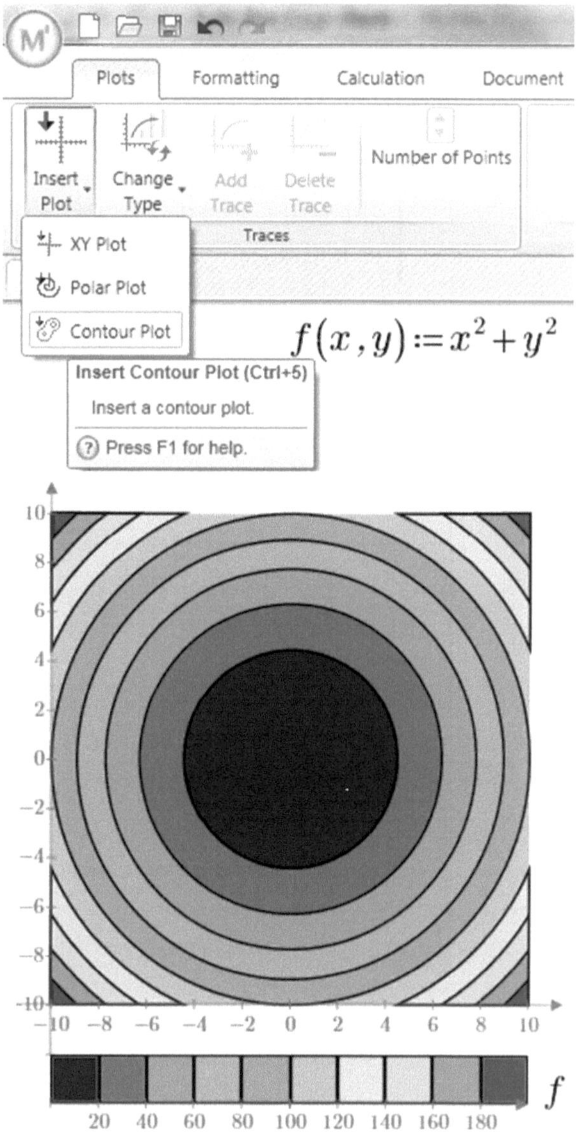

Fig. 6.86 Plotting a counter plot in Mathcad Prime

divided by age) within the range from 1 to 2 km h^{-1}. Figure 6.89 shows that we insert the *penalty* into function Cost: if the real mean speed is out of that range, the function returns not the value but the text, which is ignored in the graph.

A limitation of Mathcad 2D plotting is that variation range of the arguments must be rectangular only. By default, the range is set square having spread in values x and y from −5 to 5 and then a user varying these numbers can make it

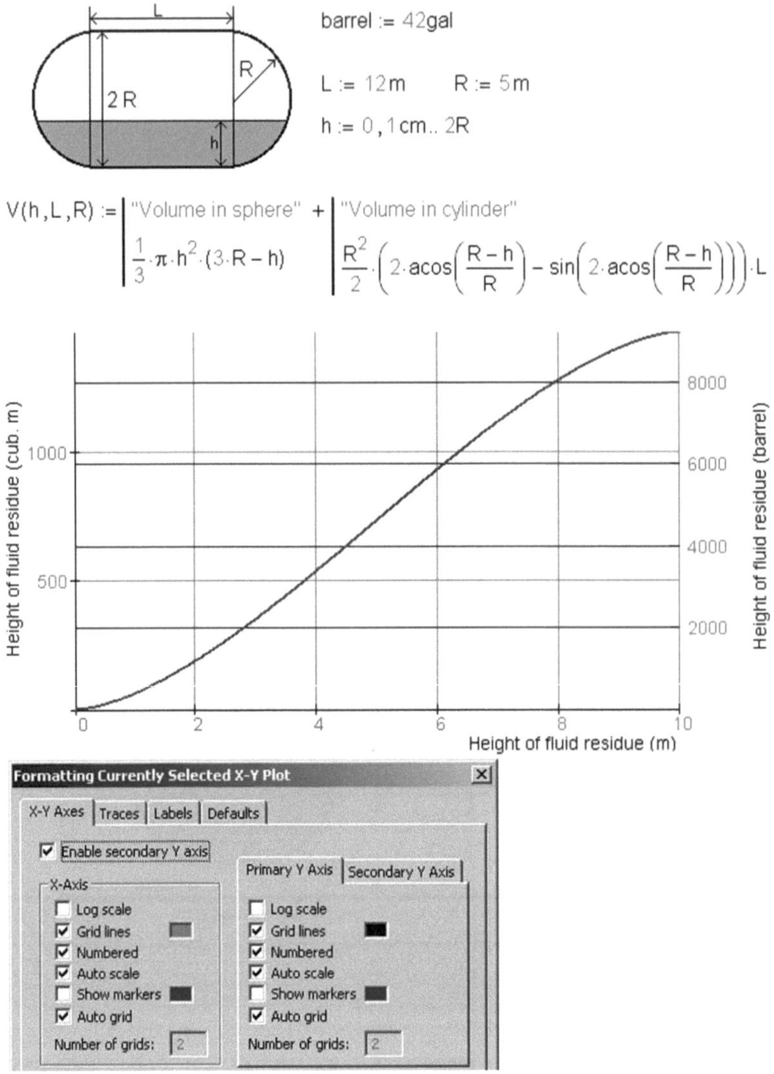

Fig. 6.87 Two y-axes

rectangular. This range can be changed via correspondent settings **QuickPlot Data** of the **3D Plot Format** dialog box but it is better to do it by function `CreateMesh` (see Fig. 6.90).

Function `CreateMesh` in Fig. 6.90 helps us, first, to draw the surface (semi-transparent) and level lines (lying on the x–y plane) within the range determined by variables x_1, x_2, y_1, and y_2 and, second, to draw isolines by *x* and *y* axes that are in essence the surfaces pared down to the lines.

$L := 12\,m \qquad R := 5\,m \qquad h := 0, 1cm .. 2R$

$$V(h,L,R) := \left| \begin{array}{l} \text{"Volume in sphere"} \\ \frac{1}{3}\cdot\pi\cdot h^2\cdot(3\cdot R - h) \end{array} \right. + \left| \begin{array}{l} \text{"Volume in cylinder"} \\ \frac{R^2}{2}\cdot\left(2\cdot acos\left(\frac{R-h}{R}\right) - sin\left(2\cdot acos\left(\frac{R-h}{R}\right)\right)\right)\cdot L \end{array} \right.$$

$$S(h,L,R) := \left| \begin{array}{l} \text{"Area in sphere"} \\ \pi\cdot\left[\sqrt{R^2 - (R-h)^2}\right]^2 \end{array} \right. + \left| \begin{array}{l} \text{"Area in cylinder"} \\ L\cdot 2\sqrt{R^2 - (R-h)^2} \end{array} \right.$$

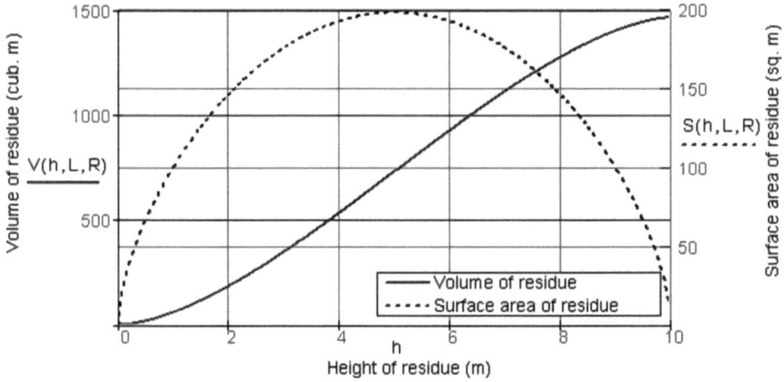

Fig. 6.88 Two y-scales

$$\begin{pmatrix} a \\ b \\ c \\ d \end{pmatrix} := \begin{pmatrix} 4064 \\ -269.4 \\ 4973 \\ -0.0000348 \end{pmatrix} \qquad Cost(Age, Run) := \frac{a + b\cdot\frac{Age}{yr} + c\cdot e^{d\cdot\frac{Run}{mi}}}{a + c}$$

$$Cost(Age, Run) := if\left(1\,\frac{km}{hr} < \frac{Run}{Age} < 2\,\frac{km}{hr}, Cost(Age, Run), \text{"Error"}\right)$$

$Run := 0, 10km .. 200000km$

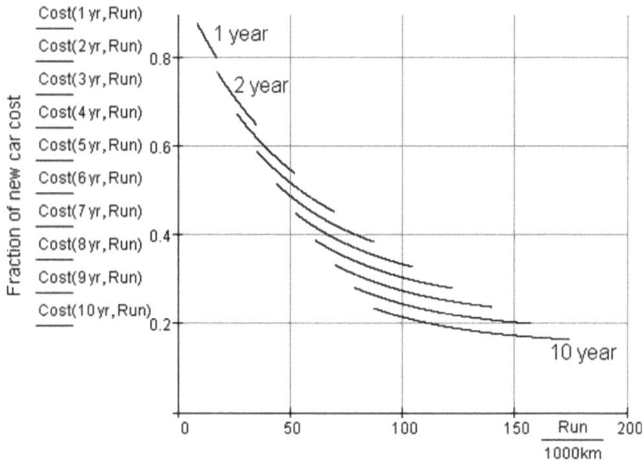

Cost(5.5years, 75000km) = 39.4 % Cost(5.5years, 100000km) = "Error"

Fig. 6.89 Plotting the function that returns a number or a text

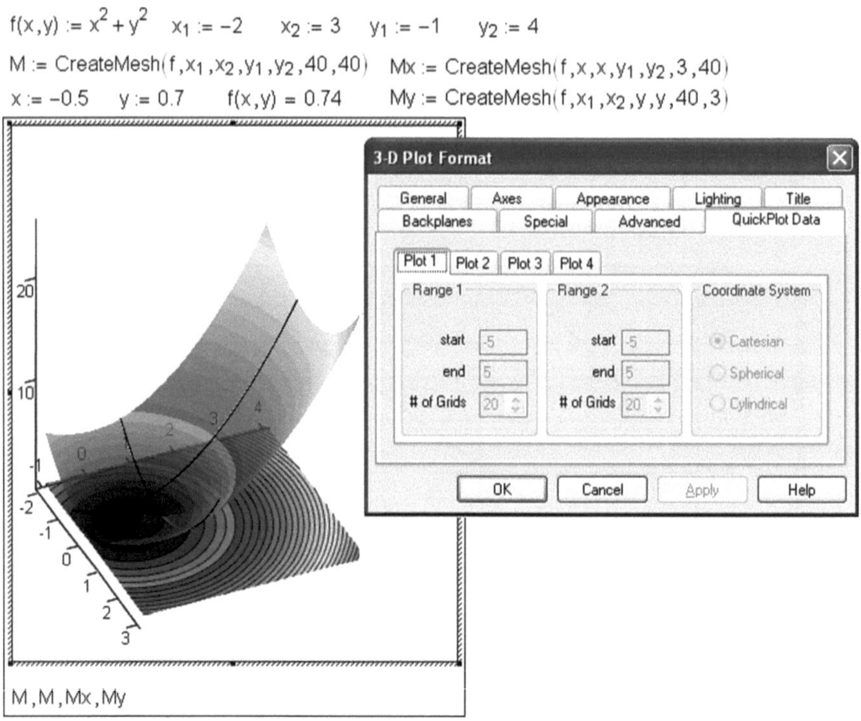

Fig. 6.90 Plotting 3-D surfaces and lines via an intermediary function

The variation range can have complex contours in the actual problems. Here is an example requiring non-rectangular variation range of the arguments in graphic solution (Fig. 6.91).

We need to design outboard tank of a plane consisting of three parts (half-sphere, cylinder, and cone), having definite volume, $V = 3\ m^3$, with the minimum surface (a typical optimization problem). This tank has three parameters: R – radius of half-sphere, cylinder and the base of cone; L – length (height) of cylinder and H – length (height) of cone. Although, a simple transformation (solving the equation of cylinder volume by variable L) brings the problem to analysis of the function of two variables – S(R, H):=...: square of the tank surface depends of the cylinder radius (R) and height of the cone (H).[31] If we try to draw level lines of this function the graph would not be created within default range of −5 to 5 for both *x* and *y*, but would be interrupted with an error message. The fact is that graph of the function having one variable is differ from that of two-dimensional: if the function y(x) is not defined in any section of the range of argument x, for example,

[31]We can choose other pair of the arguments R and L or L and H. It is possible to optimize all three variables R, L and H (two arguments allow us to make graphical analysis: to plot functions of three and more variables is difficult.

$$\begin{vmatrix}\text{"Tank"}\\ V\end{vmatrix} = \begin{vmatrix}\text{"Hemisphere"}\\ \dfrac{\frac{4}{3}\pi R^3}{2}\end{vmatrix} + \begin{vmatrix}\text{"Cylinder"}\\ \pi \cdot R^2 L\end{vmatrix} + \begin{vmatrix}\text{"Cone"}\\ \frac{1}{3}\cdot\pi\cdot R^2 H\end{vmatrix} \quad \text{solve}, L \rightarrow \frac{3\cdot V - 2\cdot\pi\cdot R^3 - \pi\cdot R^2\cdot H}{3\pi\cdot R^2}$$

$$V := 3m^3$$

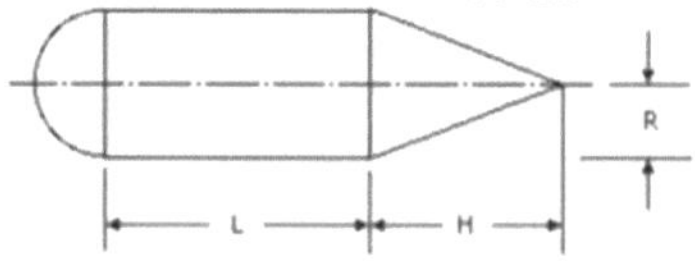

$$S(R,H) := \begin{vmatrix} L \leftarrow \dfrac{3\cdot V - 2\cdot\pi\cdot R^3 - \pi\cdot R^2\cdot H}{3\pi\cdot R^2} \\ S \leftarrow \frac{1}{2}4\pi R^2 + 2\pi\cdot R L + \pi\cdot R\sqrt{H^2 + R^2} \\ S \leftarrow \text{if}\left(L \geq 0 \wedge \dfrac{2R}{R + L + H} > 0.5, S, 0\right)\end{vmatrix}$$

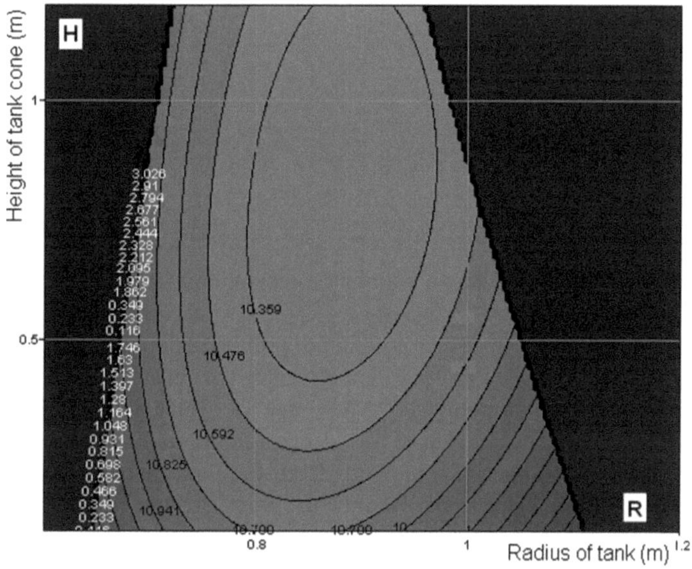

Fig. 6.91 Plotting a function of two arguments in the nonrectangular region

square root of the negative value, or returns the text constant this part will be ignored (missed). We used this property showing the cost of a used car in Fig. 6.89. Unfortunately, graph of function of two variables has no such a feature. It does not ignore complex values of the function in some sections of rectangular range but returns an error message. After that, many users give up drawing a graph.

To circumvent the limitations shown in Fig. 6.91 use the following method: If the value of L (length of the cylinder) becomes negative, function S returns not the complex number but zero. In addition, we imposed a restriction to the problem: the ratio of the tank diameter to its length must be more than 0.5. After these restrictions

$$V := 3$$

$$S(R,H) := \left| \begin{array}{l} L \leftarrow \dfrac{3 \cdot V - 2 \cdot \pi \cdot R^3 - \pi \cdot R^2 \cdot H}{3\pi \cdot R^2} \\ \dfrac{1}{2} 4\pi R^2 + 2\pi \cdot RL + \pi \cdot R\sqrt{H^2 + R^2} \end{array} \right.$$

$$\begin{pmatrix} \dfrac{\partial}{\partial R} S(R,H) = 0 \\ \dfrac{\partial}{\partial H} S(R,H) = 0 \end{pmatrix} \left| \begin{array}{l} \text{solve}, \begin{pmatrix} R \\ H \end{pmatrix} \\ \text{float}, 5 \end{array} \right. \rightarrow \begin{pmatrix} -2.2980 & 2.0554 \\ -.43888 - .76016 \cdot i & -.39257 - .67997 \cdot i \\ -.43888 + .76016 \cdot i & -.39257 + .67997 \cdot i \\ \boxed{.87776 \quad .78512} & \\ 1.1490 - 1.9901 \cdot i & -1.0276 + 1.7800 \cdot i \\ 1.1490 + 1.9901 \cdot i & -1.0276 - 1.7800 \cdot i \\ -2.2980 & 2.0554 \\ -.43888 - .76016 \cdot i & -.39257 - .67997 \cdot i \\ -.43888 + .76016 \cdot i & -.39257 + .67997 \cdot i \\ .87776 & .78512 \\ 1.1490 - 1.9901 \cdot i & -1.0276 + 1.7800 \cdot i \\ 1.1490 + 1.9901 \cdot i & -1.0276 - 1.7800 \cdot i \end{pmatrix}$$

Fig. 6.92 Problem of optimal shape of external tank

(real $L \geq 0$ and forced $2R/(R + L+H) > 0.5$) had been inserted into function S its level lines appears were drawn within non-rectangular (trapezium-shape) variation range R and H.

Figure 6.92 shows the analytical solution of problem of optimal tank, the system of two non-linear equations (partial derivatives of the function S with respect to arguments R and H) that returns 12 roots one of which ($R = 0.87776$ and $L = 0.78512$) is the coordinate of the "deepest place" in Fig. 6.91.

Figure 6.93 shows other two methods from the large number of "strategies" which we need to draw the graphs. To draw the graph of the type $f(x,y) = 0$ we can remind that it is nothing else than a zero level line of the contour graph of a function of two arguments.

This figure shows that "paying" with formats of 3D graph, turning it from surface (**Surface Plot**) to level lines (**Counter Plot**) we can solve the problem both regarding to Bernoulli's lemniscate (upper graph) and more complex closed function (bottom graph) which cannot be factorized into definite functions and plotted by the common technique.

We described only two strategies of formatting Mathcad plots (see Figs. 6.91 and 6.93). Nevertheless, there are much more of that especially in 3D plotting. Often the graphs are formatted to transport to books and articles as PrintScreen. In such cases, it is useful to insert the graph created in Mathcad to another application, for example to Paint to finish it off. We have to do that if Mathcad has not necessary formatting tools or they are difficult to access, or we do not know about them.

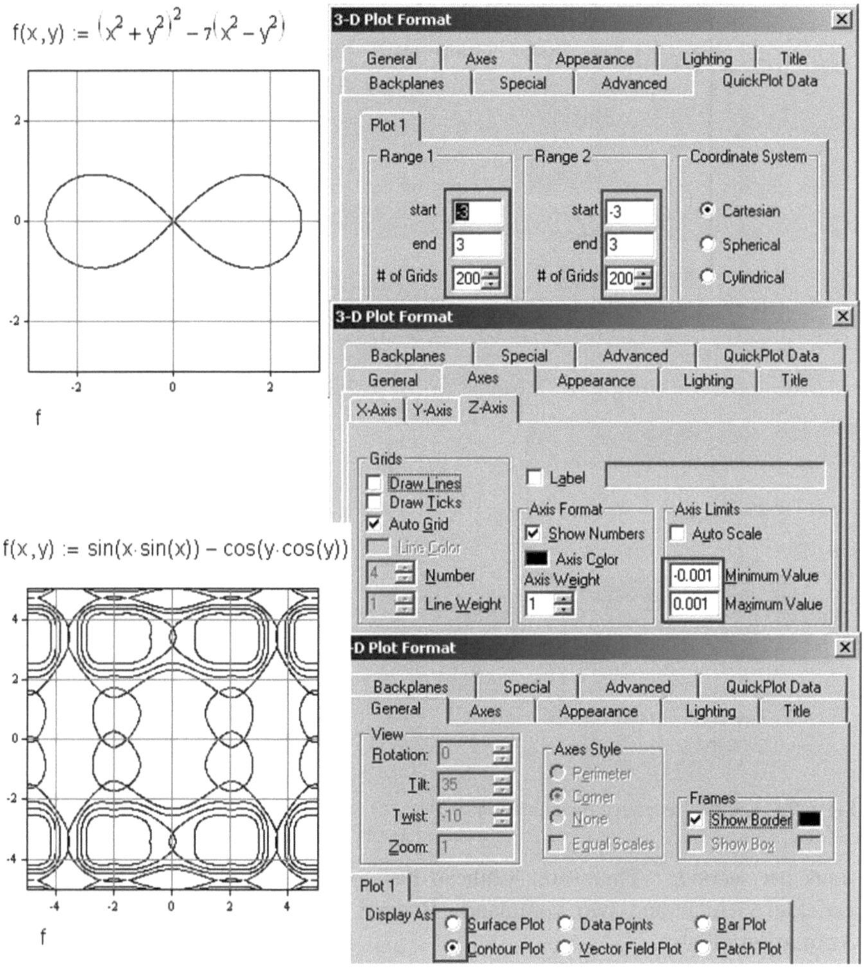

Fig. 6.93 Plot f(x, y) = 0

6.7 Animation and Pseudo-Animation

Children games "motion pictures" (i.e., "*mo*ving *vie*ws" or movies) examples provide the essential explanation of animation in Mathcad and other applications. One draws a certain picture (a "stick" man in a particular pose) on each page of a notebook changing the image smoothly from page to page. Turning the pages of such painted notebook, we shall not see the animation effect. Although, pressing and turning it quickly, releasing by a finger we can see the animation.

In Mathcad, we can change manually the value of a variable and see changes of a curve on a plot. A good animation requires the minimum picture frequency to be ten

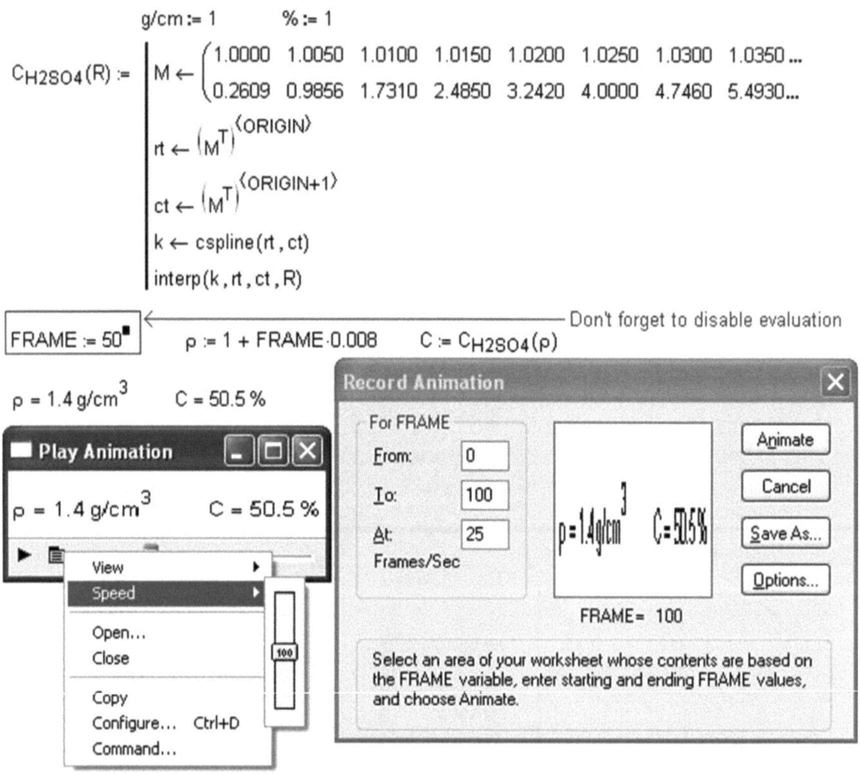

Fig. 6.94 Exe-file generated by animation

shots per second.[32] Therefore, Mathcad has special animation tools: predefined variable `FRAME` and two commands **Record** and **Playback** from the **Tools | Animation** menu.

Figure 6.94 shows the stages in creating animation in Mathcad in unusual way: to generate peculiar exe-file. The function named $C_{H_2SO_4}$ is inserted into the Mathcad worksheet, which returns the concentration of sulfuric acid depending on its density (by spline interpolation of the table values keeping in the matrix M). A user can set integer values from 1 to 100 for `FRAME` that vary the variable ρ from 1 to 1.8 (`1+ FRAME·0.008`). Variable ρ in turn changes the value of required variable `C`. Creating the animation the value of `FRAME:=` is disabled and the dialog box governs this variable: the variable range (**From** and **To**), picture frequency (**At** – shorts per second). We select the portion of worksheet to animate by dragging with the mouse. Figure 6.94 shows a dotted rectangle around two operators ρ= and `C=`. After that, clicking **Animate** makes Mathcad change the

[32]Sometime ago it was rumored that Mathcad inserted the "25th shot" to the user animation containing advertisement of the new version.

value of FRAME in specified range and draw obtained shots (pages of the notebook) for their quick displaying (turning the pages). The animation clip data (see **Play Animation** from Fig. 6.94) can be saved on the disk and played back without Mathcad. Why is it not an exe-file!? We can set the value of ρ (or C) with a slider and compute the value of variable C (or ρ).

At the end of 2003, Mathsoft Engineering & Education, Inc released Mathcad Application Server (MAS – see Sect. 6.8) for operation testing to "kill two birds with one stone". MAS allows Mathcad-based applications to be created and subsequently deployed independent from Mathcad itself (once create). MAS allow worksheets online publication and accessible both for viewing (that was done long ago) and for *calculations*. A user (of a Web browser but not Mathcad now) can change the inputs and see the result, numbers, plots, etc. on their browser window.

The guide for Mathcad Application Server states that the button **Submit** is useless in the "standard" Mathcad worksheets but necessary in the WebSheets, Mathcad worksheets converted to function as Web documents by MAS. This information may be inserted even in the **Submit Button Properties** dialog box.

Although, this information (outlined in Fig. 6.95) confuses Mathcad users and cuts a very interesting tool. First, we can manage without this button in WebSheets. After the required changes are made, we push the **Submit** button to transmit the data to the Mathcad server for computing in Mathcad installed in this server. if the inputs are changing only in Web Controls we can format this "net" interface element so that its change and pushing < Enter > results in sending the data on Mathcad server without pushing the **Submit** button which becomes unnecessary in this case.

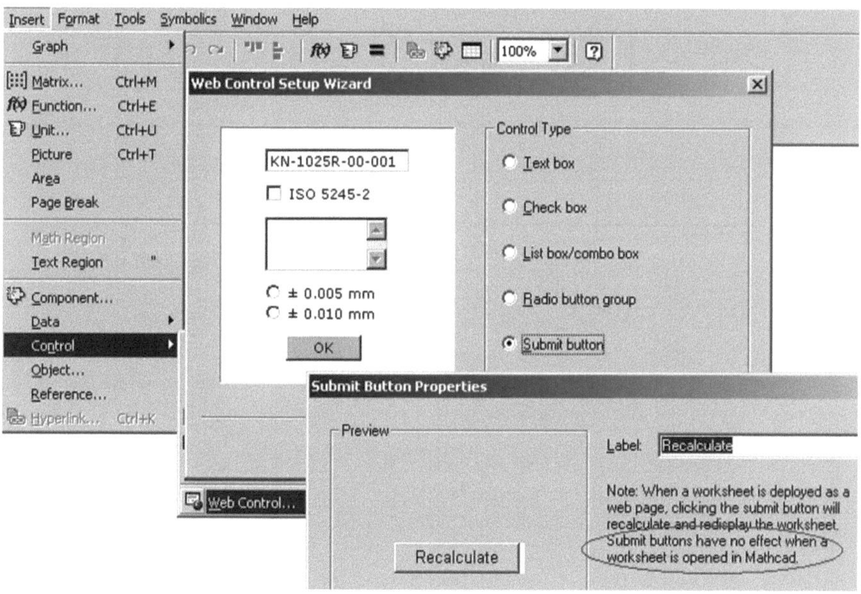

Fig. 6.95 Insert submit button

Here we can see analogies with manual mode in Mathcad (Excel, Word, BASIC, Pascal, etc.) and compare **Submit** button with < F9 > button which we press after the changes in the worksheet were made by operator := and net and standard Controls and the calculation should be done.

Second, **Submit** button can be useful both in WebSheets and in worksheets, in traditional Mathcad documents. The point is that pushing this button increase by one value of the variable connected with **Submit** (after starting Mathcad in any mode its value equals to one). This variable is a peculiar counter of pushing the **Submit** button, i.e. the number of calling to MAS. This feature of **Submit** button can be very useful in Internet and out of it: the formal variable can govern The Mathcad worksheet changing it, for example. Thus, Fig. 6.96 show Mathcad worksheet that illustrate the Newton method of numerical finding zero of the function where pushing the **Submit** button allows us to see approximation step by step.

On Mathcad server of the authors (http://twt.mpei.ac.ru/TTHB/1/Anim_Num_Met.html), there are systematic illustrations (pseudo-animation) of the following classic methods of numerical calculations based on the **Submit** button:

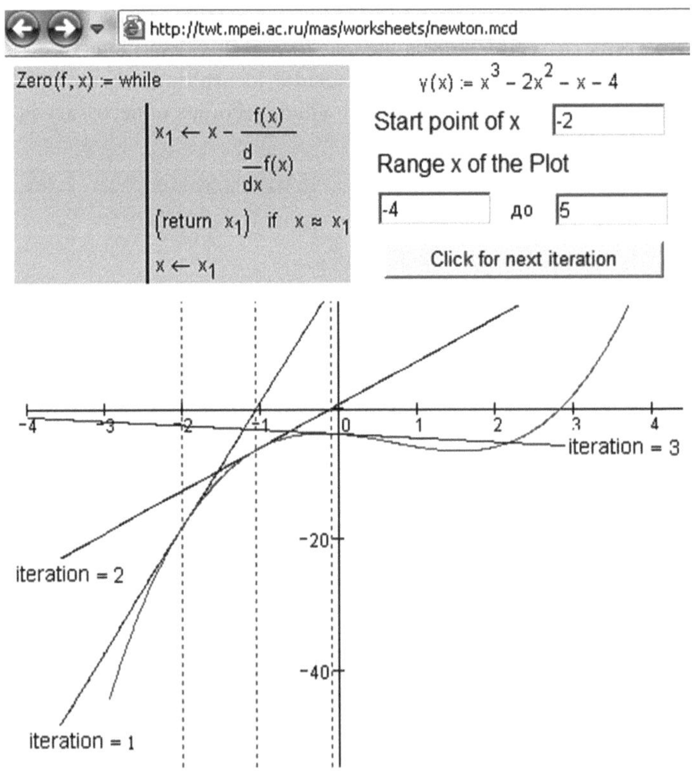

Fig. 6.96 Pseudo-animation of Mathcad

- The Newton method of finding zero of a function – http://twt.mpei.ac.ru/mas/worksheets/newton.mcd
- The Newton method of solving the system of two algebraic equations – http://twt.mpei.ac.ru/mas/worksheets/newton_2.mcd
- The method of halve in zero search of a function – http://twt.mpei.ac.ru/MAS/Worksheets/secant.mcd
- Secant method in zero search of a function – http://twt.mpei.ac.ru/MAS/Worksheets/secant.mcd
- Golden section in minimum search of a function of one argument – http://twt.mpei.ac.ru/mas/worksheets/Gold_Ratio.mcd
- Euler and Runge–Kutta methods with constant interval of solving an ordinary differential equation – http://twt.mpei.ac.ru/mas/worksheets/Euler.mcd
- Runge–Kutta method with variable interval of solving an ordinary differential equation – http://twt.mpei.ac.ru/mas/worksheets/rkadapt.mcd

6.8 Mathcad Application Server

6.8.1 Continuation of Preface

The history of using computers for engineering and technical calculations is divided into three stages (see also the preface):

- Working with computer codes
- Programming in high level languages (e.g., BASIC, C++, Fortran, Pascal etc.)
- Using mathematical packages (such as Mathcad, Maple, MatLab, Mathematica, etc.)

Comment. *The list of the packages does not contain Excel that is still the most popular application for computing. Excel stands between the programming languages and the mathematical programs.*

These listed stages (technologies) have not distinct bounders. Working, for example, in Mathcad we can call, if necessary, the user functions written in C which code has fragments in assembler. Apropos, computer codes remain in the programmable calculators still widely used in scientific and technical calculations. We should rather speak here about the tendency, which in particular shortens time of creating calculation techniques and mathematical models than about certain isolated stages of tool development for solving problems. This tendency results in excluding the programmer as additional and often muddled (in the view of applied specialist) link between a developer and a computer, increasing openness in calculations themselves when we can see both results and intermediate data along with all formulae.

We should note here one more aspect of the "person–computer interface" issue. The programming languages do not tolerate "amateurishness" of an investigator. This means, programming demands involving of an investigator (engineer) that brings skill lost in the basic profession practice. The mathematical packages, especially Mathcad, do not require such dependence. One can postpone working with Mathcad for half a year or even a year without losing the basic skills. That is, such postponed problems will be solved without help of other programmers.

The *way* of computer use may also be divided into three historical stages:

- Batch processing, when one computer worked for many users that brought their problems written on card batches and other data mediums
- Time of personal computers (PC)
- Wide-area and local networks (WAN & LAN, respectively) that, in essence, return us to batch processing on another level, more qualitative; input data and results transfer on the Internet as e-documents not written to paper (as hard copies). The calculations themselves can be parallel (cluster, distributing computing)

Again, Mathcad was created as an application *alternative* to traditional programming languages. Many specialists in particular subjects (physics, chemistry, biology, mechanical engineering and others) and students cannot use computer effectively through difficult programming languages (such as C++). Mathcad allows such users to solve wide range of scientific, engineering and educational problems without using traditional programming.

Today, the situation has notably changed. As a rule, a student starting education in his future profession knows well one of programming languages and can solve their problems with studied programming application. In this case, Mathcad may be unnecessary. Although, Mathcad has some features which not only enable "surviving" but also make it popular among experienced programmers.

The point is that Mathcad allows an effective (quick) problem definition and solution. There was such enthusiasm when high-level programming languages (Fortran, Pascal, BASIC, etc.) substituted assembly (low-level) computer codes. Exchanging higher-level computer languages for Mathcad now has removed the additional link between a computer and a specialist known as a programmer.

It is worth noting that using Mathcad a user defines and checks a mathematical model even if a user knows higher-level programming languages well. The developer group under the guidance of the authors creates and sells successfully program packages WaterSteamPro® designed for computing heat and physical properties of water and steam, the basic heat carrier in thermal and nuclear power stations, heat supply systems of cities and settlements (http://www.wsp.ru). The final version is written and compiled in Visual C++, but this project could not have been done without previous analysis of the formulae and algorithms in Mathcad (i.e. Mathcad is also a "prototyping tool").

There is no rose without a thorn. The main limitation of mathematical packages is that, as a rule, they cannot generate executable (.exe) files, which start without "progenitor" programs. In particular, that prevent from dividing people working with computer into *users* and *developers*. Those using Mathcad occupy "natural economy": work out calculation techniques only for personal use or for a circle of targeted Mathcad users. We could give such applications only to users having installed Mathcad licenses. Such users would not purchase files accessible in static form such in Adobe Acrobat® (PDF) but will try to recreate the necessary calculation.

Comment. *It is a question of small calculations, which creation and checkout time is comparable with time for search of the ready-made worksheet, installing and study of the new program. Although, large calculations hardly pave the way on the market because we can always extend and modernize one's own worksheet and not other's. Here we can draw an analogy with another "intra-Mathcad" situation. Sometimes it is better create a function instead of searching that ready-made in the labyrinth of Mathcad built-in functions. Figure 6.113 illustrates the thesis in which the cumulative interest is computed via user's algorithm and not by built-in function hidden among Finance group.*

We could sell a Mathcad worksheet to a user who does not have installed Mathcad and has not studied it with the load: installing required Mathcad version that often results in changing of Windows or even hardware. Besides, it is necessary to study Mathcad.

Mathsoft Engineering and Education and PTC, Inc have changed this abnormal situation.

First, Mathcad Explorer, free limited-version, released together with Mathcad's eighth version enables a user to open the worksheets and compute but does not save or edit them (read-only version). Mathcad Explorer is available from the Internet as free download application.

Second, tools for publication Mathcad worksheet online were developed strongly. The main Mathcad user is educational branch where the way of obtaining a result, studying calculation methods, is more important than a result itself. In particular, Mathcad 2001i was aimed at such group of users (which letter "i" means interactive).

Nevertheless all these steps were incomplete. It is better to install Mathcad itself than download rater bulky Mathcad Explorer, the limitted-version. On the other hand, we want not only to access and read Mathcad worksheets from the Internet rather html or MathML copies, but change the data inputs and see (print, save) new results.

The quite clear solution of this problem came from the Internet, also.

Mathsoft released so called *Mathcad Application Server (MAS)* at the end of year 2003.

MAS enables the solution of the following problems:

- There is no necessity to install required version of Mathcad, determine whether you have a virus, run applied mcd-files. It is enough to access the Internet and call MAS over the Internet Explorer browser (version 5.5 and later). That gives the illusion of opened Mathcad worksheet in which we can change inputs and obtain (print, save on a disk) the results. The calculation techniques (a set of formulae in tradition mathematical notation but not as a program, a feature that makes Mathcad so attractive) and intermediate data could be displayed or hidden completely or partially (sell of results but not the technique).

- New calculation techniques become accessible at once to every member of the Internet community. We should only inform the possible users of correspondent IP-address. To turn the calculations into the article we can administrate, make it paid (partially paid, reasonably paid), place banners round the references.
- Any errors (e.g., misprints, assumptions, defects), noted by developer or by a user, could be corrected at once quickly and imperceptibly for the users. It is also possible to extend and modernize a worksheet imperceptibly for users
- MAS remains the traditional option of downloading the worksheets from servers to improve or extend them (which should referenced appropriately). Thus, the site of this book contains references, pictures (pic), Mathcad files (mcd), and MAS documents, as was noted in the preface

MAS solve the problem of licensing cardinally: there is no need to buy expensive Mathcad or to palter and install an illegal copy, as well as to update or change continuously the program itself, Windows OS (Operating System), and hardware (computer); we should just call MAS. Only a developer of calculation techniques published on MAS should buy the expensive program.

It is also needless to study how to work with Mathcad worksheets. We should not be afraid to damage it: changing the inputs requires the methods that people had learnt long ago in various network forums, chats, network games.

The sections below describe how to adjust an existing Mathcad worksheet to publish it on the Internet. Described below also are the calculation techniques to make such Mathcad worksheets usable as WorkSheets.

6.8.2 *Preparation of Mathcad Worksheet for Publication Online or from WorkSheet to WebSheet*

We should rather title this section "Adjustment of Mathcad worksheet to publication..." The point is that many of the mathematical programs do not allow us to adjust the existing document, but demand that the document should be created for publication on the Internet from the start.

As a rule, the Mathcad worksheets have the structure consisting of three parts (a kind of a triptych); inputs, formulas, and results. Figure 6.97 shows such a structure in extremely simplified form: these parts are divided into three titled areas which could be operated independently – **Lock/Unlock** and (or) **Collapse/ Expand**.

The structure of a real Mathcad worksheet is usually more complex and has the fuzzy borders between these three areas: the results may be displayed just after the formula, the souse data may be duplicated in a summary table etc. If the area containing calculations itself occupies several pages it is usually collapsed during checkout to see the source data and the results at the same time. We may divide the worksheet into several areas, place the results between them and expand them in turn. Figure 6.97 shows that a user changes the values of variables a and (or) b and

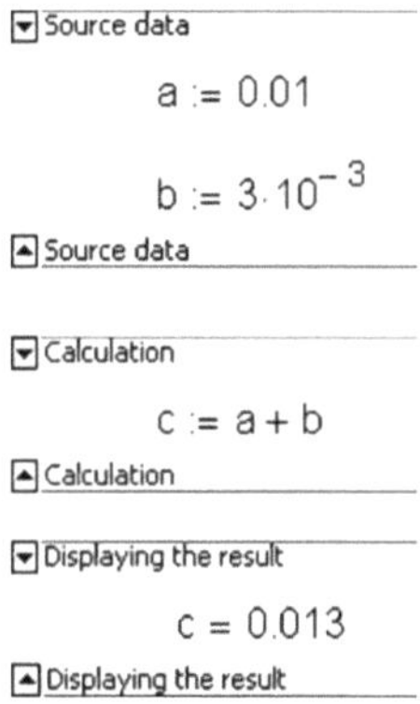

Fig. 6.97 Structure of a simple Mathcad worksheet

obtains the result, the value of c. The user may extend the worksheet (define new variables; insert formulae, graphs, etc.) In that case, the *user* turns into the *developer*. The MAS technology divides people clearly into two categories.

The first stage of preparation is to indicate assignment operators (not only $:=$ and $\equiv$ (global definition) but also so called Controls appeared in Mathcad 2001) for souse data adjustment. We should replace these operators with Web Controls. The reader knows the analogues of that by entering information on various Internet queries.

6.8.3 Web Controls: The Network Elements of the Interface

Web Controls appeared in Mathcad 11.1 Enterprise Edition and in Mathcad 12–15, see Fig. 6.98.

A Web Control distinguishes from its standard analogue. We need not write or edit the programs in form of Active Scripting (JScript and VBScript) for its operating. Such programs are not desirable on the network because they often spread the viruses. Over that reason many of the users do not insert Controls into the worksheet but use $:=$ and $\equiv$ as before. Mathcad 2001i enables us to fix such scriptable components in opening worksheet as in Word or Excel. Figure 6.99 shows **Script Security** tab from the **References** dialog box, which enables three security levels: **Low Security** – opens the worksheet with no protection against harmful scriptable components; **Medium Security** – the user can choose disabling present scriptable components or not; **High Security** – disables all present in a worksheet.

In Medium Security, opening the Mathcad worksheet containing scriptable components follows the prompt shown in Fig. 6.100.

For these reasons (threat of the viruses, increase of file size), it is recommended to use Web Controls not only in documents converted for publication online but also in standard worksheets. Besides, Web Controls are more convenient comparing

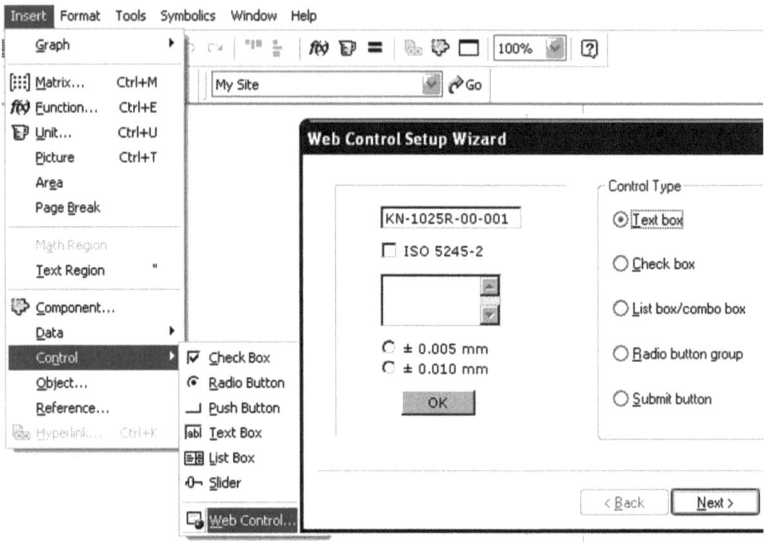

Fig. 6.98 Inserting Web Control

with "standard" Controls because setting the security level on such worksheet allow us to introduce the data into these interface elements but prevent us from "touching" the variable itself. If the worksheet contains Controls, both the access to the variable and changing its value are disabled (see Fig. 6.101).

There is one more reason that Controls with their scriptable components are useless. As we noted in the beginning of the chapter, Mathcad was created as a tool alternative to programming languages for users who cannot or do not want to program. As with Controls, we left something to come to it again: we shut the door on programming but it got in through the window.

After converting and publication on MAS the worksheet shown in Fig. 6.97 should look like that shown in Fig. 6.102.

Figure 6.102 shows that **Text boxes** substitutes for operators `a :=` and `b :=`. The **Submit** button was inserted additionally which default marking **Recalculate**. When this document is published on MAS the lines and comments `"Input Data"`, `"Answer output"`, indicating areas boarders, should not be displayed. The area `"Calculations"` could be collapsed to hide operator `c: = a + b`. The developer can hide part of the calculation online allowing a user only computing. On the contrary, he (she) can display and highlight most interesting and important parts of the worksheet, for example, for checking the method of calculation as well as inputs. Opportunely, if the variables `a` and (or) `b` from Figs. 6.97 and 6.102 should be within defined range (for example, `0 < a<3`) it will be useful to insert a kind of check-point:

```
a := if(0 < a < 3, a, "Error!")
```

that blocks computing when the user enters wrong data purposely or unwittingly.

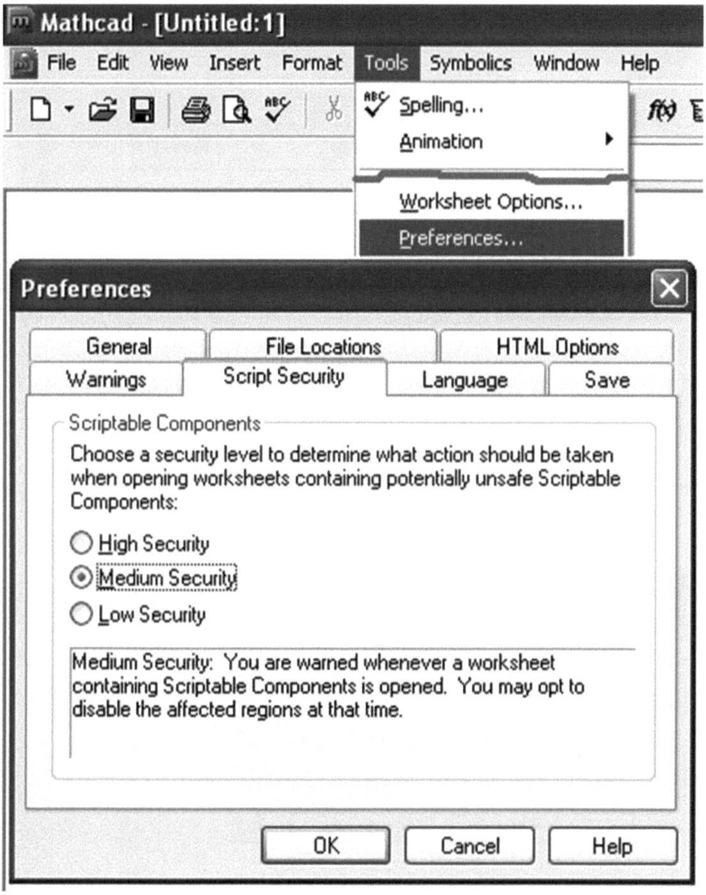

Fig. 6.99 Setting security level for work with scriptable components

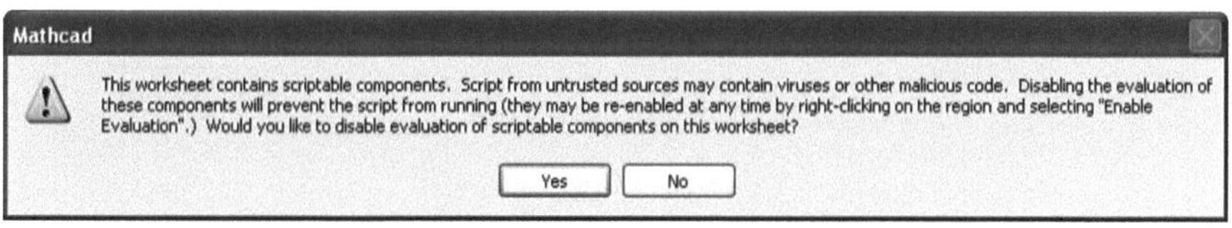

Fig. 6.100 Warning about the presence of scripts in the opened Mathcad-worksheet

Comment. *For that purpose, a user usually inserts a slider but the list of Web Controls does not contain it.*

It is better to insert such check-points above the results. That should block the result if a certain inputs combination gives wrong (unreal) result. That shows the

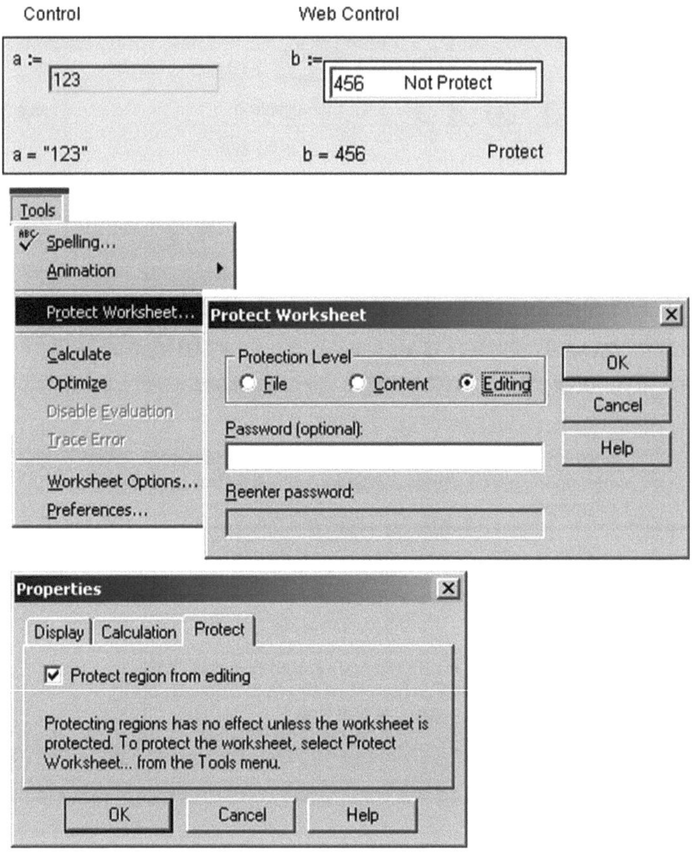

Fig. 6.101 Protect Web Controls

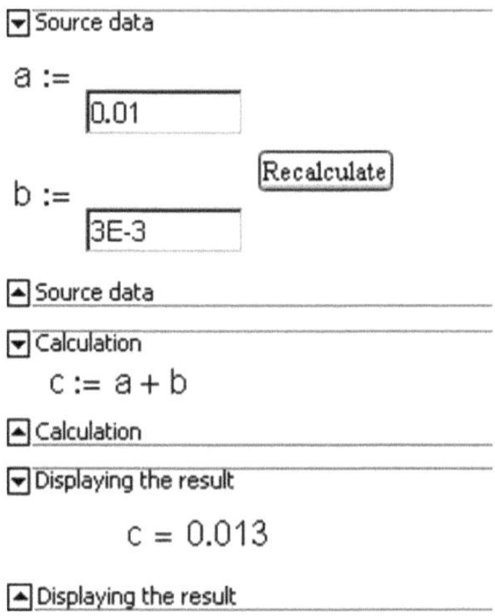

Fig. 6.102 Structure of a simple Mathcad worksheet prepared for publication on Mathcad Server

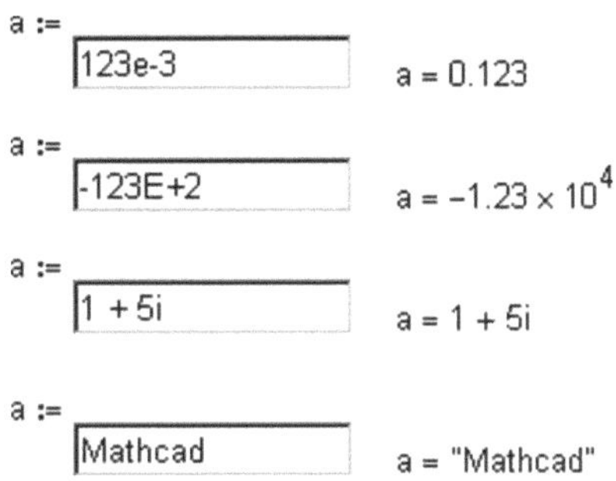

Fig. 6.103 Entering constants into text boxes

worksheet containing design output of water treatment plant for power stations at http://twt.mpei.ac.ru/mas/worksheets/Calc_Q_VPU.mcd. It returns text "Wrong source data", not a value, if the user has entered impossible data combination of designed station, for example, the station is nuclear and the fuel to be used is coal.

The name **Text box** is not quite right for Web Controls or rather it is right for the standard analog, Controls. We can enter text constants and numbers, real (as in Fig. 6.103) or complex (for example, `−5+3i`), through this network element. At that, the number should be entered in old format (DOS): `E + 3 (e + 3)` instead of 10^3, see Fig. 6.103. If entered symbols are not like a number it is understood as a text. In the standard analogue of the **Text box** any symbol file is recognized as text that could be converted into the number if needed by built-in function `str2num` (see Fig. 6.118) from the **String** category.

Tom Gudman and one of the author developed the technique of entering arrays (vectors and matrixes) and user functions through the Text box as shown in Fig. 6.104–6.106.

Figure 6.104 shows that the symbol file entered into the **Text box** is converted into the vector by the user function `Str2VMat`. That is, any symbol differing from 1, 2. . .9, the point (decimal delimiter), and minus sign can be used as the delimiter. In this case, blanks are used as the delimiters. We can change its number aligning these numbers across if, for instance, we enter some vectors of the same size through several **Text boxes** located upright.

Complicating the method shown in Fig. 6.104, it is possible to fill the matrix in by rows or by columns. If the matrix is square and its size is arbitrary we can enter it as a vector of the size n^2 and than "pack it up" the square matrix as shown in Fig. 6.105.

Another way to enter a matrix is to specify the number of rows and columns at first and then fill the matrix in through the **Text box** as the horizontal vector that converted into square matrix.

You can find the method of solving equations of the kind $A \cdot x = B$ at

http://twt.mpei.ac.ru/mas/worksheets/lsolve.mcd where the arrays A (the square matrix of the unknown quantity coefficients) and B (vector of absolute terms) are filled in by the technique shown in Figs. 6.104.and 6.105.

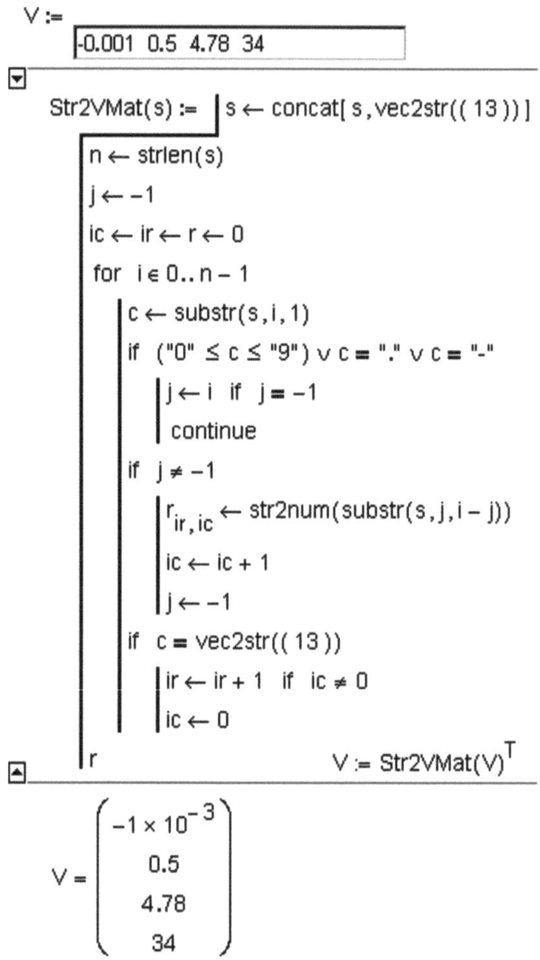

Fig. 6.104 Entering the vector elements in the text box

Comment. *The built-in function* `vec2str` *is changed in Mathcad 12. Thus, the program shown in Fig. 6.104 requires revision.*

Besides, the problem of entering formulae through the **Text box** has been partially solved, particularly user functions. Thus, Fig. 6.106 shows entering function with one argument, on which we created a plot. The algorithm of converting text into formula is rather bulky and it is not represented in Fig. 6.106, being hidden in collapsed area `"Express Compiler"`. The same algorithm could be used for entering the variable not as a constant (for example, 1.25) but as expression (`45/12 + 11`) including defined before and built-in variables (for example, `pi/4*d^2`, where `pi` is build-in and `d` defined before, i.e. the user one). If it is required to make arguments of a user-defined function, grouped in a vector, for

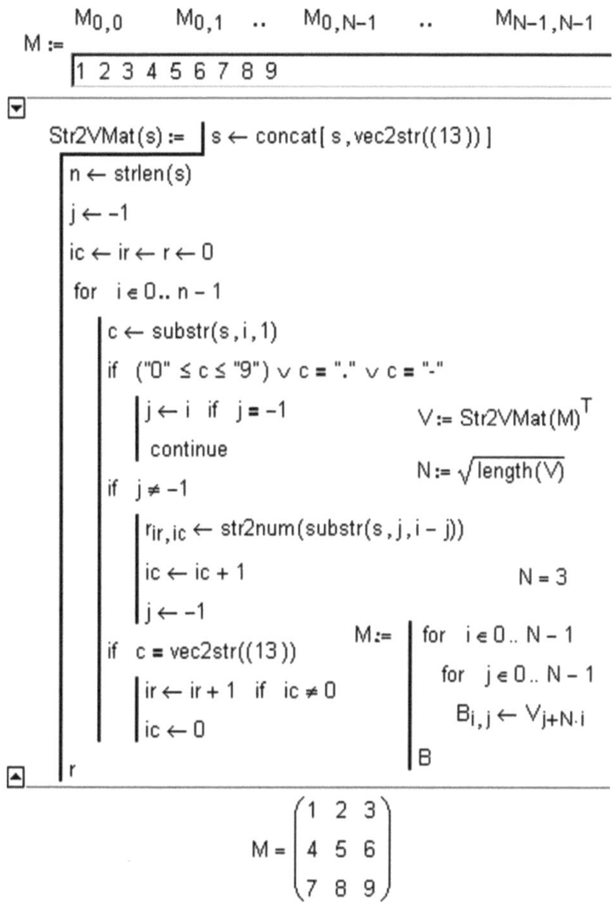

Fig. 6.105 Entering a square matrix into the text box

instance, we use the additional operator `(X):=f(X0, X1, ...)`. For example, this method was used for computing of global minimum through genetic algorithm, http://twt.mpei.ac.ru/MAS/Worksheets/Minimum.mcd.

You can see the example of work with the function entered through the **Text box** at http://twt.mpei.ac.ru/mas/worksheets/F_X_Y_Plot.mcd.

This technique is limited because we cannot apply symbol mathematics to this function and include the variables defined before into it, still the list of its arguments could be changed or extended (see the example at: http://twt.mpei.ac.ru/mas/worksheets/Fit_f_x_a_b_c.mcd.

The following symbols and functions are acceptable in the Text box for entering formulae:

- Arithmetic – + plus, − minus, * multiply, and / divide.
- Involution ^

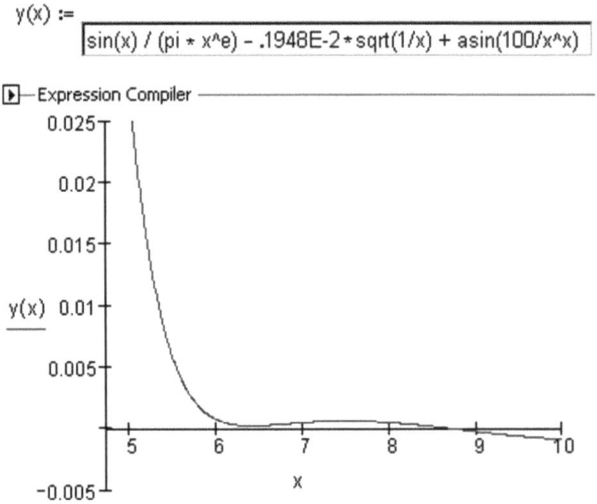

Fig. 6.106 Entering a function into a text box

- Mathematical constants – `e` is the radix of the natural logarithm; `pi` – π number. Other constants can be entered only if they are defined in the concrete worksheet.

There is another situation; to the left of the **Text box,** we can place a user-defined function with a list of required constants and variables followed by definition operator, for example `f(x,a,b,c):=`. As was noted before, entering the number we use a period as decimal delimiter and letter `e` or `E` as decimal exponent: it is possible to enter `0.001`, `1E-3,` or `10e-4`.

- Parentheses changing the order of operations: `2 + 2*2 = 6` but `(2 + 2)*2 = 8`. It is recommended that a user types the left parenthesis "`(`" followed by a right one "`)`" and then moves a cursor left to one position and enters the required expression. Thus, we can control the number of initial and close parentheses in the expression (formula).; one cannot enter brackets and braces.
- Built-in functions and operators `sqrt()` – square root, `abs()` – absolute value, `!` – factorial, `sin()` – sine, `cos()` – cosine, `tan()` – tangent, `asin()` – arcsine, `acos()` – arccosine и `atan()` – arctangent. It is possible to work with other functions if they have been defined before.

The Reader may want to use ideas of the authors on entering the formula through the **Text box**. For that, download the file Y_X_Plot.mcd from Internet description of MAS at http://twt.mpei.ac.ru/ochkov/VPU_Book_New/mas/From_WorkSheet_to_WebSheet.html and enter required changes into two functions `VARS` and `Eval` shown in Figs. 6.107 and 6.108.

Figure 6.108 shows changes in functions `VARS` and `Eval` that allow us to form the function named `lp1` ("the left part of the first equation") having seven

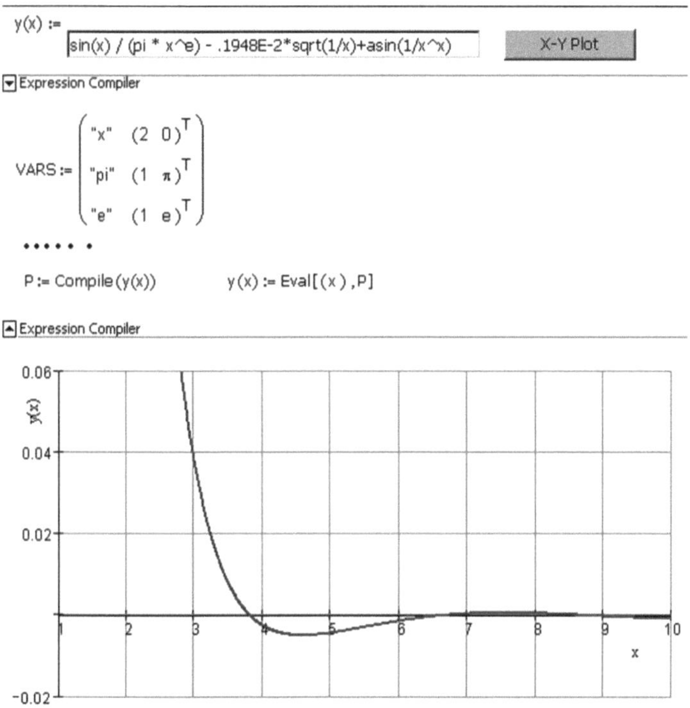

Fig. 6.107 Source functions VARS and Eval

arguments x, y, p, a, b, ang and g. These variables are used in the problem (Fig. 6.109) which suggests a user form and enter the set of three equations with three unknown quantities describing the beam equilibrium (http://twt.mpei.ac.ru/MAS/Worksheets/beam.mcd).

If we open that WebSheet all six boxes will be empty. We should type left and right parts of equations there and push the button **Ready**. That WebSheet is aimed not for computing but for checking knowledge of students (also see Sect. 6.8.8)

The very useful Mathcad feature works with physical quantities Fig. 6.110 shows that operator P:=123 mm was replaced by the **Text box** and additional definition operator (reassignment of the variable).

Dimensionless values are entered though the **Text box** into variable P, at that the comment (mm) placed to the right of the **Text box** indicates that the variable becomes dimensional after execution of the second operator P:=P·mm. Switches or **Check boxes** discussed below allows us to choose the units (Figs. 6.111 and 6.112).

Figure 6.111 shows the argument in switches (UnitP). It could be and should be hidden as shown in Fig. 6.112. For that, we call the corresponding command from pop-up menu.

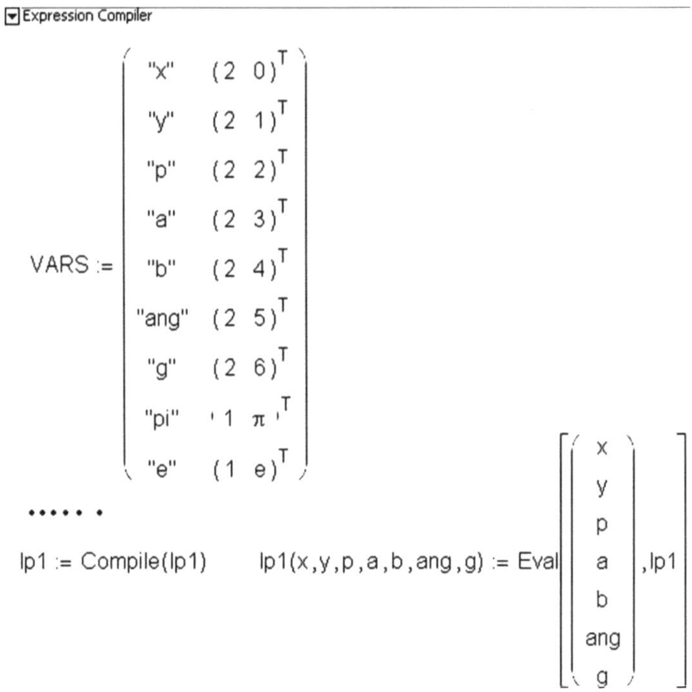

Fig. 6.108 Modified functions VARS and Eval

Earlier we used the method of choosing the unit from the list (see Fig. 1.9) to show work of operator and function `if`.

The units are often used as comments in worksheets but not as multipliers that simplify and check computing. That is connected with following:

- Many built-in functions cannot operate with dimensional arguments (see the article considering this problem at http://twt.mpei.ac.ru/ochkov/Unit_MC_MP/Unit_MC_MP.htm).
- Mathcad arrays (matrixes and vectors) operate only with dimensionless values or with quantities of the same dimension – length, pressure etc.
- Worksheets include so-called empirical formulae where the variables depend on the specific units but not on quantities that was discussed in Sect. 6.4.

If a unit (for example, `rub`) is used as a comment rather than a variable it is recommended inserting it into the calculation as a coefficient equal to one that completes evaluation operator rub := `1 C = 25` ■ ← rub. In that case, the unit becomes floating, it moves and remains in the end of the value if the number of decimal places has changed (see Fig. 6.113 showing the problem from a Chekhov's story at:

http://twt.mpei.ac.ru/mas/worksheets/Comp_procent.mcd.

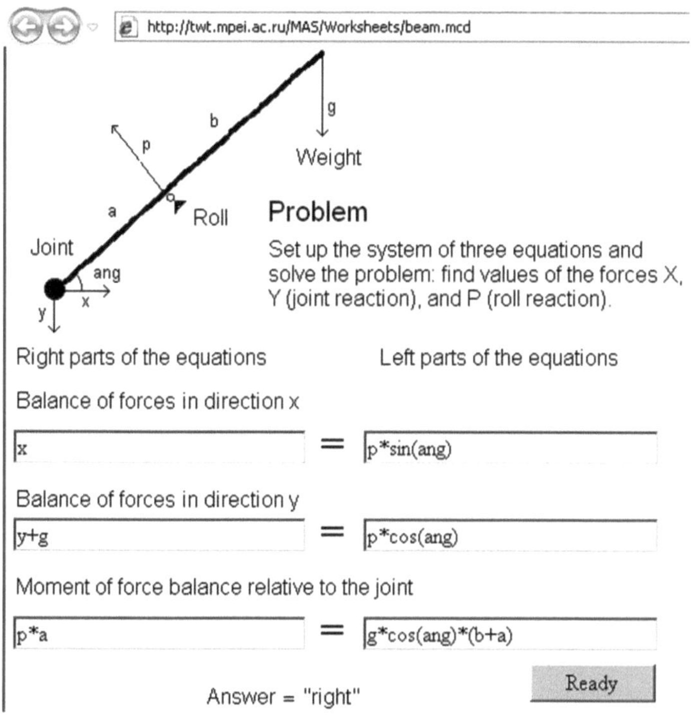

Fig. 6.109 Entering the set of equations into the text boxes

Fig. 6.110 Entering a dimensional value into a text box

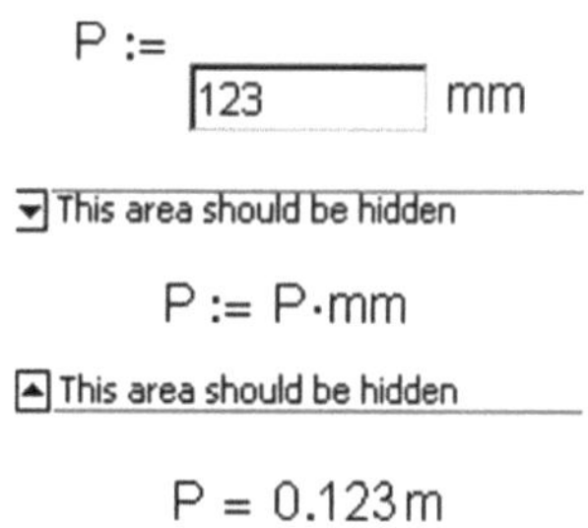

Figures 6.114–6.117 show dialog boxes to create and format Web Controls listed in Fig. 6.98: **Text box, Check box, List box/combo box** and **Radio button group**. **Submit** button was discussed previously (see Sect. 6.8.3).

If there are only a few variants of entering information the value should be selected from a preset list (by a mouse) rather than typed through a text box (from the keyboard). We did it by asking the user to choose one of three units, `m`, `mm`, and `cm`, but not type from the keyboard. If a list contain only two choices (yes or no, the alternative question) we can use a check box (entering the Boolean value), for

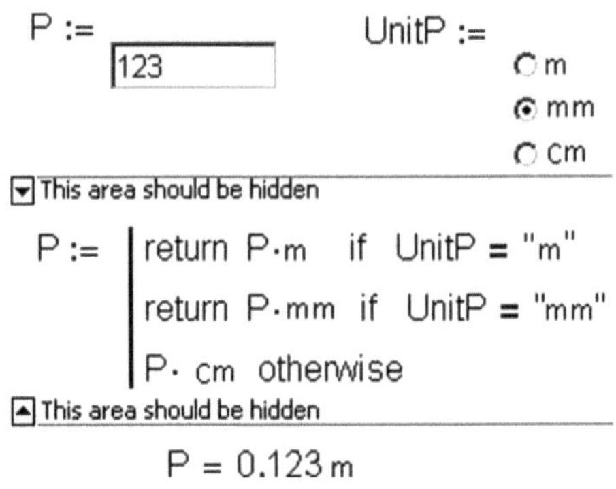

Fig. 6.111 Entering a dimensional value and choosing the unit

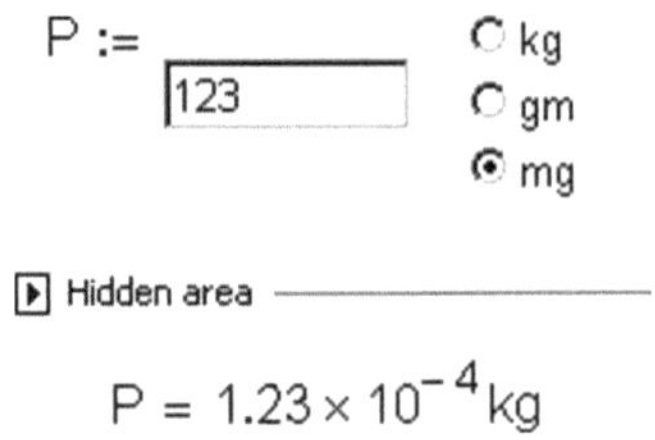

Fig. 6.112 Entering a dimensional value and choosing the unit (The area is collapsed)

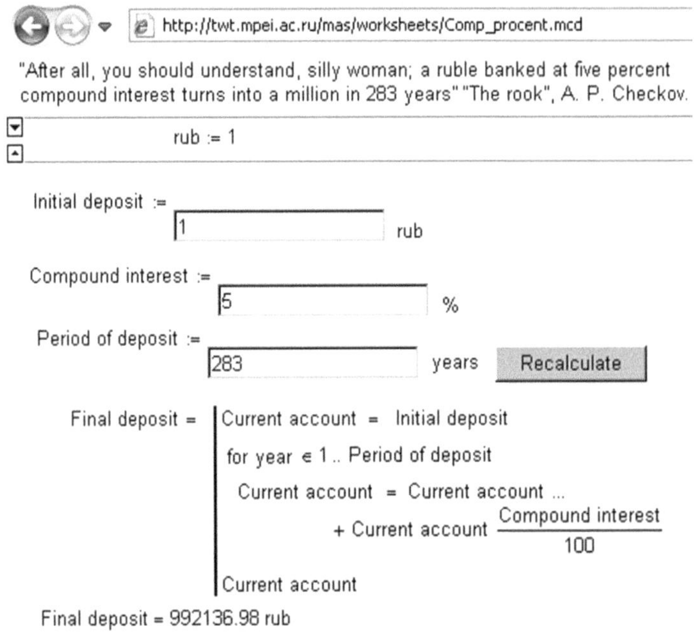

Fig. 6.113 A floating unit equal to unit

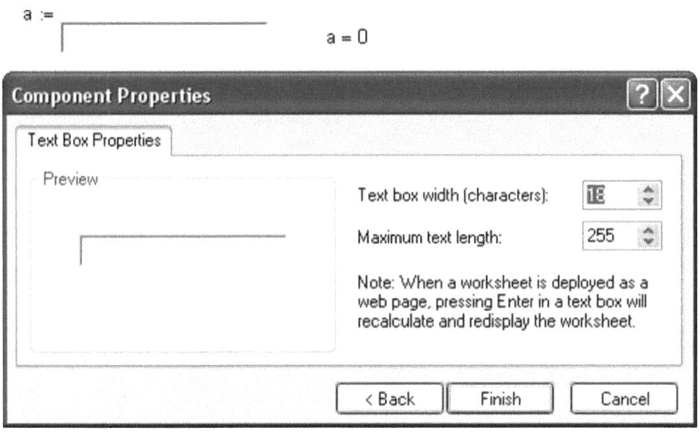

Fig. 6.114 Text box properties tab

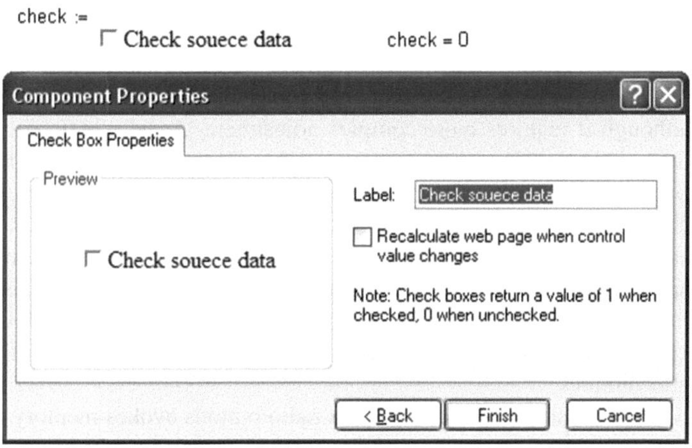

Fig. 6.115 Check box properties tab

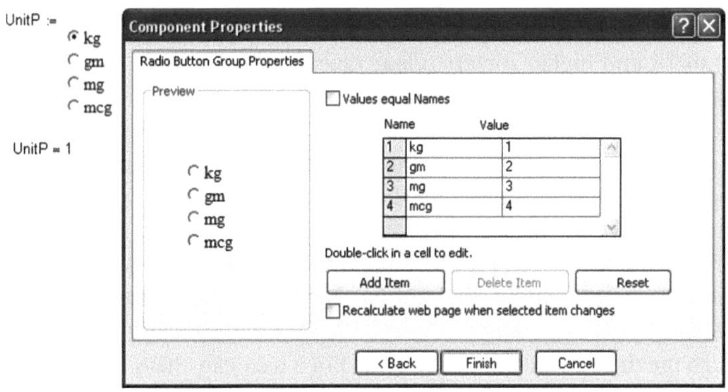

Fig. 6.116 Radio button group properties tab

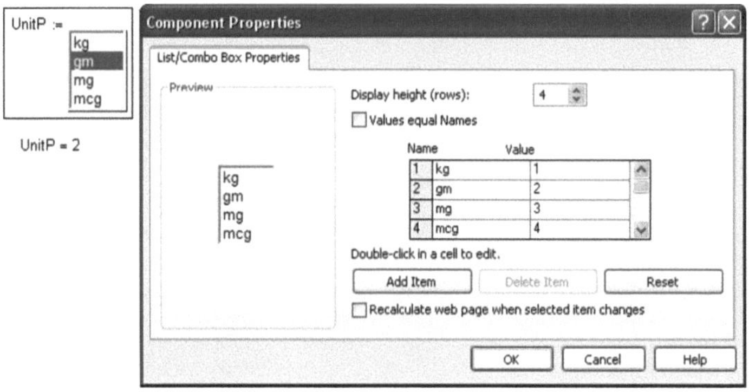

Fig. 6.117 List box properties tab

three or more choices –radio button group or/and list box. **List box** differs from **Radio button** only that it can hide a part of choices and display it by scrolling. If it is necessary to select not one but several values, we may substitute a set of check boxes for that. A check box is a universal tool which is preferable than a radio button, although it requires more complex adjustment. A radio button always has an activated value thus it may prompt a user the solution that might be false. The check boxes may be empty that gives free choice to a user without any prompts, right or false. Besides, any distance is possible between the check boxes; they can be located in any order in the worksheet but not only upright, as radio buttons. Additionally, it is possible to set rearrangement of the check boxes in every call of the document. That also excludes preferences in choice. There may be few choices in the set of check boxes that conforms to the developer idea of computing or checking techniques.

Comparing a group of check boxes and radio buttons evokes memory of an old story. During the iron curtain era, one of the authors listened to "Voice of America" and other western radio stations pushing two frequency range buttons, SW1 and SW2, of the radio set "Latvia", simultaneously. The point is that the strategy shifted the frequency range from 25 and higher meters (that was jammed by security services) to 18 and higher meters (clear range). Thus, we can choose few buttons among real, not computer radio buttons.

Speaking of radio buttons and check box sets, we shall pose a problem about `while` and `until` loops; which is more important? A set of check boxes is more important than radio buttons because, first, a check box set allows the uset to choose none button or several buttons right total number of buttons.

The dialog box shown in Fig. 6.114 and the similar (see Figs. 6.115 and 6.116) are displayed after pushing the button **Next**> in the right bottom corner of Fig. 6.98.

Through the dialog box shown in Fig. 6.114 a user can change width of a text box (it is equals to 18 in Fig. 6.114) or/and the maximum length of entered text (255 symbols). At the same time pressing <Enter> is a signal to transfer the changed data on MAS. In that case, the button Recalculate shown in Fig. 6.102 becomes unnecessary. However, it would not be an excess to insert the corresponding comments. Otherwise, a user may not see the results; a globe in the top right corner of a browser will not rotate.

There is own check box in the **Check Box Properties** tab (Fig. 6.115) which being enabling by a developer of MAS application transfer the changes in Mathcad worksheet into the MAS. There are the same check boxes in the **Radio Button Group Properties** tab (Fig. 6.116) and the **List/Combo Box Properties** (Fig. 6.117).

The check box returns zero or one to its variable (variable `check` in Fig. 6.115) depending on the state of the Web Control.

Comparing Figs. 6.116 and 6.117, we can see a counter **Display height (rows)** that fixes the number of elements displayed in a list at any given time. If this number less than the number of elements in the list they appear as a scrolling list box. If we set this number to one, the list appears as standard combo box (see Figs. 6.119 and 6.120).

We can see all the elements (buttons) in the Radio Button Group. This is the only difference between the **List box** and the **Radio Button**.

Figures 6.116 and 6.117 shows two ways of setup **Radio Button Group** and **List/Combo Box** – see the check box **Values equal Names**. In the first case (Fig. 6.116) the value connected with the Combo Box or with the List (our `unitP`) receives a text name keeping in the corresponding cell of the **Name** column; and in the second case (Fig. 6.117), the value (number or text) from the cell of the **Value** column. Unfortunately, a user cannot type there a built-in variable or an expression but only constants. In the non-Web analogues of the Radio button and the List box, such a formatting, through programming in one of the BASIC versions, in VBS (Visual Basic Script) appear in Fig. 6.118.

Note! *Assignment of the values returned by the radio button or the list box to the* ***Names*** *but not to the* ***Value*** *(compare Figs. 6.116 and 6.117) may result in errors if the names contain reserved characters, for example, degree sign (<Alt>+0176). Creating the WebSheet for temperature computing was a concern to one of the authors who struggled with this error (http://twt.mpei.ac.ru/mas/worksheets/Temperature_ Recalc.mcd) until he understood that the degree sign was losing (was used up) while transferring the data on the Web.*

As was noted above, presence of the programs (macros) inserted into Mathcad worksheet causes us to reject the Controls and replace them with the Web Controls. For this reason, the Web Control List does not contain such a convenient interface element as **Slider**. The **Slider** is a convenient tool because it allows us to set, with a mouse, the required value within defined ranges and with defined units. Nevertheless, we can simulate a slider (Fig. 6.119) formatting a list box accordingly, if we keep only one element in the box which we assign required number of the discrete values in the formatting table (in the columns **Name** and **Value**), for example, 0.1, 0.11, 0.12, . . ., 0.2 as in Fig. 6.119.

Ideally, it is possible to design such interface of a MAS user that the user will not push any keys of a keyboard; instead, he will work only with a mouse (Point-and-click).

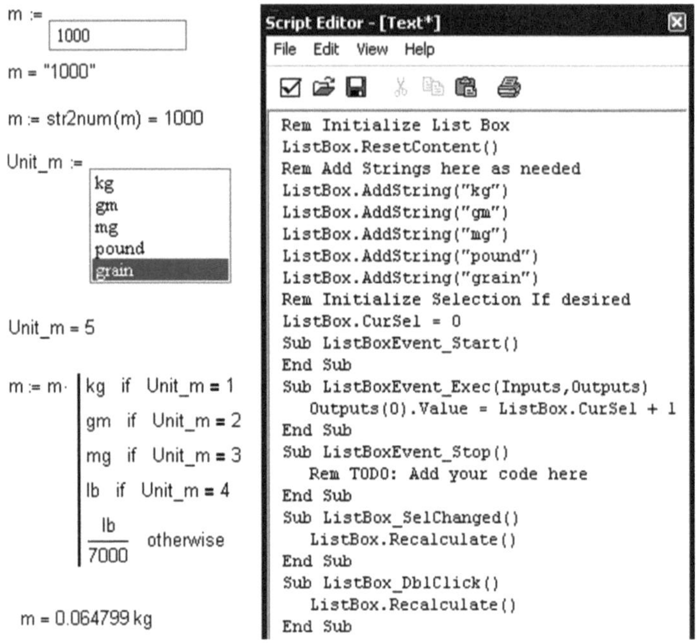

Fig. 6.118 A program working with the list box

Fig. 6.119 Simulation of a slider via the list box

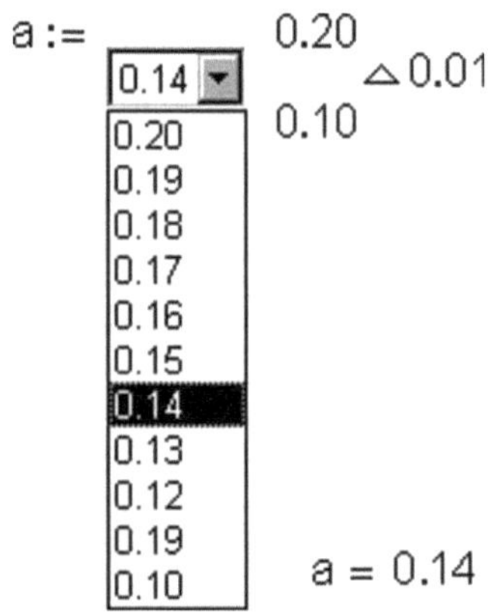

As we have previously noted, WebSheet may have not the **Submit** button. On the other hand, we may insert some of such buttons into designed document if it is bulky and the source data are entered in the different places. In that case, it is advisable to fix these places with the **Submit** button. It is a good solution to divide the document into several parts and fix the end of each one with **Submit** button. Computing of the following parts will be enabled until the user enters the required source data in the previous: check the solution and push next **Submit** button.

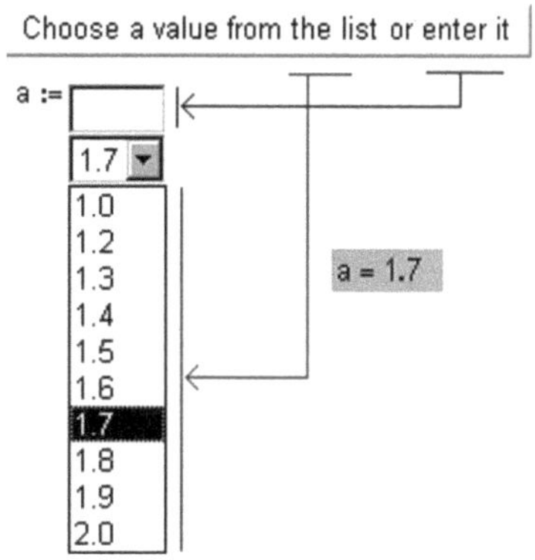

Fig. 6.120 Simulation of a Combo box via the list box and a text box

Those who designed a user interface in Visual C, Visual Basic, and Delphi, etc note that the MAS interface is rather scanty. We noted above that it has not the Combo box combining the text box and the list box.

Figure 6.120 shows a way to combine the **Text box** and the **List box** and obtain the **Combo box**.

The strategy in Fig. 6.120 is that the value in the text box is assigned to one variable and the value from the list box to another as a hidden variable. The first or the second value may be assigned to the resulting variable depending on a user choice. A sign that it was chosen from the list box but not the text box – there are no symbols in the text box. Such a box combining the text box and the list box is used in a document at http://twt.mpei.ac.ru/mas/worksheets/VPU_4_1.mcd.

A user is allowed to choose the filter diameter from the list (1.0, 1.5, 2.0, and 2.6, 3.0 or 3.4 m) or type his own value. A reader understands that the solution shown in Fig. 6.120 cannot be considered as a complete one. Thus, we still need a **Combo box**.

There are other "requirements". For example, MAS does not enable us to "suppress" some interface elements if a variable possess definite value. You will find a partial solution of this problem at http://twt.mpei.ac.ru/MAS/Worksheets/Smart_Input.mcd, as well as in Fig. 6.121.

Figure 6.121 shows the problem of converting the concentration of a solution from mass percent to titre or molarity. We use two formulae with the different sets of source data that require or, otherwise, not require entering of them through the text box. In Fig. 6.121 we suppress the text box covering its text variable with operator ■ = with varying length, highlighted in black. Here we use such Mathcad feature as overlapping regions (see Sect. 6.5). The check boxes of the titre (the concentration of solution expressed in mg l^{-1}) and molarity (the concentration of solution expressed in mol l^{-1}) are set; therefore, their changes are transformed on the server right away that results in opening or closing corresponding text boxes of density entrance or/and mass of the solute. Figure 6.121 shows three computing situations; the upper part is the WebSheet state after it being open on the Web; the middle part is computing of the titre; and the bottom part is computing of the

Fig. 6.121 Disabling the text boxes

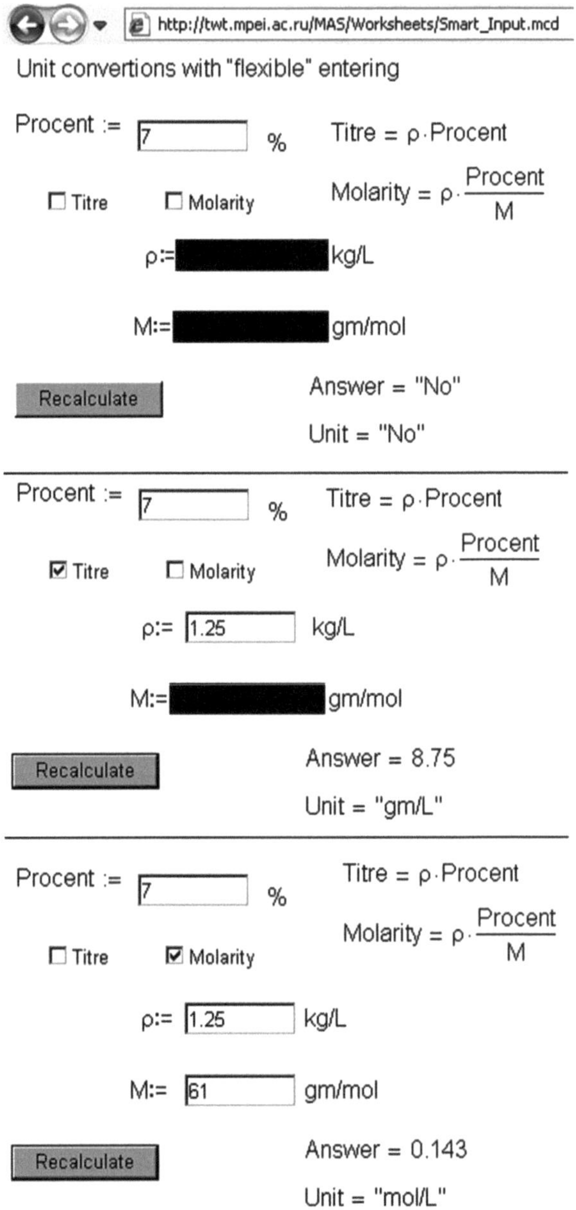

molarity. There is also the fourth state; a user enables two check boxes accidentally or intentionally that correspond to initial state of computing.

Discussing limitations of Web Controls, we should note that Mathsoft did not create Controls but are standard tools for data transfer on the Internet, rather weak for the present time. Controls are half-finished products that require to be improved. We can achieve such improvements through programming only, through a script (see above.) There is a magic circle and the problem is where we can tear it!

6.8.4 Comments in the WebSheets

As was noted above, we should not hesitate creating a WebSheet and insert enough comments, text and graphic. They do not affect computing but help viewers and even the developer to understand it (see Sect. 6.3).

Converting a worksheet for Web distribution, we should examine this commenting rule. The point is that a bulky document decelerates the Web and loads a server with superfluous operations. For this reason, it is worthwhile to place the main comments, the creation history, the detailed description, instructions etc, into a separate non-Mathcad file, for example, html-file created in Word or FrontPage. The Mathcad worksheet itself should contain short comments and references to the "bulky", additional comments that the user may open in a new browser window. For the same reason, the user should avoid inserting too many complex plots. It is better to "freeze" static plots that do not change with the source data varying and insert them as pictures.

When publishing a document on the Internet, we should think of making it polylingual. That is, we may duplicate all comments in some languages, in Russian, English, Spanish, etc. Another way is to allow a user to choose the language (Fig. 6.122).

The solution shown in Fig. 6.122 is the following:

We enter the comments into the vectors and their elements are displayed depending on the chosen language (the first language corresponds to the first elements of vectors, the second language – to the second elements etc). There is the third solution t allowing a user to open some variants of a worksheet (several files) with variables and comments written in different languages, for example V_{input} and $V_{вх}$. The WebSheet at: http://twt.mpei.ac.ru/mas/worksheets/Tests/Test_Cone.mcd and http://twt.mpei.ac.ru/mas/worksheets/Tests/Test_Cone_Eng.mcd were created using this method.

Comment. *It is possible that the server application will "feel" what browser, Russian or English, "touches" it and open the Russian or English variant of the WebSheet accordingly.*

There is another solution shown in Fig. 6.123. Information representing also a part of Mathcad worksheet could be turned into the graphic files with different names (in Fig. 6.123, their names are Counter_Flow.gif and Direct_Flow.gif). Such pictures we can read from a disk and insert in any part of a worksheet following the calculation logic. The reader will find a virtual laboratory work concerning heat mass exchange at http://twt.mpei.ac.ru/MAS/Worksheets/Solodov/Heat%20exchanger_Design.mcd in which

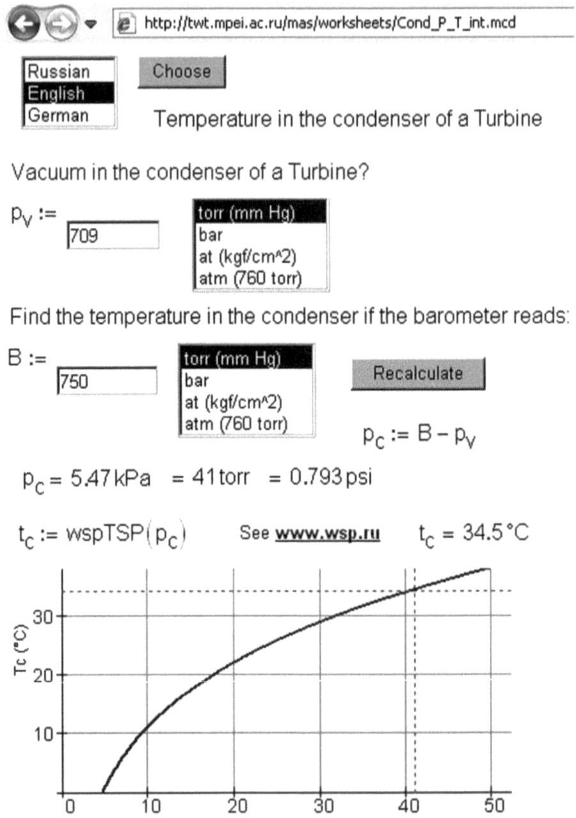

Fig. 6.122 A multilanguage web sheet

the student choice of the heat exchanger (direct flow or counter flow) opens a proper apparatus scheme.

Finally, a word about letters is in order. When opening MAS applications on the Internet, we can face the problem of Cyrillic characters representing both in operators and in text comments. These problems are widespread – representing of characters in wrong coding ("abracadabra") – the characters are not placed in a visible region but fall down behind the following operator (the common problem for the browser sets not only for Cyrillic characters). The "abracadabra" problem may be solved partly by addition of such notice inserted as a picture at the top; "If coding of the Cyrillic characters is wrong change it pushing Encoding from the View menu". It is advantageous for users to substitute variable names (pictures) for text comments in which reserved characters are inserted after pushing < Ctrl> + <Shift>+<j>. We should remember that their number must be minimal, and it is better to type some blank characters before and after such comments. We should debug a ready-made WebSheet in computers, which monitors differ in size and

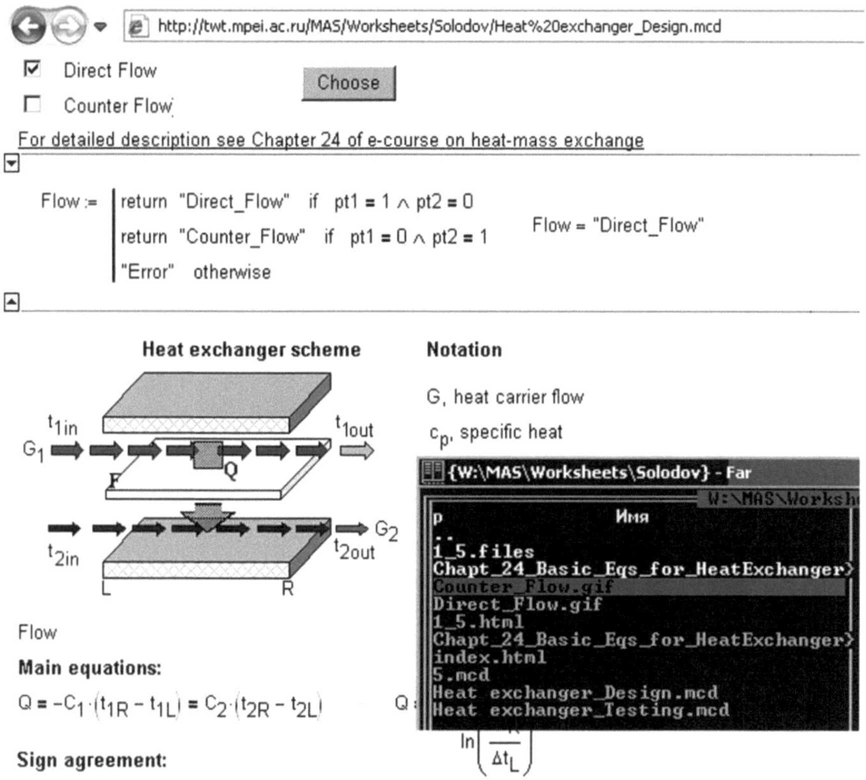

Fig. 6.123 Interactive change of the information in the web sheet

settings and in different browsers (Netscape, Mozilla etc., not only in Internet Explorer).

6.8.5 *Inserting Other Applications*

We are able to insert fragments of other applications, Excel, MATLAB, LabView, into a Mathcad worksheet and exchange data with these programs. The insertion of fragments of Excel often facilitates the entry and representation of tables (vectors and matrixes, see Fig. 6.27) or for bringing its feature sets.

The list of such applications depends on the programs installed on the computer (worksheets) or on the server (WebSheets, Fig. 6.124 shows the dialog tab of inserting components into the Mathcad).

We may publish the Mathcad worksheets containing such components on the MAS only if we have solved license problems with developers of corresponding

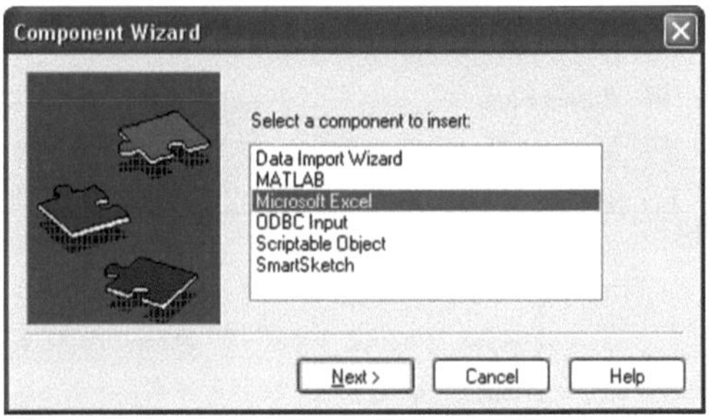

Fig. 6.124 Insert components into Mathcad worksheet

applications (such as if the program license is for use on a workstation and not for installing on the server and publishing on the Internet).

6.8.6 Names of Variables and Functions

When preparing Mathcad worksheets for Web publication we can pose a problem that the name of a function or a variable working in the worksheet has failed while we open the document on the Internet. Usually this situation occurs if we use Cyrillic characters in the names. It is not advisable to use Cyrillic characters in the "working" names of variables, functions, files and even in comments (see Sect. 6.8.4). Nevertheless, we have to use them, basically in special calculations in which one or another object has got customary Russian names. Some of them have their Latin analogues others have not. If we need Russian characters, it is a good idea to smooth away the "Russianness" of these variables as was noted above. The authors noted that variables ч (hour), мг−экв (milligram-equivalents), etc. fail while one opens a Mathcad websheet — they became red and clicking on them displays the error message "This variable or function is undefined". We can "cure" such variables starting their names from blank. They will have the same look (but will appear shift slightly to the right) and the document will work again. It is possible to "cure" such variable by changing its name via **Edit | Replace** clicking **Find in Math Region**.

We should remember that Mathcad really supports only Greek and Latin letters. Our strategies to "heal" reddening variable names that contain reserved characters (blanks) are non-documented. Therefore, this technique may not work in next Mathcad versions. If we cannot do without Russian characters, then a user may cover Latin names with the pictures of Russian characters, and MAS allows us to remove any restriction from the variable names. Thus, Fig. 6.125 shows computation of a temperature difference in which the variables contains "tops", - and ˜, subscripts, `in` and `out`, `superscripts`, 2 and ξ, located one over another without horizontal shift.

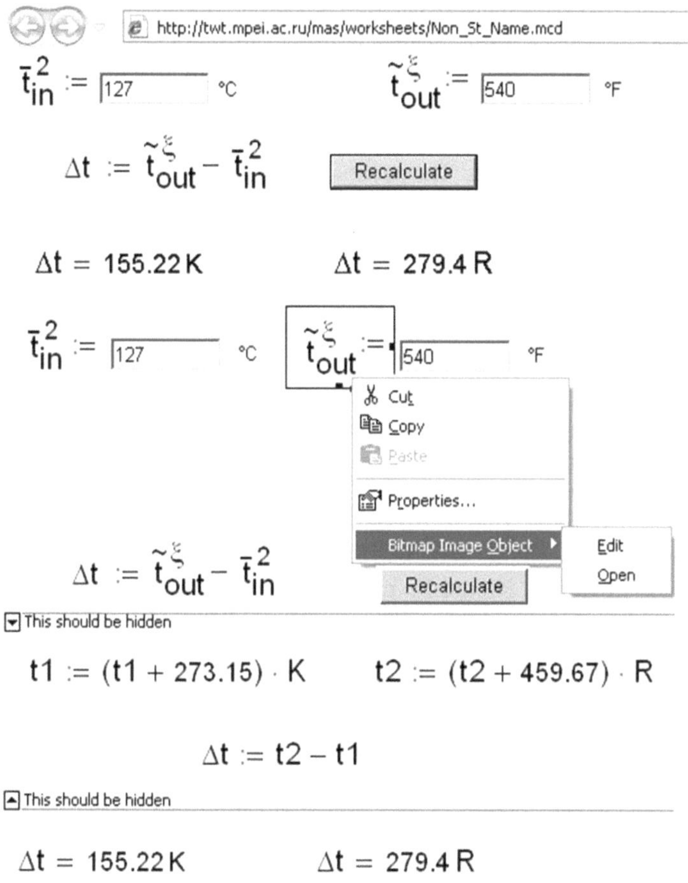

Fig. 6.125 Optional variable names on MAS

The secret in Fig. 6.125 is that the real names of `t1` and `t2` are invisible and a user view pictures of the variables and the formulae. Such "trick" does not work in a Worksheet but is applicable to a WebSheet. As mentioned in Sect. 6.2.2, the document requires minimum comments if the variable names are conventional for the specialists – it is sufficient to name the document and refer to the description in HTML. Possibly, this recommendation of naming convention, if implemented, is superfluous (or reduntant) because the variable names speak for themselves and for the document.

6.8.7 Problem of Extensional Source Data

When working with Mathcad worksheets or WebSheets, we may encounter problem of entering a unintensional change of array source data (see the example at http://twt.mpei.ac.ru/mas/worksheets/Exp_Steam_K_300_240_LMZ.mcd in which a user should view and change values in dozens of text boxes before pushing the **Recalculate** button). Entry of such data via Web Controls is time-consuming and inconvenient. Working in individual computers (not on the Network) may solve this problem saving the data to disk and reading it into the arrays via the READPRN function and other commands. This technique is difficult for work on the MAS and we will suggest another one. Before a user works on the MAS, for instance, one prepares new source data in electronic tables (Fig. 6.126).

After preparing the data sets (for instance, changing data of the previous connection with MAS which saved in the user archive) we may copy them (pressing the key-sequence < Ctrl> + <Insert > followed by < Shift> + <Insert>) to the long text box of the Mathcad WebSheet, one text box or several of them if the length of the file exceeds 255 symbols. After that, we press on the **Submit** button and the calculation operates with new source data. If the results are also extensional, we recommend representing them as tables not as scalar values. Copying tables from the WebSheet to transfer it to the user Mathcad worksheet for analysis and redesign is an easy task. That is, we should remember that scrolling of extensional arrays does not work on the Network (besides, we cannot display nested arrays on the Network).Such data should be prepared for display beforehand: insert some tables, expand nested arrays, etc.

Suggested solution for archiving and entering extensional data in a WebSheet is integrating such data into MAS. Many non-network Windows applications for calculation have the special commands such as **New [Project], Open [Project]** and **Save [Project]** in the **File** menu. These special commands allow us to work with an archive-copy of source data, see Fig. 6.127 showing these commands (the menu and dialog windows of a Window application that save data concerning separate projects to a disk in files Sample1.rod, Sample2.rod ... Sample8.rod).

Converting worksheets to functions as Web documents, we should choose start values of the source data. As a rule, WebSheet developers follow the tradition of Mathcad with its basic interface operator := and creates it in such a way that the

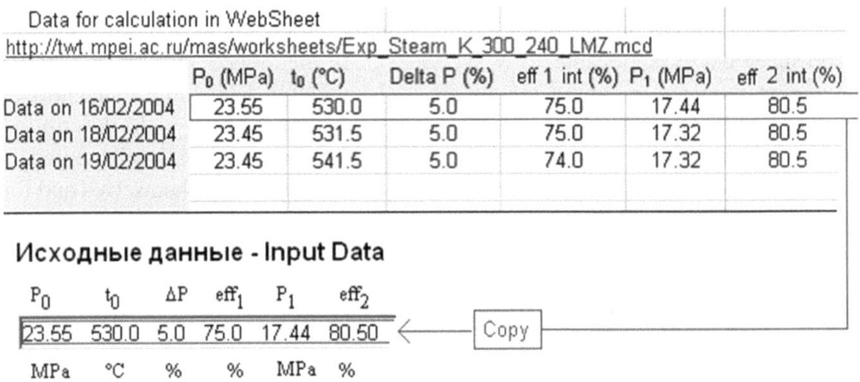

Fig. 6.126 Source data table for MAS (shows minor part of data)

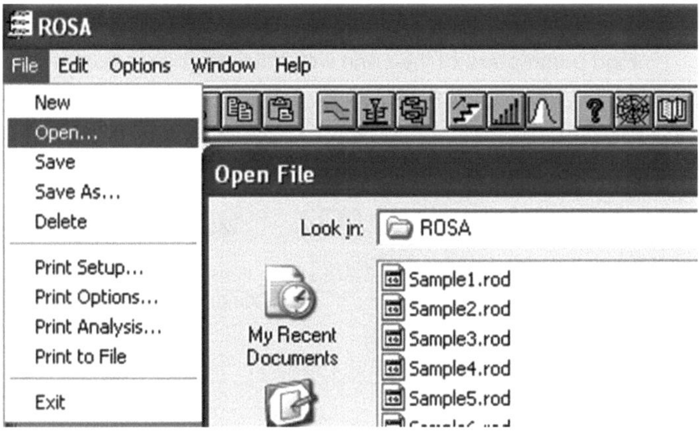

Fig. 6.127 Menu and the dialog box for creating, opening, and saving calculations

document contains common initial (baseline) data that could be edited by a user. On the one hand, it is convenient to be able to view a document immediately and do not change anything there or change just some initial data. On the other hand, it may result in errors if a user have not changed the required data and obtains wrong solutions. Therefore, it is recommended that the developer put a dash ("–") into a text box but does not type standard values. This will disable calculations and thus demand that a user enters the required data. For this same reason, it is advisable to use check boxes instead of list boxes or the radio buttons in creating data choices. As previously mentioned, we must select at least one choice in a check box ("clean page for calculation").

6.8.8 Knowledge Checking Via MAS

The main object of MAS is the solution of practical problems. User may enter initial data via Web Controls, push the **Recalculate** button and obtain a result. However, it is possible to set Web Controls for knowledge checking if we "turn it inside out", such document is shown in Fig. 6.109. It is acceptable to generate initial data and ask a user (a trainee) to enter the required formulae. The results of calculations using these formulae are validated with the true solutions. We can check knowledge of corresponding process, equipment or technology on the base of comparisons. Web Controls implemented in education may provide practical training as self-control. That is, it is necessary to give the theoretical footing and references prompting a user to the right answer in the document itself or in the referenced worksheets. The example of such WebSheet is available at http://twt.mpei.ac.ru/mas/worksheets/Tests/Test_Cone.mcd, which asks to define the height and volume of the right circular cone by given geometry (Fig. 6.128).

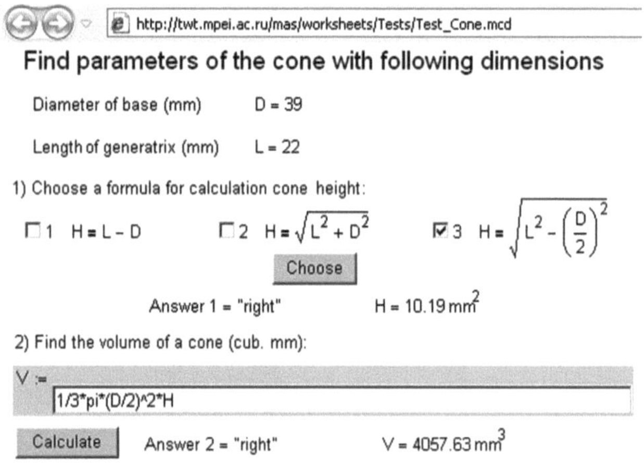

Fig. 6.128 The test on the MAS: choosing a formula and entering an expression

In Fig. 6.128, point-1 suggests traditional choice the right solution from the list containing also some wrong solutions (this technique is obsolete) – Point 2 asks to enter solution but not choose. That is, we can enter both variables (`D`, `H`, and `pi`) and numerical constants.

The WebSheet at http://twt.mpei.ac.ru/MAS/Worksheets/Chem/Chem_R_N26t.mcd shown in Fig. 6.129 asks a student to solve a chemical problem on computer not by entering numerical constants as shown in Fig. 6.128 but by composing the required equation. That is, Mathcad with its procedures of finding a solution allows us to change method in one or another branch. Thus, in formal school education Mathcad is applied to the solution of similar problems. Having the method of finding a solution, we can forget about "primitive" calculations and turn to the essence of problem, to solve equation for conservation of mass.

We should solve the following problems in such documents (Figs. 6.128 and 6.129):

- It is necessary to generate data within the reasonable range.
- The initial data should be chosen from quite extensional table or created via true or pseudo random number generator excluding wrong data.
- Initial data may have different units that encourages a student to study some sections of metrology.
- A trainee can choose a formula from the list (height of the cone, Fig. 6.128) or enter it in a text box (volume of the cone). These methods have their advantages and limitations. Choice of solution implies wrong variants. It is controversial point in methodology: a student may remember a wrong result from the list but not the right one. Entrance of a formula in the text box is rather inconvenient: we should enter the formula in the text mode basing on "computer" but not mathematical way of operator notation. There are some advantages in this method — a user studies some techniques of work with computer and special science at the same time.

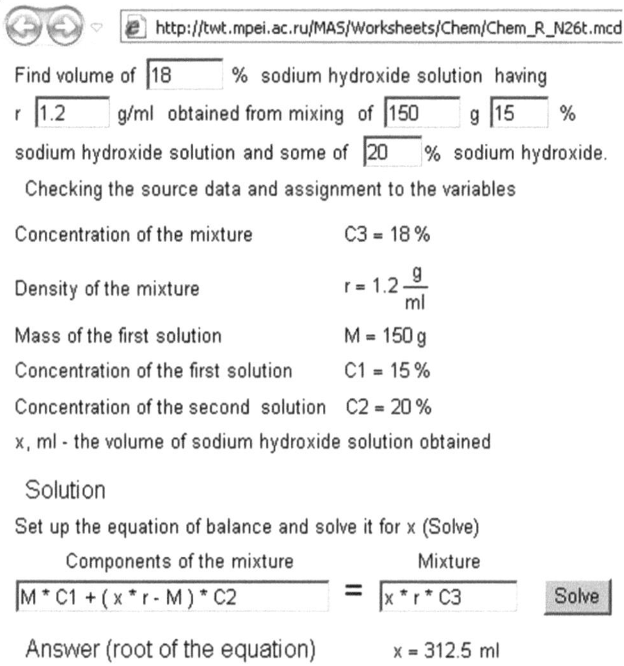

Fig. 6.129 The test on the MAS: setting up and solving the equation

- Comparing numerical answers of the user including those entered via formulae with true results we should take into consideration accuracy of mathematical operations. For instance, the user may enter p equal to 3.14 that differ from the actual value of this fundamental mathematical constant. A solution of this problem is to display not the message of the kind "Right/Wrong" but two values, user's result and true result analyzed by a person.

6.8.9 *Access to Calculations Via Password*

The solution of the problem of access to the MAS websheets with traditional technologies used on the Internet with the following method:

Registration, password, etc.

These technologies are applicable to the entire site or to separate calculations worksheets. For instance, Fig. 6.130 shows calculations of the temperature in the condenser of a turbine. The solution is created in such a way that result is displayed only with proper initial data: at Pv = 0.95 bar and B = 750 torr (mm Hg) the result will be Pc = 4.99 kPa tc = 32.8°C etc, but the calculations are disabled with other data sets. The user may view the WebSheet but not calculate. To make calculations,

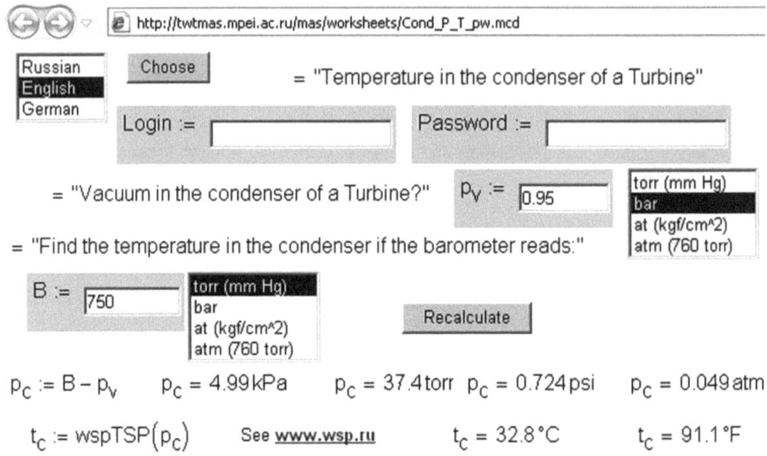

Fig. 6.130 A WebSheet containing the password

the user with name **Login** should get the password from the developer and enter it in the corresponding text box.

Entered registration name and the password are compared with (operators or functions `if`) that keeping in the program and if they agree the calculation is enabled. Therefore, here is what we should do to make a Mathcad document accessible on the Internet:

1. Edit Mathcad worksheet in Mathcad 11.2 Enterprise Edition or Mathcad 12–15 (all versions). Insert Web Controls instead of assignment operators excluding or changing superfluous comments, excluding the formatting of some results, etc. If the developer has no such versions he(she) may, for instance, highlight these operators and ask the pertinent MAS administrator to make required changes (see Figs. 6.97 and 6.102)
2. Send this document to MAS administrator, for instance, to the authors of this book. It is better to place this WebSheet on two MAS, for instance, also in Mathcad libraries. Therefore, you can see the word "double" many a time at http://www.vpu.ru/mas. This means that the corresponding calculation is a duplicated existing in the mirror server. Of course, it is possible to acquire MAS for individual use.
3. After installing the WebSheet on the MAS check it and work it out to some iterations limit (correspondence with the server administrator).

The Mathcad worksheets converted using this method may also work in individual computers that have Mathcad installed. It is possible to protect the documents as shown in Fig. 6.34) Therefore, we stick a kind of protective film with openings in the Web Controls that protect documents from intentional or unintentional damage and from curious look to the collapsed areas.

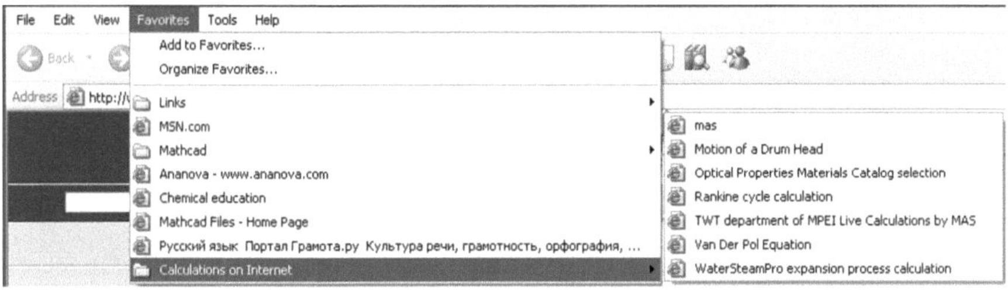

Fig. 6.131 Creating a list of references to the Internet calculations

Fig. 6.132 The site on thermal conductivity of metals

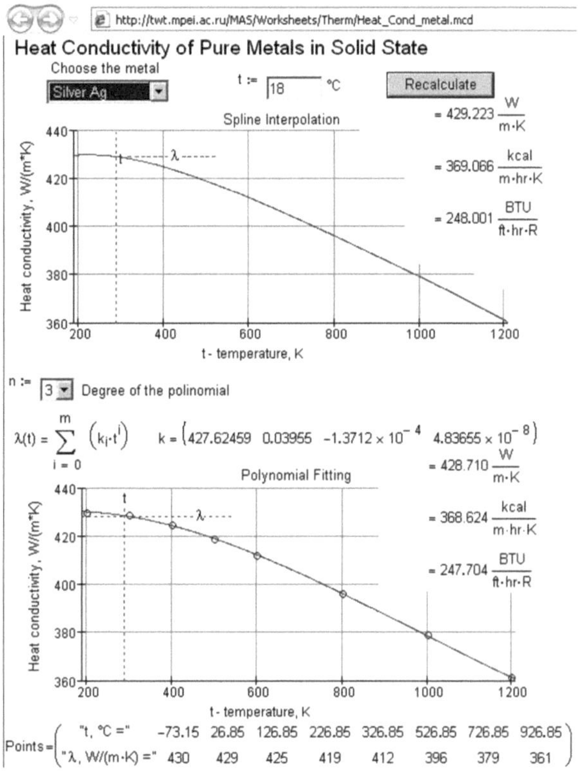

The only difference between such WebSheet and WorkSheet is recalculation begins after we have change the Web Controls but not after pushing the **Submit** button (see Fig. 6.95). In that case, we can switch to manual mode and ask to press **F9**, not the **Submit** button.

Readers can add interesting WebSheets to favorites list via the **Favorites** menu of their Internet Explorer browsers as shown in Fig. 6.131.

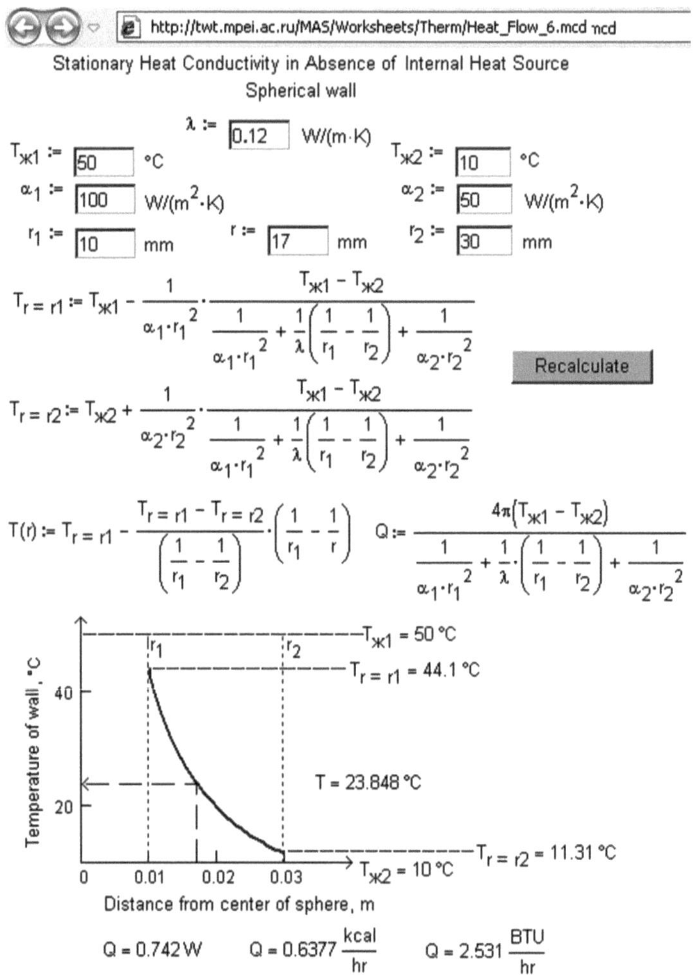

Fig. 6.133 The site containing thermal conductivity formulae

Authors hope that readers will expand the list of problems shown in Fig. 6.131. MAS may also be useful in creating the Internet versions of reference books and encyclopedias, inserting calculations in their Internet copies.

Many of the properties of metals (and non-metals) including thermal conductivity are functions of temperature. Figure 6.132 shows the WebSheet containing detailed information on thermal conductivity of metals.

This WebSheet has three features that give additional information:

1. A user may choose both metal and temperature.
2. The results are displayed numerically and on the plot that allow us to view changing of thermal conductivity with temperature

3. The site shows values of coefficients of the polynomials which smooth the data tables. A user can define exponents of the polynomials and transfer the interpolation formula obtained not only to Mathcad but also to other application (such as spreadsheets).

Besides tables, the formulae for calculations serve as reference information. The MAS technology allows us to "animate" such formulae. Figure 6.133 shows the WebSheet enabling us to both, compute the temperature inside of the wall and view temperature distribution in the wall depending on changing conditions of heat exchange.

The list of sites "animated" by the MAS technology is extensive and increasing. Students help the authors in converting reference books into the Internet sites. Therefore, the students combine usefulness (studying basis of informatics on examples of power engineering) and pleasure (to see own work on the Internet).

Chapter 7
Problems

1. Show by computer means, that all tangents, drawn to the starting points of kinetic curves for an irreversible first-order reaction with rate constant k, intersect time axis at the same point. What is the physical meaning of the abscissa of the intersection point?
2. Reaction $2A + B \to C$ has first order for each reactant. Find an analytic time-dependence of the concentration of each reactant under condition $C_{A_0} \neq C_{B_0}$.
3. Reaction proceeds according to the following stoichiometry $-\,2A + 3B \to P\,-$ and has first order for each reactant. Deduce an integral form of the kinetic equation.
4. For each of the reactions of the first, second and third orders: $A \to P$, $A + B \to P$ and $A + B + C \to P$ (initial concentration of all the reactants are equal) find, how many times slower comparing to the half-life time is the transformation rate of 99% reached.
5. Some reaction has an order of '-1' for an initial substance. Deduce an integral form of the kinetic equation for a time-dependence of a reactant. Obtain expressions for a half-life time and for the overall reaction time.
6. Investigated sample of a rock contained 100 mg of uranium and 14 mg of lead. Half-life time of uranium is 4.5×10^9 years, and the total decay of 238 g of uranium yields in 206 g of lead. Define the age of the rock. Assume that when formed the rock contained no lead. Also neglect the presence of the intermediate decay products considering that they decay much faster than uranium.
7. Fifty percent of the initial quantity of a radioactive substance decayed in 30 days. In what time will only 1% of the initial quantity be left? Assume that decay proceeds according to the first-order regularities.
8. Third-order one-way reaction follows the scheme $A + 2B \to$ *Products*. Rate constant equals to 0.005 $L^2\,mol^{-2}\,s^{-1}$. Initial reactants' concentrations equal to $A_0 = 1$ mol L^{-1} and $B_0 = 2$ mol L^{-1}. Build kinetic curves $A(t)$ and $B(t)$ based on the results of the analytic solution of the direct kinetic problem. What are the concentrations of the reactants 500 s after the start of the reaction? Assume that kinetics of the reaction fully follows its stoichiometry.

V.I. Korobov and V.F. Ochkov, *Chemical Kinetics with Mathcad and Maple*,
DOI 10.1007/978-3-7091-0531-3_7, © Springer-Verlag/Wien 2011

9. Build a plot of the change in the quantity of each substance, involved in the process of radioactive decay:

$$^{239}_{92}U \xrightarrow{\tau_{1/2}=23.5\text{ min}} {}^{239}_{93}Np \xrightarrow{\tau_{1/2}=235\,days} {}^{239}_{94}Pu.$$

Initial uranium's weight is 2.0212 mg.

10. Bismuth's radioactive decay

$$^{210}_{83}Bi \xrightarrow{k_1} {}^{210}_{84}Po \xrightarrow{k_2} {}^{206}_{82}Pb,$$

proceeds according to the regularities of the successive first-order reaction with rate constants $k_1 = 0.144$ days^{-1} and $k_2 = 5.102 \times 10^{-3}$ days^{-1}. Initial bismuth's weight is 1.0000 g. Define: (a) initial mass of polonium, formed during the decay; (b) time required to reach the maximum of polonium's concentration; (c) composition of the reactive mixture (weight %) 10 days after the start of the reaction.

11. In a successive reaction, consisting of to irreversible first-order stages, half-life time of an initial reactant equals to the time from the start of the reaction to the moment intermediate's concentration reaches its maximum. What rate constants' ratios make such situation possible? Initial concentrations of the intermediate and the product are zero.

12. In a first-order one-way successive reaction, consisting of two elementary stages, 80% transformation of the reactant A took 72.5 min. Maximum concentration of the intermediate was reached in 55.5 min. Determine the rate constants for each stage.

13. Cyclopentadiene (A_1) reacts with butadiene (A_2) to form vinyl dicycloheptene (A_3) according to the regularities of the second-order reaction. Intermediate A_3 transforms into tetrahydroindene (first-order reaction). Under initial concentrations $c_{1(0)} = 1$ mol L^{-1} and $c_{2(0)} = 11$ mol L^{-1} half of the initial cyclopentadiene reacted in 2.3 h. By that moment concentration of A_3 had been increasing and after some time t_{max} it has reached its maximum value of 0.8 mol L^{-1}. Calculate the rate constants for the given successive reaction. What is the value of t_{max}?

14. Let's picture, that reaction is a subsequence of two irreversible first-order stages with equal rate constants. Deduce the equations of kinetic curves for all reactants analytically.

15. It has been determined that birotation of alumo benzene camphor in CCl_4 under 293.16 K proceeds in two successive first-order stages with rate constants $k_1 = 9.01 \times 10^{-3}$ s^{-1}, $k_2 = 1.91 \times 10^{-3}$ s^{-1}. According to experiment, under initial concentration of the reactant equal to 10 mmol L^{-1} intermediate's concentration reaches its maximum value of 6.59 mmol L^{-1} 218 s after the start of reaction. Model kinetic curves for the participants of that reaction and define, whether the modelling results agree with the experimental data.

16. In a parallel first-order reaction time of 1/3 conversion of reactant A equals to 40 min. Yield of substances B and D equals correspondingly to 32 and 16%. Determine the rate constants k_1, k_2, k_3. Kinetic scheme of the reaction is as follows:

$$B \xrightarrow{k_1} A \xrightarrow{k_2} C$$
$$\downarrow k_3 \quad .$$
$$D$$

17. Successive reaction $A \rightarrow B \rightarrow P$ proceeds in solution. Compose MathCAD or Maple worksheet, showing that very different forms of the time-dependencies of an optical density of reactive mixture are possible depending on the ratio of light extinction molar coefficients $\varepsilon_A, \varepsilon_B, \varepsilon_C$.
18. Solve direct kinetic problem for a second-order reaction $A + B \rightarrow Products$ numerically and analytically. Initial concentrations A_0 and B_0 are 0.50 and 0.25 mol L^{-1} correspondingly. Rate constant equals to 0.05 L $mol^{-1}s^{-1}$.
19. In a third-order one-way reaction $A + B + D \rightarrow Products$ initial reactants' concentration are equal: $A_0 = B_0 = D_0 = 1$ mol L^{-1}. Rate constant is 0.05 L^2 mol^{-2} s^{-1}. Build kinetic curve for reagent A. Compare results of analytic and numeric solutions.
20. In a reversible reaction

$$A \xrightarrow{k_1} B, B \xrightarrow{k_{-1}} A,$$

initial concentrations of the reactants are $[A]_0 = 0.4$ and $[B]_0 = 0.3$ mol L^{-1}. Rate constants are $k_1 = 0.1$ и k_{-1}=0.03. Calculate and build kinetic curves for A and B. Determine the equilibrium time for the following cases: (a) forward and reverse reactions have first order; (b) forward and reverse reactions have second order; (c) forward reaction has first order, reverse – second order; (d) forward reaction has second order, reverse – first order.

21. Consider one of the kinetic models of data flows in science. For example, when analysing a number of scientific publications Y in a certain field, one assumes that the rate of increase $dY(t)/dt$ is proportional to the current level of publications number:

$$\frac{dY(t)}{dt} = kY(t).$$

Solution of the equation defines the number of publications depending on time and has a form of $Y(t) = Y_0\exp(kt)$, where k is a constant, which characterises replies to publications in a particular field on the average ($k > 0$), Y_0 is a value, which characterises some initial level of the science development. However, exponential increase of function $Y(t)$ in time does not continue, as it is influenced by a number of various limiting factors. Therefore, mechanism of the change in the number of publications is given by a differential equation:

$$\frac{dY(t)}{dt} = kY(t)(Y_{\max} - Y(t)),$$

where Y_{max} is a maximum possible value of $Y(t)$. Calculate the dependence $Y(t)$ for $k = 0.1$, $Y_0 = 0.01$, $Y_{max} = 1$, which will satisfy the equation above. Compare results of analytic and numeric solutions. Determine the value of conventional time t_{max}, corresponding to a maximum rate of increase in the number of scientific papers.

22. Analyse kinetics of your progress in the cafeteria queue during the lunch break, using the following model (*I. Leenson*). Cashier's rate is constant and equals to two persons per minute; people join queue uniformly and either stand at the end of the queue, if they didn't find acquaintances there, or stick to an acquaintance in the queue, if they found one. Possibility of a person being an acquaintance is constant and equals to 0.01. What are the possible options of your progress in the queue?
23. Some reactions follows the mechanism:

$$A \xrightarrow{k_1} M + C;$$
$$M + C \xrightarrow{k_{-1}} A;$$
$$M + B \xrightarrow{k_2} D.$$

Here A, B, C, D – are the stable participants of the reaction, M – is an unstable intermediate. Rate constants are k_1 =0.007; $k_{-1} = 0.07$; $k_2 = 0.28$; initial concentrations are $[A]_0 = [B]_0 = 1$, $[C]_0 = [D]_0 = [M]_0 = 0$. Find a numeric solution for the set of differential equations, corresponding to this mechanism. Build time-dependence of the formation rate of the final product D. Afterwards, verify the applicability of the quasistationery principle: deduce expression for the formation rate of D assuming that $d[M]/dt = 0$, then put the values of the necessary parameters into the obtained expression and build a graphic time-dependence of the formation rate of D. Compare two plots.

24. In some reaction stable products R and S are formed from the stable reactants A and B. Mechanism of this processes includes formation of the intermediate X:

$$A + B \underset{k_2=1/10}{\overset{k_1=1/10}{\rightleftarrows}} X$$
$$X + B \xrightarrow{k_3=10} R + S.$$

Make and solve numerically a set of differential equations for the given kinetic scheme relatively to the current concentration of all reaction's participants. Build corresponding kinetic curves. Deduce kinetic equation for the formation rate of the product R using quasistationery approximation. Define, how justified is the application of the quasistationery principle to the given mechanism.

25. Mechanism of some fermentative reaction is as follows:

$$E + S \underset{k_{-1}}{\overset{k_1}{\rightleftarrows}} ES \underset{k_{-2}}{\overset{k_2}{\rightleftarrows}} E + P,$$

where E is a ferment; S is a substrate; ES is a ferment–substrate complex; P is a final product. Using quasistationery approximation, express a formation rate of the product through the current concentrations of the ferment, substrate and product.

26. For a reaction that follows the overall scheme $NO_2Cl \rightarrow NO_2 + \frac{1}{2}Cl_2$, next mechanism has been suggested:

    (I) $NO_2Cl \rightarrow NO_2 + Cl^* \ (k_1)$,
    (II) $NO_2Cl + Cl^* \rightarrow NO_2 + Cl_2 \ (k_2)$.

    Deduce an expression for the rate of NO_2Cl decomposition, using the method of quasistationery concentrations.

27. Conversion of para-hydrogen into ortho-hydrogen proceeds according to the following mechanism (M is an inert particle):

    (I) $n - H_2 + M \rightarrow H^* + H^* + M \quad (k_1)$,
    (II) $H^* + n - H_2 \rightarrow o - H_2 + H^* \quad (k_2)$,
    (III) $H^* + H^* \rightarrow n - H_2(k_3)$.

    Using the method of quasistationery concentrations, deduce an expression for the rate of para-hydrogen conversion.

28. Thermolysis of a hydrocarbon R_2 proceeds according to the following mechanism:

    (I) $R_2 \rightarrow 2R^* \quad (k_1)$,
    (II) $R_2 + R^* \rightarrow P_B + R^{/*} \quad (k_2)$,
    (III) $R^{/*} \rightarrow P_A + R^* \quad (k_3)$,
    (IV) $2R^* \rightarrow P_A + P_B \quad (k_4)$,

    where R_2, P_A, P_B are the stable hydrocarbons, R^* and $R^{/*}$ are radicals. Define the dependence of the decomposition rate of an initial hydrocarbon on its current concentration. Use quasistationery approximation.

29. In a one-way parallel reaction

    (I) $A \xrightarrow{k_1} P_1$,
    (II) $A \xrightarrow{k_2} P_2$.

    stage (I) follows the regularities of the first-order reaction, and stage (II) has a second order. Calculate kinetic curves of a reactant A under its initial concentration 1 mol L^{-1} and rate constants $k_1 = 0.1$ min^{-1} and $k_2 = 0.005$ L mol^{-1} min^{-1}. Determine the current concentration of A 30 min after the start of the reaction. Provide both analytic and numeric solutions. Compare results.

30. Radiolysis rate of A in reaction $A \rightarrow B$ is proportional to a given dimentionless variable – radiation density D and the concentration of the initial reactant. Decay rate is given by the equation

    $$dC_A(t)/dt = -k_0 D(t) C_A(t).$$

    Radiation density also changes in time as follows:

    $$D(t) = D_0 e^{-k_1 t}.$$

    Let $k_0 = 0.001$ s^{-1}; $k_1 = 0.001$ s^{-1}; $D_0 = 1$; initial concentration of A equals to 1 mol L^{-1}. Calculate time-dependence of the current concentration of A. Compare results of analytic and numeric solutions. Define the value of the current concentration of the product in 2,000 s.

31. Radiolysis rate of substances A and B in a successive first-order reaction

$$A \xrightarrow{k_1} B \xrightarrow{k_2} P$$

is proportional to the dimensionless radiation density D. D is a function of time according to the expression

$$D(t) = D_0 \exp(-kt)$$

Initial A and B concentrations were 1 and 0 mol L^{-1} correspondingly; rate constants of individual stages were $k_1 = 0.01\ s^{-1}$, $k_2 = 0.00005\ s^{-1}$. Radiation parameters are: D_0=1, $k = 0.00005\ s^{-1}$. Calculate kinetic curve for an intermediate.

32. Decomposition of dichlorosilane proceeds according to the following mechanism:
(I) $SiH_2Cl_2 \Leftrightarrow SiHCl + HCl, \quad (k_1, k_{-1})$
(II) $SiHCl + SiH_2Cl_2 \Leftrightarrow HCl_2SiSiH_2Cl, \quad (k_2, k_{-2})$
(III) $HCl_2SiSiH_2Cl \Leftrightarrow HCl_2SiSiCl + H_2, \quad (k_3, k_{-3})$
(IV) $HCl_2SiSiCl \Leftrightarrow HClSiSiCl_2, \quad (k_4, k_{-4})$
(V) $HClSiSiCl_2 \Leftrightarrow SiHCl + SiCl_2, \quad (k_5, k_{-5})$

Build mathematic model of the process in the form of full set of differential equations. Then, assuming that $SiHCl, HCl_2SiSiH_2Cl, HCl_2SiSiCl, HClSiSiCl_2$ are unstable intermediates, make a reduced set of differential equations. Based on your results, explain why decomposition of SiH_2Cl_2 is inhibited by hydrogen chloride.

33. Order of monomolecular thermal decomposition of a substance A equals to 2, and decomposition of A under UV irradiation has zero order. If a system is irradiated with UV light, stages from above proceed in parallel. Calculate the kinetic curve for a reactant: (a) based on the analytic solution of the direct problem; (b) numerically. Present graphic information that illustrates your results. Assume $C_{A_0} = 1\,\text{mol}\,L^{-1}$; $k_1 = 0.001\,\text{mol}^{-1}\,L\,s^{-1}$; $k_2 = 0.0002\,\text{mol}\,L^{-1}\,s^{-1}$. Determine the concentration of A 1,500 s after the start of the reaction.

34. Thermal decomposition of N_2O_5 is a multistep process with a summary scheme $N_2O_5 \rightarrow 2NO_2 + 1/2O_2$. Total of elementary stages and their kinetic parameters are given in the table.

Stage	Elementary reaction	Preexponential factor	Activation energy, kcal mol^{-1}
I	$N_2O_5 \longrightarrow NO_2 + NO_3$	$1.3 \times 10^{16}\ s^{-1}$	19.3
II	$NO_2 + NO_3 \longrightarrow N_2O_5$	$1.8 \times 10^{9}\ L\ mol^{-1}\ s^{-1}$	0
III	$NO_2 + NO_3 \longrightarrow NO + O_2 + NO_2$	$2.3 \times 10^{8}\ L\ mol^{-1}\ s^{-1}$	4.4
IV	$NO + NO_3 \longrightarrow 2NO$	$6.0 \times 10^{10}\ L\ mol^{-1}\ s^{-1}$	1.4

Initial concentration of N_2O_5 is $[N_2O_5]_0 = 0.040\,\mathrm{mol\,L^{-1}}$. Build kinetic curves for all participants of the reaction: (a) under $T = 450$ K; (b) under $T = 600$ K. Further analyse kinetic scheme using quasistationery principle, assuming that NO and NO_3 are unstable intermediates. Calculate kinetic curves for the stable participants of the reaction based on the cut mathematic model that does not include the concentrations of the intermediates. Compare results of both calculations and make a conclusion about the applicability of the quasistationery principle for a given kinetic scheme.

35. First-order reaction proceeds with rate constant k in an isothermal continuous stirred-tank reactor with volume V. Reactant with concentration C_0 is continuously added to the reactor with a speed w, and the reactive mixture is removed with the same speed. Using interface elements **Control** (or **Web Control**) create a document, which will allow to calculate the concentration of unreacted initial reactant when it exits reactor at any time based on the values of V, k, C_0 and w, specified by the user. Show that in a particular moment in time concentration on the exit of the reactor will become constant.
36. A following mechanism has been proposed for a thermal decomposition of acetaldehyde:

$$CH_3CHO \xrightarrow{k_a} \bullet CH_3 + \bullet CHO,$$

$$\bullet\, CH_3 + CH_3CHO \xrightarrow{k_b} CH_4 + \bullet CH_2CHO,$$

$$\bullet\, CH_2CHO \xrightarrow{k_c} CO + \bullet CH_3,$$

$$\bullet\, CH_3 + \bullet CH_3 \xrightarrow{k_d} CH_3CH_3.$$

Find values of rate constants for elementary stages of the process for a chosen temperature in the Internet (e.g. http://www.nist.gov) and solve direct kinetic problem, assuming arbitrary value for an initial concentration of the reactant. Verify the applicability of the quasistationery principle for a given kinetic scheme.

37. Process of ionisation of the molecules of a gas proceeds in gas phase under irradiation; q cations and the same number of anions is formed in 1 s in a given volume. Because ions carrying opposite charges recombine back to molecules, their number decreases. Assume that rate of ions recombination is proportional to the square of their current quantity n with a constant of proportionality α. Determine the time-dependence of the ions number n and illustrate it, assuming $q = 2{,}000$, $\alpha = 1 \times 10^{-8}$. In a moment in time $t = 0$ quantity of ions $n_0 = 0$.
38. Model kinetics of gossips' spread in the population. Assume there are A citizens in a town. In some moment in time $t = 0$ one of them found out a news. The speed of that news spreading can be characterized as a derivative

$dN(t)/dt$, where $N(t)$ is a number of citizens, who had found out about the news by the moment in time $t > 0$. Assume the speed of news spread is proportional to the number of contacts of those, who already knows the news and those, who doesn't know about it yet, with a constant of proportionality (rate constant) a. Calculate the dependence $N(t)$ for $A = 60{,}000$ and $a = 0.0000003$. Deduce and expression for a time t_{max}, corresponding to a maximum speed of gossips' spread. What chemical reactions are analogous to the given kinetic model?

39. Consider one of the kinetic models for the spread of epidemic. Let N to be a number of citizens in a town; I is a number of the ones, who felt ill; G is a number of healthy citizens, who can potentially fall ill; E is a number of people, who has already been down and developed immunity to this decease. Corresponding set of differential equations is as follows:

$$\frac{dG}{dt} = -k_1 GI\,; \quad \frac{dI}{dt} = k_1 GI - k_2 I\,; \quad \frac{dE}{dt} = k_2 I.$$

Specify acceptable values of the parameters k_1, k_2, G_0, I_0, E_0 and calculate the dynamics of the epidemics spread. Illustrate a well-known epidemiologic theorem about the threshold character of the epidemics: show that infection does not become an epidemic, if the condition $G < k_2/k_1$ is satisfied.

40. Fifty contagious patients appeared in a town with 20,000 citizens caused epidemic. Increase in the number of ill per day is proportional to a product of a number of healthy citizens (haven't been down yet) and a number of ill. It is necessary to trace the dynamics of epidemic spread, e.g. to define, in what way the number of ill changes daily. In Basic environment the problem is solved as follows:

```
For t=0 to 12 rem "Dynamics of epidemic spread for 12
days"
Ill(t+1) = Pr*Ill(t)*Healthy(t)
Healthy(t+1)=Healthy(t)-Pr*Ill(t)*Healthy(t)
Next
```

(Pr here is a constant of effectiveness of a preventive control). Translate given fragment to the "language" of MathCAD or Maple and carry out the calculation of the dynamics of the epidemic spread. Also solve the problem using build-in integrators. Compare the results and explain the possible reasons of their mismatch.

41. Reaction proceeds according to the mechanism:

(I) $A + X \longrightarrow R + Y \quad (k_1 = 0.10)$,
(II) $B + Y \longrightarrow S + X \quad (k_2 = 0.05)$.

Calculate and build kinetic curves for all participants of the reaction for $[A_0] = 1$; $[B_0] = 0.02$.

42. Scheme of the reaction, rate equations for each stage and the kinetic parameters are as follows:

(I)	$A + B \rightarrow 2A$	$r_1 = k_1 C_A C_B$	$k_1 = 0.01\ \text{cm}^3\ \text{mmol}^{-1}\ \text{h}^{-1}$
(II)	$A + B \rightarrow A + C$	$r_2 = k_2 C_A {C_B}^2$	$k_2 = 0.001\ \text{cm}^6\ \text{mmol}^{-2}\ \text{h}^{-1}$
(III)	$A + C \rightarrow 2C$	$r_3 = k_3 C_A C_C$	$k_3 = 0.001\ \text{cm}^3\ \text{mmol}^{-1}\ \text{h}^{-1}$
(IV)	$A \rightarrow C$	$r_4 = k_4 C_A$	$k_4 = 0.1\ \text{h}^{-1}$

Deduce the set of differential equations for a given kinetic scheme and provide numeric calculation of the kinetic curves for all participants of the reaction.

43. Investigation of the combustion reaction requires the information about the concentrations of hydrogen atoms and hydroxyl radicals. Information about the following reactions is available:

(I) $H + NO_2 \longrightarrow OH + NO \quad k_1 = 2.9 \times 10^{10} dm^3\, mol^{-1}\, s^{-1}$;
(II) $OH + OH \longrightarrow H_2O + O \quad k_2 = 1.55 \times 10^{9} dm^3\, mol^{-1}\, s^{-1}$;
(III) $O + OH \longrightarrow O_2 + H \quad k_3 = 1.1 \times 10^{10} dm^3\, mol^{-1}\, s^{-1}$.

Based on the values for the initial concentrations of hydrogen atoms and NO_2 molecules (4.5×10^{-10} and 5.6×10^{-10} mol cm^{-3} correspondingly) calculate and build time-dependences of the concentrations of O, O_2 and OH in the range 0–10 ns. Determine, how the chosen algorithm of solving the set of differential equations influences the final results of the calculation.

44. Thermal decomposition of the hydroperoxide $ROOH$ proceeds according to the scheme:

$$ROOH \longrightarrow RO^\bullet + OH^\bullet \quad k_1 = 10^{-4} s^{-1};$$

$$OH^\bullet + ROOH \longrightarrow H_2O + RO_2^\bullet \quad k_2 = 10^5 M^{-1} s^{-1};$$

$$RO^\bullet + ROOH \longrightarrow ROH + RO_2^\bullet \quad k_3 = 10^4 M^{-1} s^{-1};$$

$$RO_2^\bullet + RO_2^\bullet \longrightarrow stable\ products \quad k_4 = 10^7 M^{-1} s^{-1}.$$

Make mathematic model for a process and calculate kinetic curves for all given substances for $[ROOH]_0 = 0.1$ M. Show that for given values of rate constants application of the quasistationery principle is justified. Determine the values of the constant concentrations of the radicals and the value of the constant rate of the decomposition of $ROOH$.

45. Deduce analytic expressions for the time-dependences of the concentrations of substances A, B, C and D for given reactions' mechanisms using operational or the classic matrix method of solving sets of differential equations. Assume following numeric values for the rate constants (s^{-1}): $k_1 = 1/2$; $k_2 = 1/4$; $k_3 = 1/5$; $k_4 = 1/10$. Assume also that before the start reactive system consisted only of reactant A with the concentration 1 mol L^{-1}.

No	Kinetic scheme	No	Kinetic scheme
1.	$B \underset{k_4}{\overset{k_1}{\rightleftarrows}} A \xrightarrow{k_2} C \xrightarrow{k_3} D$	7.	$A \underset{k_4}{\overset{k_1}{\rightleftarrows}} B \xrightarrow{k_2} C \xrightarrow{k_3} D$
2.	$B \xleftarrow{k_{-1}} A \underset{k_4}{\overset{k_2}{\rightleftarrows}} C \xrightarrow{k_3} D$	8.	$A \xleftarrow{k_{-1}} B \underset{k_4}{\overset{k_2}{\rightleftarrows}} C \xrightarrow{k_3} D$
3.	$B \underset{k_4}{\overset{k_1}{\rightleftarrows}} A \xrightarrow{k_2} C \xrightarrow{k_3} 2D$	9.	$A \underset{k_4}{\overset{k_1}{\rightleftarrows}} B \xrightarrow{k_2} C \xrightarrow{k_3} 2D$
4.	$B \xleftarrow{k_{-1}} A \underset{k_4}{\overset{k_2}{\rightleftarrows}} C \xrightarrow{k_3} 2D$	10.	$A \xleftarrow{k_{-1}} B \underset{k_4}{\overset{k_2}{\rightleftarrows}} C \xrightarrow{k_3} 2D$
5.	$A \xrightarrow{k_1} B \xrightarrow{k_2} C \xrightarrow{k_3} D \xrightarrow{k_4} E$	11.	$A \xrightarrow{k_1} B \xrightarrow{k_2} C \underset{k_4}{\overset{k_3}{\rightleftarrows}} D$
6.	$\begin{array}{l} A \xrightarrow{k_1} B \xrightarrow{k_2} C \xrightarrow{k_4} E \\ \qquad\;\; \downarrow k_3 \\ \qquad\;\; D \end{array}$	12.	$\begin{array}{l} A \xrightarrow{k_1} B \xrightarrow{k_2} C \xrightarrow{k_4} E \\ \qquad\;\; \downarrow k_3 \\ \qquad\;\; D \end{array}$

Instructions. In order to prevent too bulky expressions it is recommended to assign rate constants numeric values before deducing kinetic equations. Don't use the numbers with floating point. For example, it is recommended to assign $k_1 := 1/2$, and not $k_1 := 0.5$.

46. Solve (analytically) a problem about dissolving the salt in a pool. Pool's bottom is covered with a layer of caked salt. To wash the pool it was filled with water. Deduce an expression for the concentration of the salt as a function of time t and distance from the pool's bottom x. Apply mathematic approach of the Laplace transformation to solve the problem.
47. When analysing Lotka–Volterra equations (Sect. 3.4) it was assumed that food resources are unlimited. How would the dynamics of populations look if one assumes that the quantity of food decreases in time according to the regularities of: (a) zero-order reaction; (b) first-order reaction?
48. Apparently, a number of factors affecting the development of the populations of the predators and their preys in reality id much larger, than is considered by *Lotka–Volterra* equations. Assume that the hunters invaded this isolated system. Part of them hunts the prey and constantly "kills" some quantity of the preys with speed r_1. Others kill the predators with speed r_2. Investigate, how the interference of the hunters influences the dynamics of the development of both populations.
49. Consider one of the simplest models of the development of the population of the organisms, according to which birth rate is proportional to the population size P with a rate constant k_1, and death rate is proportional to the square of the population size with a rate k_2. Determine, how the population size changes for $k_1 = 1, k_2 = 0.0001$, initial population size is $P = 10$ individuals. What would be the overall population size after a very long time? First solve the problem analytically: obtain expressions for the dependence $P(t)$ and also expression for the size of the stable population $P(t)_{t\to\infty}$. Then perform numeric calculation. Compare results of analytic and numeric calculations.
50. Dynamics of development of the population of certain types on animals is described by the differential equation

$$\frac{dP}{dt} = k_1 P^2 - k_2 P,$$

and unlike the model in the previous problem, birth rate is proportional to the square of the current population size, and death rate is proportional to the population size. Let there are two independent populations in the moment in time $t = 0$, and each of them develops in its closed econiche. Initial size of the population P_0 equals 30, for the second one $P_0 = 70$. Constants of proportionality equal correspondingly 0.001 and 0.040 conventional units. Model dynamics of development for both populations in the range 0–20 conventional units of time. Compare results of numeric and analytic compare. What conclusions can be made based on the obtained results?

51. Modify the *Frank-Kamenetsyi* model for the oxidation of higher hydrocarbons (see Sect. 3.4) assuming that initial substance burns out. Perform corresponding calculations.
52. Self-oscillations of the surface concentrations are a rather widespread phenomenon in a heterogeneous catalysis. Following mechanism has been proposed for an oxidation reaction of hydrogen on the platinum catalyst:

    (1) $O_2 + 2[\,] \rightarrow 2[O]$,
    (2) $H_2 + 2[\,] \rightarrow 2[H]$,
    (3) $[O] + 2[H] \rightarrow H_2O + 3[O]$,
    (4) $H_2 + [O] \rightarrow H_2O + [\,]$,

    where [ ] is a surface vacancy. If one assumes that an activation energy of stages (3) and (4) depends on the degree of the coverage of the catalyst surface by the oxygen atoms $[O]$, e.g. E_{a3} and E_{a4} are described with the linearity ratios

    $$E_{a3} = E^0_{a3} + \alpha_3[O]\ ;\ \ E_{a4} = E^0_{a4} + \alpha_4[O],$$

    then solution for the set of kinetic equations for a given scheme can reveal the presence of the self-oscillations in a particular range of parameters' values.

    Make a set of differential equations, which would describe the time-change of surface concentrations for atomic oxygen $[O]$, atomic hydrogen $[H]$ and surface vacancies []. Let the surface concentrations of the molecular forms O_2, H_2, H_2O are constants. Set the physically acceptable values for the standard activation energies E^0_{a3}, E^0_{a4} and coefficients α_3, α_4, chose the values for the effective rate constants for all stages in the process in order to demonstrate a solution of the set, indicative of the presence of self-oscillations in a system. Present the necessary graphic information (kinetic curves, phase trajectories).
53. Apart from *Lotka–Volterra* equations there are many well-known models for populations' development. Model of coexistence of the competitive species (*Holling–Tenner*) is one of them. Speed of change of the population size of the preys dx/dt in this model is a sum of three values, first of which is a speed of reproduction in the absence of predators rx; second reflects the influence of the intraspecific competition for food and equals to $(-rx \times x/K)$. The third value reflects the influence of predators (under assumption that a predator kills a prey

only when being hungry) and equals $\omega yx/(D + x)$. Speed of change of the predators' population dy/dt is built under assumption that preys meet predators rarely. If J preys are needed to maintain life of one predator, then a population of x preys can supply food for x/J predators. In this case a set of equations corresponding to Holling-Tenner's model is as follows:

$$dx/dt = r(1 - x/K)x - \omega \cdot yx/(D + x),$$
$$dy/dt = s \cdot (1 - Jy/x)y,$$

where r, s, K, D, J – *const* (>0). Model dynamics of the evolution of the population size under the following values of parameters: $r = 1$, $K = 7$, $\omega = 1$, $D = 1$, $s = 0{,}2$, $J = 0.5$, and also under the following initial conditions: $x_0 = 1$, $y_0 = 1$. Present the time-dependence curves for a number of preys and predators, and build a phase portrait of the system.

54. Mechanism of the reaction is given below (see Table):

(1)	$A + X \rightarrow 2X$	$k_1 = 0.4$ L mol^{-1} s^{-1}
(2)	$2X \rightarrow A + X$	$k_{-1} = 0.35$ L mol^{-1} s^{-1}
(3)	$X + Y \rightarrow 2Y$	$k_2 = 0.15$ L mol^{-1} s^{-1}
(4)	$2Y \rightarrow X + Y$	$k_{-2} = 0.28$ L mol^{-1} s^{-1}
(5)	$Y + Z \rightarrow 2Z$	$k_3 = 0.15$ L mol^{-1} s^{-1}
(6)	$2Z \rightarrow Y + Z$	$k_{-3} = 0.20$ L mol^{-1} s^{-1}
(7)	$X + Z \rightarrow X + B$	$k_4 = 0.1$ L mol^{-1} s^{-1}
(8)	$X + B \rightarrow X + Z$	$k_{-4} = 0.13$ L mol^{-1} s^{-1}

Initial concentrations of the reactants are (mol L^{-1}): $A_0 = 1$; $X_0 = 0.5$; $Y_0 = 0.3$; $Z_0 = 1$, $B_0 = 0.5$. Calculate and build all possible kinetic curves for the reaction's participants, as well as phase trajectories. Investigate the influence of the initial concentration of reactant B on the character of the oscillations of other participants' concentrations.

55. During a spontaneous cooling of a matter, heated earlier up to temperature T_0, down to temperature of the environment T_s speed of temperature change equals to the difference between its current temperature T in a moment in time t and the temperature of the environment with a constant of proportionality k. Deduce an expression that describes a time-dependence of a cooling matter's temperature. Let cooling proceeds from $t_0 = 100°C$ to $t_s = 20°C$ with a constant $k = 0.01$ s^{-1}. Present kinetic curves of cooling based on both numeric and analytic solutions.

56. In a consecutive reaction $A \xrightarrow{k_1} B \xrightarrow{k_2} C$ change of intermediate's B concentration is described by expression

$$r = \frac{dC_B(t)}{dt} = k_1 C_A(t) - k_2 C_B(t) = k_{0,1} \cdot e^{-E_1/RT} \cdot C_A(t) - k_{0,2} \cdot e^{-E_2/RT} \cdot C_B(t).$$

Measurements under four different concentrations of A and B and different temperatures gave the following results (see Table):

C_A, mol L^{-1}	C_B, mol L^{-1}	T (K)	r (mol L^{-1} s^{-1})
1	1	300	$+1.966 \times 10^{-3}$
1	2	310	−0.0975
2	1	320	+0.2435
1	1	330	−0.06219

Calculate the parameters of Arrhenius equation $k_{0,1}$, E_1, $k_{0,2}$, E_2.

57. Temperature-dependence of a rate constant is described by Arrhenius equation. Define the temperature, under which this dependence has a bend. Illustrate the results of analytic calculations with a correspondent plot.
58. Temperature-dependences of the rate constants of two first-order reactions are described by the equations

$$k_1 = 2.54 \times 10^{15} \exp\left(-\frac{31600}{RT}\right), \quad k_2 = 1.53 \times 10^{17} \exp\left(-\frac{42000}{RT}\right).$$

Under what temperature should one expect the rates of two reactions to become equal? Activation energies are in cal mol^{-1}.

59. Let substance X transforms into substances A and B through two parallel reactions with the rate constants

$$k_A = 10^{15} \exp\left(-\frac{30000}{RT}\right), \quad k_B = 10^{15} \exp\left(-\frac{20000}{RT}\right).$$

Determine: (a) under what temperature are both final products formed with equal rates? (b) under what temperature does substance A form 10 times faster than substance B; (c) under what temperature is a rate of A formation equal to 0.1 of the B's formation rate? Activation energies are in cal mol^{-1}.

60. First-order reaction $A \rightarrow$ *Products* with initial reactant's concentration $A_0 = 2$ mol L^{-1} starts at $T_0 = 293$ K. At the same time the temperature of the reactive mixture is being changed according to $T(t) = T_0 + bt$, where parameter b equals to 0.5 K s^{-1}. Temperature-dependence of the rate constant follows Arrhenius equation with a preexponential factor 4.1025×10^7 s^{-1} and activation energy $E_a = 50$ kJ mol^{-1}. Calculate numerically kinetic curve for reactant A. Compare obtained curve with a curve for $A(t)$ under isothermal conditions at 293 K.
61. Solve previous problem under a condition that a temperature of reactive mixture changes in time according to $T = T_0 + t\sin(t)$ (time is given in seconds).
62. Phenomenon of a *thermal autocatalysis* could be one of the reasons for an unstable operation of the reactor. The simplest model to give thermokinetic oscillations in a well-stirred continuous reactor is

$$B_0 \rightarrow B \rightarrow P + Q.$$

Substance B gets into continuous stirred-tank reactor, in which a monomolecular exothermal decomposition reaction proceeds. The heat is eliminated thought the wall of reactor. Reaction's kinetics is described by two differential equations relative to B's concentration and the temperature T inside reactor:

$$\frac{d[B]}{dt} = k([B_0] - [B]) - A\exp\left(-\frac{E_a}{RT}\right)\cdot[B];$$
$$\frac{dT}{dt} = \frac{1}{c_p\rho}\cdot\left[QA\exp\left(-\frac{E_a}{RT}\right)\cdot[B] + h(T_0 - T)\right],$$

where $[B_O]$ is a coerced concentration on the reactor's entrance; T_O is a temperature of reactor's wall; k is a coefficient of the speed with which reactive mixture is renewed in reactor; h is a coefficient of the heat exchange rate; c_p is an isobaric heat capacity; ρ is density; E_a and A are activation energy and preexponential factor correspondingly. In such system warming up accelerates the reaction, which leads to the exhaustion of substance B in reactor and the decrease in its rate. Further concentration of B increases, and cycle repeats. Try to choose such values for the parameters, so that solution of the set of differential equations would show the periodic character of the time-dependence of the temperature of reactive mixture.

63. Kinetic data for a one-way reaction

$$C_4H_4O_4Br_2 \xrightarrow{k} C_4H_3O_4Br + HBr,$$

are given in the table:

t (min)	0	10	20	30	40	50	60
C (mol L^{-1})	0.511	0.377	0.274	0.202	0.148	0.108	0.080

where C is a current concentration of $C_4H_4O_4Br_2$. Calculate the rate constant assuming reaction has a first order for the reactant.

64. Calculate the initial rate of the second-order reaction $A + B \rightarrow X$, build kinetic curves for participating substances and determine the rate constant. Initial concentrations are $A_0 = 0.00323$ and $B_0 = 0.002508$ mol L^{-1}, and the concentration of product X changes in time as follows:

t (min)	3.65	7.65	15.05	26	47.60	90.60
$X\cdot10^{-5}$ (mol L^{-1})	56.0	95.3	142.0	180.0	211.7	236.7

65. Determine an order and a rate constant of the decomposition reaction of the malonic acid vapour into methane and carbon dioxide based on the change of the overall pressure of the mixture:

t (min)	0	10	20	30	50	75	100	150	∞
p (kPa)	24.7	29.5	33.9	37.5	43.7	49.5	53.7	58.9	64.5

66. Reaction profile was investigated through the change in the concentration of the reactant A under two temperatures T_1 and T_2. Determine the reaction's

order for a substance A, rate constants under T_1 and T_2 and an activation energy based on the following data:

t (min)	0	10	15	20	25	30
C_A (300°C), mol L^{-1}	1.000	0.898	0.855	0.806	0.769	0.735
C_A (600°C), mol L^{-1}	1.000	0.746	0.658	0.592	0.541	0.490

67. Acetone vapour decompose under 777 K according to scheme:

$$CH_3COCH_3 \longrightarrow C_2H_4 + H_2 + CO.$$

It is easy to follow the reaction path by measuring an overall pressure P in a static reactor. Experimental data is given in the table (t – time from the reaction start):

t (°C)	0	390	777	1,195	3,255
P (mmHg)	312	408	488	562	779

Determine the rate constant and the half-life time of acetone under given temperature.

68. Thermal decomposition of cyclobutane under 383°C is a parallel first-order reaction:

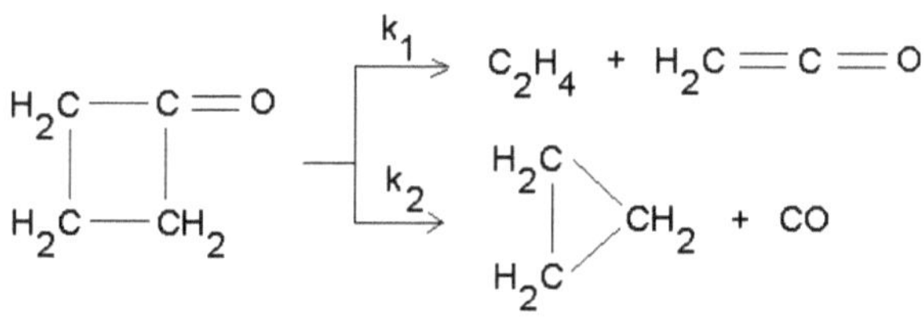

The following kinetic data was obtained under the initial concentration of C_4H_6O equal to 6.5×10^{-5} mol L^{-1}:

t (min)	0.5	1.0	3.0	6.0
$[C_2H_4]\cdot 10^5$ (mol L^{-1})	0.31	0.68	1.53	2.63
$[C_3H_6]\cdot 10^7$ (mol L^{-1})	0.21	0.47	1.24	2.20

Determine the rate constants k_1 and k_2.

69. Reaction $A + B + C \longrightarrow Products$ was investigated by changing initial rates r_0 for different combinations of the initial reactants' concentrations C_{i_0}. The following results were obtained:

C_{A_0}, mol L^{-1}	0.20	0.60	0.20	0.60
C_{B_0}, mol L^{-1}	0.30	030	0.90	0.30
C_{C_0}, mol L^{-1}	0.15	0.15	0.15	0.45
r_0, mol $(L \times s)^{-1}$	0.60×10^{-2}	1.81×10^{-2}	5.38×10^{-2}	1.81×10^{-2}

Determine the partial reaction orders and the rate constant.

70. Recombination reaction of iodine atoms in gas phase in the presence of argon has been investigated. The following rates of molecular iodine formation

according to reaction $I + I + Ar \longrightarrow I_2 + Ar$ have been found for the different concentrations of I and Ar atoms:

$[I]_0$, (mol L^{-1})		1.0×10^{-5}	2.0×10^{-5}	4.0×10^{-5}	5.0×10^{-5}
$(d[I_2]/dt)_0$, mol $L^{-1}s^{-1}$	(a)	8.70×10^{-4}	3.48×10^{-3}	1.39×10^{-2}	3.13×10^{-4}
	(b)	4.35×10^{-3}	1.74×10^{-2}	6.96×10^{-2}	1.57×10^{-1}
	(c)	8.69×10^{-3}	3.47×10^{-2}	3.48×10^{-1}	3.13×10^{-1}

Argon concentrations were:
(a) 1.00×10^{-3} mol L^{-1}; (b) 5.00×10^{-3} mol L^{-1}; (c) 10.00×10^{-3} mol L^{-1}.

Determine the reaction orders for iodine and argon atoms, as well as the rate constant.

71. Let's consider a hydrolysis reaction of a methyl halogenide CH_3X in an alkaline aqueous solution. The process could be described by the following scheme:

$$\underset{(a-x)}{CH_3Hal} + H_2O \xrightarrow{k_1} CH_3OH + H^+ + X^-,$$

$$\underset{(a-x)}{CH_3Hal} + \underset{(b-x)}{OH^-} \xrightarrow{k_2} CH_3OH + X^-.$$

If x is a change in the concentration of halogenide in time t, then current rate of reaction is described by the differential equation

$$\frac{dx}{dt} = k_1(a - x) + k_2(a - x)(b - x).$$

Prove that part ω of initial substance's molecules, which had reacted by the moment t, equals to

$$\omega = \frac{e^{\alpha t} - 1}{e^{\alpha t} - \beta},$$

where $\alpha = k_1 + k_2(b - a)$; $\beta = k_2 a/(k_1 + k_2 b)$; a and b are initial concentrations of CH_3Hal and OH^- correspondingly.

Find k_1 and k_2 for a hydrolysis reaction of methyl halogenides based on the following data:

1. CH_3Cl	t, (min)	0	2	30	65	109	214	1,160
$a = 35.3$; $b = 102.5$	$\omega \cdot 100$	0	0.6	8.8	18.0	27.7	45.8	93.0
2. CH_3Br	t, (min)	0	200	500	900	1,400	2,000	3,000
$a = 57.2$; $b = 105.0$	$\omega \cdot 100$	0	15.5	32.9	49.1	62.7	72.5	84.4

Concentrations a and b are given in mmol L^{-1}.

72. Reaction of triphenylmethyl chloride with methanol is an example of a third-order reaction:

$$(C_6H_5)_3CCl + CH_3OH \rightarrow (C_6H_5)_3COCH_3 + HCl.$$

There is a discrepancy between the process' kinetics and its stoichiometry: one more molecule of ethanol participates in an elementary act as a catalyst. If we designate A and B as current concentrations of triphenylmethyl chloride and methanol correspondingly, and x as a current concentration of one of the products, then kinetic equation for a reaction could be given as follows

$$dx(t)/dt = k(C_{A_0} - x(t))(C_{B_0} - x(t))^2$$

Determine a rate constant k based on the following experimental data:

t (min)	168	174	418	426	444	1,150	1,440	1,510	1,660
$x \cdot 10^2$(mol L^{-1})	0.91	1.10	1.81	1.89	2.07	3.18	3.34	3.45	3.54

Initial concentrations of the reactants were $C_{A_0} = 0.106\,\text{mol}\,\text{L}^{-1}$, $C_{B_0} = 0.054\,\text{mol}\,\text{L}^{-1}$

73. Reaction between C_3H_6 and HCl in the presence of catalyst (phosphoric acid) can be formally considered as a reverse second-order reaction

$$C_3H_6 + HCl \underset{k_2}{\overset{k_1}{\rightleftarrows}} C_3H_7Cl.$$

Data on the accumulation of the reaction's product in time (x is a current partial pressure of C_3H_7Cl) is given below:

t (min)	3	7	10	20	30	40	60	100	140	170
x (mmHg)	3	13	30	60	78	87	97	110	115	116

Initial partial pressures of the reactants were equal to 200 mmHg. When the reaction started, product was not present in the reactive mixture. Determine the constants k_1 and k_2.

74. Solve the inverse kinetic problem of calculating the rate constants for each stage of the reaction using given kinetic data for the current concentrations (mol L^{-1}) of the participants of a multistage reaction in time (s), as well as the kinetic scheme of the process.

Variant 1.

Kinetic scheme:

(I) $A + B \xrightarrow{k_1} 2C,$

(II) $A + B \xrightarrow{k_2} 2D,$

(III) $C + D \xrightarrow{k_3} 2E.$

Experimental data:

t	0	20	46	66	82	108	254
C_A	0.100	0.081	0.064	0.054	0.048	0.040	0.017
C_B	0.150	0.131	0.114	0.104	0.098	0.090	0.067
C_C	0.000	0.026	0.050	0.063	0.071	0.082	0.108
C_D	0.000	0.012	0.022	0.028	0.032	0.036	0.046
C_E	0.000	0.000	0.000	0.001	0.001	0.002	0.011
t	300	402	464	588	690	776	1,000
C_A	0.014	0.009	0.007	0.004	0.003	0.002	0.001
C_B	0.064	0.058	0.057	0.054	0.053	0.052	0.051
C_C	0.112	0.115	0.115	0.114	0.113	0.111	0.107
C_D	0.047	0.046	0.046	0.042	0.040	0.037	0.032
C_E	0.014	0.022	0.027	0.036	0.042	0.048	0.060

Variant 2.
Kinetic scheme:
(I) $A + B \xrightarrow{k_1} B + C$
(II) $A + B + C \underset{k_3}{\overset{k_2}{\rightleftarrows}} 2D$
Experimental data:

t	0	0.1	0.2	0.3	0.4	0.5	0.6
C_A	2	1.441	1.094	0.867	0.707	0.588	0.500
C_B	2	1.728	1.568	1.474	1.416	1.381	1.361
C_C	0	0.015	0.043	0.083	0.128	0.175	0.222
C_D	0	0.544	0.863	1.053	1.168	1.237	1.278
t	0.8	1	1.6	2	3	4	5
C_A	0.377	0.296	0.169	0.126	0.072	0.048	0.035
C_B	1.344	1.345	1.383	1.415	1.486	1.543	1.587
C_C	0.312	0.394	0.597	0.703	0.900	1.037	1.139
C_D	1.312	1.309	1.234	1.171	1.028	0.915	0.826

Variant 3.
Kinetic scheme:
(I, II) $A \underset{k_2}{\overset{k_1}{\rightleftarrows}} 2B + C$
(III, IV) $B \underset{k_4}{\overset{k_3}{\rightleftarrows}} D$
Initial concentrations of the reactants (mol L^{-1}): $C_{A_0} = 1; C_{B_0} = 0$.
Experimental data:

t	0	10	20	30	40	50	70
C_C	0	0.398	0.634	0.772	0.853	0.902	0.952
C_D	0	0.068	0.217	0.395	0.574	0.74	1.019
t	90	100	120	140	160	180	200
C_C	0.974	0.98	0.988	0.992	0.995	0.996	0.997
C_D	1.227	1.308	1.433	1.522	1.584	1.627	1.656

Variant 4.
Kinetic scheme:

(I) $A \xrightarrow{k_1} C$
(II) $B \xrightarrow{k_2} 2C + D$

(III) $A + B \xrightarrow{k_3} E$
(IV) $C + D \xrightarrow{k_4} E$
Experimental data:

t	0.0	0.6	4.8	10.2	21.0	25.2	30.0
C_A	0.5	0.464	0.281	0.153	0.049	0.032	0.019
C_B	0.600	0.566	0.379	0.237	0.096	0.070	0.046
C_C	0.000	0.086	0.535	0.856	1.052	1.069	1.065
C_D	0.000	0.031	0.169	0.249	0.211	0.186	0.147
t	35.4	40.2	42.0	47.4	50.4	55.8	60.0
C_A	0.011	0.007	0.005	0.003	0.002	0.001	0.000
C_B	0.031	0.020	0.019	0.011	0.010	0.006	0.004
C_C	1.06	1.045	1.045	1.029	1.027	1.013	1.010
C_D	0.119	0.089	0.087	0.061	0.057	0.044	0.029

Variant 5.
Kinetic scheme:
(I) $A \xrightarrow{k_1} B + C$
(II) $C \xrightarrow{k_2} D$
(III) $B + D \xrightarrow{k_3} E$
Experimental data:

t	0	2	4	7	12	15	20
C_A	2.000	1.289	0.833	0.430	0.143	0.076	0.026
C_B	0	0.711	1.166	1.567	1.843	1.900	1.925
C_C	0	0.691	1.108	1.416	1.527	1.500	1.372
C_D	0	0.010	0.069	0.150	0.316	0.410	0.554
C_E	0	0	0.001	0.002	0.012	0.025	0.050
t	30	50	60	70	96	149	200
C_A	0.002	0.000	0.000	0.000	0.000	0.000	0.000
C_B	1.882	1.702	1.618	1.526	1.315	0.994	0.791
C_C	1.103	0.673	0.548	0.433	0.234	0.066	0.021
C_D	0.779	1.027	1.070	1.095	1.081	0.924	0.769
C_E	0.115	0.298	0.380	0.470	0.685	1.010	1.212

Variant 6.
Kinetic scheme:
(I, II) $A + 2B \underset{k_2}{\overset{k_1}{\rightleftarrows}} 2C$
(III) $C \xrightarrow{k_3} D$
Initial concentrations of the reactants (mol L^{-1}): $C_{A_0} = C_{B_0} = 1$.
Experimental data:

t	0	5	10	15	20	25	30
C_C	0	0.23	0.32	0.375	0.403	0.413	0.414
C_D	0	0.011	0.037	0.07	0.106	0.144	0.183
t	35	40	45	50	60	80	100
C_C	0.407	0.396	0.382	0.367	0.334	0.267	0.21
C_D	0.221	0.259	0.296	0.331	0.396	0.509	0.598

Variant 7.

Kinetic scheme:

(I) $A + B \xrightarrow{k_1} C$
(II) $A + B \xrightarrow{k_2} D$
(III) $D \xrightarrow{k_3} 2E$
(IV) $C \xrightarrow{k_4} E$

Experimental data:

t	0	0.12	0.2	0.4	0.6	0.84	1.04
C_A	1.00	0.63	0.51	0.36	0.28	0.23	0.20
C_B	0.90	0.53	0.41	0.26	0.18	0.13	0.10
C_C	0.00	0.04	0.06	0.07	0.08	0.09	0.09
C_D	0.00	0.32	0.41	0.51	0.53	0.53	0.51
t	1.8	2.00	2.20	2.44	3.04	3.52	4.00
C_A	0.15	0.14	0.14	0.13	0.12	0.11	0.11
C_B	0.05	0.04	0.04	0.03	0.02	0.01	0.01
C_C	0.09	0.09	0.09	0.09	0.09	0.08	0.08
C_D	0.41	0.39	0.36	0.33	0.27	0.23	0.19

Variant 8.

Kinetic scheme:

(I) $A \xrightarrow{k_1} C$
(II) $B \xrightarrow{k_2} C$
(III) $A + B \xrightarrow{k_3} D$
(IV) $C + D \xrightarrow{k_4} E$

Experimental data:

t	0.000	0.038	0.054	0.129	0.210	0.360	0.540
C_A	0.230	0.158	0.111	0.084	0.047	0.024	0.015
C_B	0.120	0.056	0.033	0.014	0.006	0.002	0.002
C_C	0.000	0.025	0.034	0.060	0.070	0.092	0.088
C_D	0.000	0.061	0.082	0.097	0.102	0.096	0.091
t	0.780	0.900	1.050	1.350	2.040	2.610	3.000
C_A	0.006	0.004	0.002	0.001	0.000	0.000	0.000
C_B	0.001	0.001	0.000	0.000	0.000	0.000	0.000
C_C	0.092	0.088	0.081	0.077	0.064	0.056	0.051
C_D	0.078	0.076	0.069	0.062	0.045	0.039	0.033

75. A mechanism of the decomposition of aluminium chloride $AlCl_3$ in the presence of molecular hydrogen has been suggested. A list of stages and their kinetic parameters (rate constants of direct and reverse reactions k_f and k_r) are given in a Table:

No	*Elementary stage*	k_f (1,300 K, 0.1 bar)	k_r (1,300 K, 0.1 bar)
1	$AlCl_3 \leftrightarrow AlCl_2 + Cl$	Unknown	3.92×10^{19} cm^3 mol^{-1} s^{-1}
2	$AlCl_2 \leftrightarrow AlCl + Cl$	1.54×10^{6} s^{-1}	5.86×10^{17} cm^3 mol^{-1} s^{-1}
3	$AlHCl \leftrightarrow AlCl + H$	5.87×10^{9} s^{-1}	9.78×10^{15} cm^3 mol^{-1} s^{-1}
4	$H_2 \leftrightarrow 2H$	1.43×10^{-8} s^{-1}	1.13×10^{9} cm^3 mol^{-1} s^{-1}

(continued)

5	$HCl \leftrightarrow H + Cl$	1.64×10^{-7} s^{-1}	3.91×10^{10} cm^3 mol^{-1} s^{-1}
6	$AlHCl_2 \leftrightarrow AlCl + HCl$	1.77×10^{5} s^{-1}	4.64×10^{12} cm^3 mol^{-1} s^{-1}
7	$AlCl_3 + H \leftrightarrow AlCl_2 + HCl$	3.29×10^{10} cm^3 mol^{-1} s^{-1}	1.28×10^{11} cm^3 mol^{-1} s^{-1}
8	$AlCl_2 + Cl_2 \leftrightarrow AlCl_3 + Cl$	2.20×10^{12} cm^3 mol^{-1} s^{-1}	9.45×10^{3} cm^3 mol^{-1} s^{-1}
9	$AlHCl_2 + H \leftrightarrow AlCl_2 + H_2$	3.42×10^{13} cm^3 mol^{-1} s^{-1}	7.13×10^{9} cm^3 mol^{-1} s^{-1}
10	$AlHCl_2 + H \leftrightarrow AlHCl + HCl$	1.26×10^{10} cm^3 mol^{-1} s^{-1}	1.99×10^{11} cm^3 mol^{-1} s^{-1}
11	$AlHCl_2 + Cl \leftrightarrow AlCl_2 + HCl$	7.50×10^{13} cm^3 mol^{-1} s^{-1}	5.18×10^{9} cm^3 mol^{-1} s^{-1}
12	$AlHCl + Cl_2 \leftrightarrow AlHCl_2 + Cl$	2.20×10^{12} cm^3 mol^{-1} s^{-1}	2.33×10^{3} cm^3 mol^{-1} s^{-1}
13	$AlCl + HCl \leftrightarrow AlCl_2 + H$	3.07×10^{8} cm^3 mol^{-1} s^{-1}	1.92×10^{14} cm^3 mol^{-1} s^{-1}
14	$AlCl + Cl_2 \leftrightarrow AlCl_2 + Cl$	5.18×10^{13} cm^3 mol^{-1} s^{-1}	5.42×10^{11} cm^3 mol^{-1} s^{-1}
15	$AlHCl + H \leftrightarrow AlCl + H_2$	1.00×10^{13} cm^3 mol^{-1} s^{-1}	2.11×10^{2} cm^3 mol^{-1} s^{-1}
16	$AlHCl + Cl \leftrightarrow AlCl + HCl$	1.00×10^{13} cm^3 mol^{-1} s^{-1}	7.00×10^{1} cm^3 mol^{-1} s^{-1}
17	$Cl_2 + H \leftrightarrow Cl + HCl$	1.08×10^{14} cm^3 mol^{-1} s^{-1}	1.81×10^{6} cm^3 mol^{-1} s^{-1}
18	$H_2 + Cl \leftrightarrow HCl + H$	6.94×10^{12} cm^3 mol^{-1} s^{-1}	2.30×10^{12} cm^3 mol^{-1} s^{-1}

Reaction pathway was investigated by measuring the current concentration C of aluminium (III) chloride. At the initial moment in time composition of the reactive mixture was 0.997 mol rates of hydrogen and 0.003 mol rates of $AlCl_3$ under the overall pressure in the system equal to 0.1 bar. Conditions of the reaction are isochoric-isothermal (T = 1,300 K). The following experimental results were obtained:

t (s)	0.02	0.06	0.10	0.22	0.32	0.50	0.64	0.74	0.88	0.98
$C \cdot 10^9$ (mol cm^{-3})	2.548	2.398	2.308	2.172	2.045	1.944	1.892	1.870	1.849	1.831
t, s	1.12	1.24	1.34	1.40	1.48	1.62	1.74	1.86	1.90	2.00
$C \cdot 10^9$ (mol cm^{-3})	1.822	1.825	1.814	1.813	1.806	1.812	1.803	1.802	1.794	1.800

Calculate the unknown rate constant using these data.

76. Reaction between CF_3CFHO_2 and NO was investigated. Peroxy radical CF_3CFHO_2 is formed in troposphere as a result of reaction between CF_3CFH_2, OH-radicals and molecular oxygen. By-turn a fluorine-containing substance CF_3CFH_2 is a main component of a marketable product HFC-134a, which is a substituent for Freon, used in automobile systems of climate-control. It is important to know how intensively HFC-134a decomposes in a troposphere, because at higher layers of the atmosphere it can participate in the processes, contributing to exhaustion of the ozone layer of Earth.

Kinetic model, suggested for chemical transformations in troposphere involving HFC-134a, includes a set of the following elementary reactions:

Stage	*Elementary reaction*	*Rate constant*
1	$Cl + CF_3CFH_2 \rightarrow CF_3CFH + HCl$	2.7×10^{-15} cm^{3}molecules^{-1}s^{-1}
2	$CF_3CFH + O_2 \rightarrow CF_3CFHO_2$	2.1×10^{-12} cm^{3}molecules^{-1}s^{-1}
3	$2CF_3CFHO_2 \rightarrow 2CF_3CFHO + O_2$	5.0×10^{-12} cm^{3}molecules^{-1}s^{-1}
4	$CF_3CFHO_2 + NO \rightarrow CF_3CFHO + NO_2$	Unknown
5	$CF_3CFHO \rightarrow CF_3 + HCOF$	5.0×10^{4} s^{-1}
6	$CF_3 + O_2 \rightarrow CF_3O_2$	6.0×10^{-12} cm^{3}molecules^{-1}s^{-1}
7	$CF_3O_2 + NO \rightarrow CF_3O + NO_2$	1.6×10^{-11} cm^{3}molecules^{-1}s^{-1}
8	$2CF_3O_2 \rightarrow 2CF_3O + O_2$	1.8×10^{-12} cm^{3}molecules^{-1}s^{-1}
9	$CF_3O + NO \rightarrow CF_2O + FNO$	4.7×10^{-11} cm^{3}molecules^{-1}s^{-1}
10	$CF_3O_2 + CF_3O \rightarrow CF_3OOOCF_3$	1.4×10^{-11} cm^{3}molecules^{-1}s^{-1}

An experiment was performed in order to verify the model. In this experiment decomposition of CF_3CFH_2 was initiated by atomic chlorine, formed during Cl_2 photolysis. Further produced radical CF_3CFH reacts with O_2 to form peroxy radical CF_3CFHO_2, which was a reactant in stage 4 together with oxide NO. The product of this stage CF_3CFHO participated in a further chain of transformations. Reaction path was investigated by recording relative concentration X of a final product – nitrogen dioxide NO_2. Results of the experiment are given in a Table:

t, s	9.127×10^{-5}	1.838×10^{-4}	5.421×10^{-4}	9.743×10^{-4}	1.405×10^{-3}	2.501×10^{-3}	3.498×10^{-3}
X	2.057×10^{-2}	6.294×10^{-2}	2.173×10^{-2}	9.070×10^{-2}	0.1415	0.3860	0.4742
t, s	4.861×10^{-3}	5.604×10^{-3}	6.171×10^{-3}	6.916×10^{-3}	7.300×10^{-3}	8.181×10^{-3}	8.614×10^{-3}
X	0.6630	0.6544	0.7222	0.7354	0.7449	0.7812	0.8635
t, s	1.004×10^{-2}	1.188×10^{-2}	1.540×10^{-2}	1.678×10^{-2}	1.798×10^{-2}	1.915×10^{-2}	2.918×10^{-2}
X	0.8900	0.9128	0.9475	0.9655	1.0030	1.0000	1.0000

The following initial concentrations of the reactants (molecules cm^{-3}) correspond to the given data: $[Cl]_0 = 4.32 \times 10^{13}$; $[CF_3CFH_2]_0 = 2.67 \times 10^{17}$; $[NO]_0 = 7.26 \times 10^{13}$; $[O_2]_0 = 2.3 \times 10^{16}$. It is assumed that time, needed to form the given quantity of atomic chlorine, is negligibly small comparing to the time of experiment. Calculate rate constant k_4 based on the experimental data above.

Instructions. Relative concentration of nitrogen dioxide X is related to the absolute value of its current concentration as follows:

$$X = 1 - [NO]/[NO]_{max}.$$

77. Investigated reaction between CH_2ClOCO_2 proceeds parallel to the monomolecular decomposition of CH_2ClO according to the total scheme:

$$CH_2ClO + O_2 \rightarrow HC(O)Cl + HO_2,$$

$$CH_2ClO \rightarrow HCO + HCl.$$

Initial chlorine-containing radical CH_2ClO was formed as a result of methyl chloride CH_3Cl oxidation. Participants of the process were involved in the following chain of chemical transformations (Table).

The process was initiated by atomic chlorine, formed as a result of Cl_2 photo dissociation. Dimensionless concentration of HCl was measured in the experiment in different moments in time.

Reaction	Rate constant
(1) $Cl + CH_3Cl \rightarrow CH_2Cl + HCl$	5.6×10^{-13} cm^3 molecules^{-1} s^{-1}
(2) $CH_2Cl + O_2 \rightarrow CH_2ClO_2$	2.05×10^{-12} cm^3 molecules^{-1} s^{-1}
(3) $2CH_2ClO_2 \rightarrow 2CH_2ClO + O_2$	3.39×10^{-12} cm^3 molecules^{-1} s^{-1}
(4) $CH_2ClO + O_2 \rightarrow HCOCl + HO_2$	Unknown
(5) $CH_2ClO \rightarrow HCO + HCl$	Unknown
(6) $Cl + CH_2ClO_2 \rightarrow CH_2ClO + ClO$	7.7×10^{-11} cm^3 molecules^{-1} s^{-1}
(7) $Cl + CH_2ClO_2 \rightarrow HCl + CHClO_2$	7.4×10^{-11} cm^3 molecules^{-1} s^{-1}

(*continued*)

Reaction	Rate constant
(8) $ClO + CH_2ClO_2 \rightarrow CH_2ClO + ClOO$	1.7×10^{-12} cm^3 molecules^{-1} s^{-1}
(9) $2HO_2 \rightarrow H_2O_2 + O_2$	1.6×10^{-12} cm^3 molecules^{-1} s^{-1}
(10) $HCO + O_2 \rightarrow CO + HO_2$	6.6×10^{-12} cm^3 molecules^{-1} s^{-1}
(11) $Cl + HO_2 \rightarrow HCl + O_2$	3.1×10^{-11} cm^3 molecules^{-1} s^{-1}
(12) $Cl + HO_2 \rightarrow OH + ClO$	9.4×10^{-12} cm^3 molecules^{-1} s^{-1}
(13) $CH_2ClO_2 + HO_2 \rightarrow CH_2ClOOH + O_2$	2.4×10^{-12} cm^3 molecules^{-1} s^{-1}
(14) $CH_2ClO_2 + HO_2 \rightarrow CH_2ClO + HO_3$	2.4×10^{-12} cm^3 molecules^{-1} s^{-1}
(15) $CH_2Cl + Cl_2 \rightarrow CH_2Cl_2 + Cl$	3.0×10^{-13} cm^3 molecules^{-1} s^{-1}
(16) $OH + CH_3Cl \rightarrow CH_2Cl + H_2O$	4.1×10^{-14} cm^3 molecules^{-1} s^{-1}
(17) $OH + Cl_2 \rightarrow HOCl + Cl$	7.4×10^{-14} cm^3 molecules^{-1} s^{-1}
(18) $HCO + Cl_2 \rightarrow HCOCl + Cl$	5.6×10^{-12} cm^3 molecules^{-1} s^{-1}

The following experimental data was obtained as a result:

$t\,10^3$, s	0.385	0.688	0.748	0.990	1.172	1.476	1.657	3.477	4.265
[NO_2]	0.401	0.406	0.445	0.465	0.480	0.482	0.516	0.593	0.634
$t\,10^3$, s	5.600	8.210	8.877	9.909	10.395	11.062	12.640	13.369	14.401
[NO_2]	0.698	0.755	0.784	0.787	0.809	0.821	0.856	0.861	0.875
$t\,10^3$, s	15.919	16.586	17.619	19.561	21.444	23.326	25.876	27.810	29.823
[NO_2]	0.883	0.910	0.905	0.957	0.932	0.939	0.991	0.996	1.001

Determine rate constants k_4 and k_5, if the reactants' concentrations in the beginning of the reaction were (molecules cm^{-3}): $[Cl]_0 = 2.00 \times 10^{13}$; $[Cl_2]_0 = 1.66 \times 10^{17}$; $[CH_3Cl]_0 = 2.76 \times 10^{17}$; $[O_2] = 6.52 \times 10^{17}$.

78. Lead gets inside the human body from the environment through respiratory apparatus, as well as with food and water. From lungs and digestive system lead gets into blood, and then it quickly spreads into liver and kidneys from there. Moreover, other soft tissues and bones absorb it. Lead is removed from the body with urine, sweat and via accumulation in hair and nails. All this could be illustrated by the following kinetic scheme:

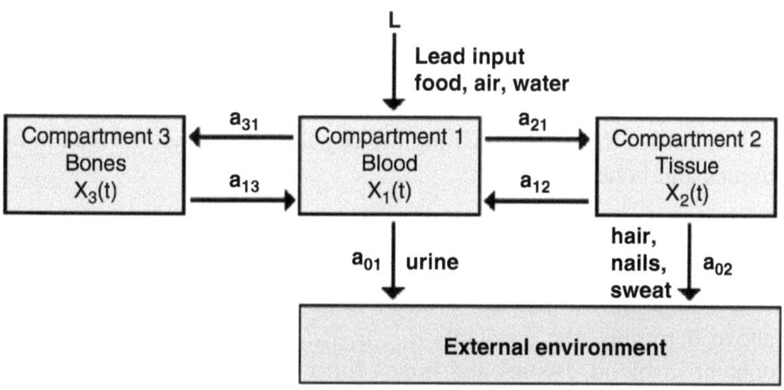

Based on this scheme one can deduce a mathematic model, describing dynamics of lead transfer in a living body. Let *L(t)* is a speed of lead input into the body with air, water and food (*microgram day*$^{-1}$); $x_1(t)$ is a time-dependence of lead quantity in blood (*microgram day*$^{-1}$); $x_2(t)$ is a function, describing the change in lead quality in tissues (*microgram day*$^{-1}$); $x_3(t)$ is a function, describing the change in lead quality in bones (microgram day^{-1}). Mathematic model can be given as a set of equations:

$$dx_1(t)/dt = L(t) - a_{21}x_1(t) - a_{31}x_1(t) - a_{01}x_1(t) - a_{02}x_2(t) + a_{12}x_2(t) + a_{13}x_3(t);$$

$$dx_2(t)/dt = a_{21}x_1(t) - a_{12}x_2(t) - a_{02}x_2(t);$$

$$dx_3(t)/dt = a_{31}x_1(t) - a_{13}x_3(t).$$

Here a_{01}, a_{02}, a_{12}, a_{13}, a_{21}, a_{31} are constant coefficients, referring to the rate constants of the separate stage in lead transfer. Each constant has a dimension of days^{-1}. Let's assume that a citizen of an ecologically pure area came for a 60 days visit to an industrial city. Here he was consuming 49 *micrograms* of lead daily with water, air and food. During his stay in the city a citizen underwent thorough medical examination every 10 days. Lead content in his blood, tissues and bones has been measured each time. After leaving a city a citizen was examined for the next 2 month by the local GP. Summary of the analysis results are given in a Table.

Time (days)	Lead content (μg)		
	In blood	In tissues	In bones
10	403.0	21.4	8.3
20	683.2	69.7	29.8
30	877.1	128.5	60.5
40	1010.4	188.4	97.4
50	1101.6	244.2	138.7
60	1160.8	293.6	182.9
70	802.3	314.3	220.7
80	550.0	301.2	246.7
90	374.6	271.1	264.4
100	253.2	234.3	276.4
110	169.8	197.0	284.5
120	112.6	162.1	289.8

It is known that before visiting the city this citizen had no lead in his blood, tissues and bones at all, e.g.:

$$x_1(0) = x_2(0) = x_3(0) = 0.$$

It is also known that $a_{01} = 0.021$ days^{-1}, $a_{02} = 0.016$ days^{-1}. Using data above determine the constants a_{12}, a_{13}, a_{21}, a_{31}. Create plots, showing lead content in blood, tissues and bones for a 1-year period from the moment a citizen came to a city.

79. $PbCl_2$ melt was electrolysed. Cathode area was 40 m^2. Six experiments under different values of the intensity of current I and quantity of electricity Q were performed. Mass of lead formed on cathode was registered. The following results were obtained:

I, A	2.5	1.5	1.0	0.6	0.25	0.075
Q, F	4,825	12,000	7,308	9,600	7,504	17,020
m_{Pb}, g	5	11.5	7.0	9.25	6.5	10.5

Build dependencies of lead's current-yield and speed of its precipitation on the current density.

80. Electrolysis of $AgNO_3$ solution with a silver anode proceeded for 2 h. During this time current in a chain changed as $I(\tau) = (a + \tau)^{-1}$, where τ is time, *h*. Calculate the increase in the mass of anode during electrolysis, if speed of anode dissolution in the end of electrolysis was equal to 4.48×10^{-4} g s^{-1}.
81. Copper content in some alloy of mass m = 22.1234 g was analysed using coulometry. The following data was obtained by registering the intensity of current in different moments in time under some fixed value for an electrode's potential:

t, (min)	0	2	4	6	8	10
I, (mA)	501.0	339.4	238.1	163.3	110.6	71.6
t, (min)	12	14	16	18	60	120
I, (mA)	47.6	31.2	18.4	11.6	0.00	0.00

Current in a cell changed in time according to a law: $I(t) = A\cdot\exp(-Bt)$. Determine fraction of cooper in an alloy. Assume that under given conditions copper is the only electrochemically active component.

82. Dependence of overvoltage η on the current density i for a two-electron cathode reaction under 298.2 K is as follows:

$i\cdot10^4$ (A cm^2)	2	10	20	50	80	120	160	200
η, (V)	−0.212	−0.258	−0.277	−0.307	−0.315	−0.326	−0.334	−0.341

Determine the coefficients in Tafel's equation and kinetic parameters of an electrochemical reaction – exchange current and transport factor α. Assume that an electrode process proceeds according to a theory of delayed discharge.

83. Polarisation curve of some electrode process is described by an equation of the theory of delayed discharge. Build plots, which illustrate the change of the form of the total polarisation curve and partial curves of anode and cathode processes depending on the change of exchange current i_0 and transport factor α.
84. Consider the applicability of *genfit* function to determine the parameters of an electrode process with a delayed stage of charge transfer based on the polarisation measurements. Firstly, create a data array for the dependence of current density i on the overvoltage η, using the equation of the theory of delayed discharge. Set acceptable values for the parameters, e.g. exchange current $i_0 = 1.4 \times 10^{-6}$ A cm^{-2}, transport factor $\alpha = 0.5$, number of electrons $n = 1$ in a range of overvoltage from −0.100 V to +0.100 V with a step size of 1 mV. Further, solve an inverse problem: process a data array using *genfit* in order to calculate i_0, α, n.

85. Determine constants a and b in an empiric equation $N_k = a \cdot e^{-b/\eta^2}$, connecting a number of 3D embryos Nk to an overvoltage during cathode precipitation of mercury on the platinum surface, using the following data:

$1/\eta^2$, (V^{-2})	227.8	215.1	202.3	191.5	180.8	172.3	163.5
lg N_k	0.389	0.834	1.111	1.332	1.555	1.945	2.140

Apply function *genfit*. Compare the results of non-linear approximation to the results, obtained by linearization of the data.

86. Current ran through an electrochemical cell during 2π hours. Current changed according to a law $I(t) = 2+\sin(t)$, where t is time, hours. Calculate the quantity of electricity that passed through the chain for a period of time, corresponding to the change in current from its maximum to its minimum value.

87. Consider kinetic model of a phasic zinc ionisation

(I) $Zn - e \underset{k_{-1}}{\overset{k_1}{\rightleftarrows}} Zn^+,$

(II) $Zn^+ - e \xrightarrow{k_2} Zn^{2+},$

Assuming that each of the electron transfer stages follows the regularities of a delayed discharge, determine a theoretical value of a Tafel's slope of polarisation characteristics depending on a number of a delayed stage.

88. Irreversible process

$$Peroxy\ benzoic\ acid + 2e + H^+ \longrightarrow Benzoic\ acid + OH^-,$$

proceeds in $NaOH$ solution of 0.1 mol L^{-1} concentration on a silver electrode, covered with a thin layer of Ag_2O. Tafel's straight line with a slope of 38 ± 2 mV was found for that process. It is assumed that kinetic scheme of the process could be considered as a set of stages:

$$A + e \underset{k_{-1}}{\overset{k_1}{\rightleftarrows}} B \quad (a);\ B + e \underset{k_{-2}}{\overset{k_2}{\rightleftarrows}} C \quad (b)$$

Kinetic equation for an overall process is as follows:

$$i = 2 \cdot \frac{k_1 k_3 c_A \exp\frac{-(2-\alpha_1-\alpha_2)FE}{RT} - k_{-4}k_{-2}c_C \exp\frac{(\alpha_1+\alpha_2)FE}{RT}}{k_3 \exp\left[-\frac{(1-\alpha_1)FE}{RT}\right] + k_{-2}\exp\frac{\alpha_1 FE}{RT}},$$

where C_A, C_C are the concentrations of an initial substance and a final product. Is it reasonable to assume that peroxy benzoic acid reduction is a one-stage process? Which one of the stages is irreversible? Consider the following possibilities:
(1) reactions (a) and (b) are irreversible, e.g. k_{-1} and k_{-2} equal to zero; (2) $k_{-1} = 0$, and reaction (b) is irreversible; (3) reaction (a) is reversible, $k_{-2} = 0$, and $k_2 \approx k_{-1}$; (4) reaction (a) is reversible, $k_{-2} = 0$, and $k_2 \ll k_{-1}$. Determine, which of the possibilities does the experimentally defined Tafel's slope correspond to.

89. The following dependence of a transition time of a current density was obtained for some electrode process using chronopotentiometry.

i_k, (mA cm^{-2})	2	5	8	9	14
τ, (s)	60.84	9.73	3.80	3.00	1.24

Create a document, confirming the delay in a diffusion stage.

90. When considering a problem of non-stationery diffusion in a galvanostatic chain shorting it was assumed that during the whole time of the electrolysis a current, passing through a working electrode, was direct, e.g. one of the boundary conditions was a direct diffusion current of an oxidised form of an active substance. Consider a case, when diffusion current is a periodic function of time, e.g. sinusoidal. Then a boundary condition, characterising an electrolysis regimen, is as follows

$$\left(\frac{\partial C_{Ox}(x,t)}{\partial x}\right)_{x=0} = \frac{i(t)}{nFD_{Ox}} = \frac{\sin(\omega t)}{nFD_{Ox}}$$

Here ω is a circular frequency. Deduce analytic expression, describing distribution of an oxidised form in time and depending on the distance to an electrode.

91. Solve a problem of non-stationary diffusion to a spherical electrode under galvanostatic chain shortening, using equation of the second Fick's law in spherical coordinates. Use your solution to process experimental data, obtained by investigating a process, controlled by a mass-transfer stage, using galvanostatic method on a spherical electrode. The data show the following dependence of a transition time τ on the current density i:

i, (mA cm^{-2})	0.05	0.075	0.10	0,20
τ, (s)	9	1.35	0.61	0.11

Diffusion constant of a reactant D_{Ox} equals to 1×10^{-5} cm^2 s^{-1}, electrode process is a one-electron process. Determine concentration of a reactant C_{Ox} and radii of electrode r_0.

Instructions. Correct solution gives the following relationship between the current density and the transition time:

$$\frac{nFD_{Ox}C_{Ox}}{ir_0} = 1 - \exp\left(\frac{D_{Ox}\tau}{r_0{}^2}\right) erfc\left[\frac{(D_{Ox}\tau)^{1/2}}{r_0}\right].$$

Still, this equation is difficult to apply in practice. Prove, that under a condition, when $D_{Ox}\tau/r_0^2$ is large, it could be deduced to an approximate, but easy to use in practice equation:

$$i\tau^{1/2} = \frac{nF\sqrt{\pi D_{Ox}}C_{Ox}}{2} + \frac{\pi nFD_{Ox}C_{Ox}\tau^{1/2}}{4r_0}.$$

92. In 1950s *Gerisher* and *Filstich* suggested potentiostatic method to investigate the processes of electrochemical kinetics. The idea beyond the method is that electrode potential is dramatically changed from its equilibrium value φ_p to some value φ and measure the resulting current depending on time. As a rough approximation one can assume that diffusion rate is inversely proportional to $\sqrt{t}$, and a rate of discharge-ionisation stage does not depend on time. Then if one decreases the time of electrolysis, rate of diffusion stage would increase, and if the ions' discharge was instant, then under $t \to 0$ one could expect an infinite value of current. Under a finite discharge rate current under $t \to 0$ is not infinite, and has a specific value, defined by a rate of the slowest stage, which could be exactly a discharge stage. Correspondingly, by analysing dependence $i(t)$, one can obtain quantitative information about kinetic parameters of discharge-ionisation stage.

A general equation of mixed kinetics is as follows

$$i = i_0\left\{\frac{C_{Ox}{}^s}{C_{Ox}{}^0}\exp\left(\frac{\alpha nF\eta}{RT}\right) - \frac{C_R{}^s}{C_R{}^0}\exp\left[-\frac{(1-\alpha)nF\eta}{RT}\right]\right\},$$

where surface concentrations C^s_{Ox} and C^s_R are in this case functions of time. Therefore, to obtain theoretic dependence $i(t)$, it is necessary to solve the corresponding diffusion problem relatively to $C_{Ox}(x,t)$ and $C_R(x,t)$, and then put obtained values for $C_{Ox}(0,t)$, $C_R(0,t)$ into an equation of mixed kinetics. Initial and boundary conditions are: under $t = 0$ $C_{Ox}(x,t) = C^0_{Ox}$; $C_R(x,t) = C^0_R$; under $t > 0$ i $x = \infty$ $C_{Ox}(\infty,t) = C^0_{Ox}; C_R(\infty,t) = C^0_R$. Sum of flows of the substances through electrode surface equals zero:

$$D_{Ox}\left(\frac{\partial C_{Ox}(x,t)}{\partial t}\right)_{x=0} + D_R\left(\frac{\partial C_R(x,t)}{\partial t}\right) = 0.$$

Under conditions, when process is simultaneously controlled by diffusion and discharge-ionisation stage

$$D_{Ox}\left(\frac{\partial C_{Ox}(x,t)}{\partial t}\right)_{x=0} = k_1C_{Ox}(0,t) - k_2C_R(0,t),$$

where formal rate constants k_1 and k_2 are given by the equations:

$$k_1 = \frac{i_0}{nFC_{Ox}{}^0}\exp\left[\frac{\alpha nF\eta}{RT}\right]; \quad k_2 = \frac{i_0}{nFC_R{}^0}\exp\left[\frac{-(1-\alpha)nF\eta}{RT}\right].$$

Kambara and *Tachi* first solved this problem in 1952. Deduce the dependence $i(t)$ applying methods of computational mathematics.

93. Blood cells, moving inside a capillary blood vessel, release oxygen, which spreads from there into surrounding tissues due to diffusion. Tissue could be described as a concentric cylinder of radii R_I with a blood vessel of radii R_0 ($R_0 < R_1$) in the middle. Concentration of oxygen in a capillary space (including capillary walls) equals C_0 mol L^{-1}. Oxygen flow inside the tissue (dC/dr under $r = R_0$) is known and equals 2 mol L^{-1} m^{-1}. Tissue spends oxygen on the life support necessities with speed M mol L^{-1} s^{-1}. Deduce analytic expression for a stationery oxygen concentration in tissue as a function of distance to a capillary wall. Assume that oxygen diffusion constant in tissue is $D = 2$ m^2 s^{-1}, $M = 1$ mol L^{-1} s^{-1}, $C_0 = 3$ mol L^{-1}.

Bibliography

1. S. Benson, *The Foundation of Chemical Kinetics* (McGraw-Hill Company, New York, 1960)
2. F. Daniels, R. Alberty, *Physical Chemistry* (Wiley, Hoboken, NJ, 1965)
3. N. Emanuel, D. Knorre, *Chemical Kinetics (Homogeneous Reactions)* (Vysshaja Shkola, Moscow, 1984) (in Russian)
4. P. Atkins, *Physical Chemistry* (Oxford University Press, Oxford, 1998)
5. M. Swihart, Constructing Reaction Mechanisms, Chapter 5, in *Modelling of Chemical Reactions*, ed. by R. Carr. Comprehensive Chemical Kinetics, vol. 42 (Elsevier, Netherlands, 2007), pp. 189–246
6. M. Swihart. Applied Chemical Kinetics (Lecture Notes), http://www.eng.buffalo.edu/Courses/ce561/LecNotes.html
7. H. Bittrich, D. Haberland, G. Just, *Methoden chemisch-kinetischer Berechnungen* (VEB Deutscher Verlag fur Grundstoffindustrie, Leipzig, 1979)
8. J. Noggle, *Physical Chemistry Using Mathcad* (Pike Creek Publishing Company, Petersburg, IN, 1997)
9. D. Skoog, *Introduction to Analytical Chemistry: Using Mathcad* (Henry Holt and Company, New York, 1997)
10. G. Calzaferri. Skript und MathCad Programme fuer die Vorlesung PC IIA (Kinetik), http://iacrs1.unibe.ch/kinetik/startdcb.html
11. K. Krasnov, N. Vorob'ev, *Physical Chemistry. Vol. 2. Electrochemistry. Physical Chemistry and Catalysis* (Vysshaja Shkola, Moscow, 1995). in Russian
12. A. Kubasov, *Chemical Kinetics and Catalysis. Part 1* (Moscow University Press, Moscow, 2004). http://www.chemnet.ru/rus/teaching/kubasov/welcome1.html, (in Russian)
13. A. Kubasov, *Chemical Kinetics and Catalysis. Part 2* (Moscow University Press, Moscow, 2005). http://www.chem.msu.su/rus/teaching/kubasov/1-3.pdf, (in Russian)
14. Z. Galus, *Teoretyczne Podstawy Elektroanalizy Chemicznej* (Panstwowe Wydawnictwo Naukowe, Poland, 1971). in Polish
15. B. Damaskin, O. Petrij, *Introduction in Electrochemical Kinetics* (Vysshaja Shkola, Moscow, 1975). in Russian
16. E. Moelwyn-Hughes, *The Chemical Statics and Kinetics of Solutions* (Academic, London, 1971)
17. D. Frank-Kamenetskii, *Diffusion and Heat Transfer in Chemical Kinetics* (Springer, Heidelberg, 1959)
18. N. Emanuel, *Experimental Methods of Chemical Kinetics* (Vysshaja Shkola, Moscow, 1971) (in Russian)
19. N. Semenov, *Chain Reactions* (Nauka, Moscow, 1986) (in Russian)
20. G. Riznichenko, *Mathematical Models in Biology* (Izhevsk, Moscow, 2002). in Russian
21. H. Adidharma, V. Temyanko, *Mathcad for Chemical Engineers* (Trafford Publishing, Bloomington, IN, 2007)
22. V. I. Korobov. Laplace Transform in Physical Chemistry and Electrochemistry, http://www.exponenta.ru/educat/systemat/korobov/laplas/main.asp, (in Russian)

V.I. Korobov and V.F. Ochkov, *Chemical Kinetics with Mathcad and Maple*,
DOI 10.1007/978-3-7091-0531-3, © Springer-Verlag/Wien 2011

23. J. Andraos, A Streamlined Approach to Solving Simple and Complex Kinetic System Analytically. J. Chem. Educ. **76**(11), 1578–1583 (1999)
24. V. I. Korobov, E. Khren. Chemical Kinetics Tasks on the MAS, http://twt.mpei.ac.ru/TTHB/1/ChimKin.html, (in Russian)
25. Chemical Kinetics (http://www.chm.davidson.edu/ChemistryApplets/kinetics/)
26. P. Keusch. Datenanalyse mit Microsoft Excel – Chemische Kinetik, http://www.uni-regensburg.de/Fakultaeten/nat_Fak_IV/Organische_Chemie/Didaktik/Keusch/excel.htm
27. A. F. Bopp. A Chemical Kinetics Application of Mathcad, http://bluehawk.monmouth.edu/~tzielins/mathcad/ABopp/doc001.htm
28. J. A. Pojman. Studying Chemical Dynamics with Numerical Experiments, http://bluehawk.monmouth.edu/~tzielins/mathcad/JPojman/doc001.htm
29. R. Poshusta. Oscillating Chemical Reactions, http://www.sci.wsu.edu/idea/OscilChem/index.html
30. J. Garcia. ABC Chemical Kinetics, http://jchemed.chem.wisc.edu/JCEDLib/SymMath/collection/article.php?id=27
31. I. Kudrjashov, G. Karetnikov, *The Collection of Examples and Problems in Physical Chemistry* (Vysshaja Shkola, Moscow, 1991) (in Russian)
32. N. Fok, M. Melnikov, *The Collection of Problems in Chemical Kinetics* (Vysshaja Shkola, Moscow, 1982) (in Russian)
33. L. Labowitz, J. Arents, *Problem and Solutions Physical Chemistry* (Academic, London, 1969)
34. F. Kukoz et al., *The Collection of Problems in Theoretical Electrochemistry* (Vysshaja Shkola, Moscow, 1982) (in Russian)
35. I. Leenzon. Chemical Kinetics: Control Questions and Tasks, http://www.chem.msu.su/rus/teaching/leenson/zadaniya/welcome.htm, (in Russian)
36. V. F. Ochkov, *Mathcad 14 for Students, Engineers and Designers* (BHV Publisher House, Saint-Petersburg, 2007). in Russian
37. D. Kir'anov, *Mathcad 13* (BHV Publisher House, Saint-Petersburg, 2006) (in Russian)
38. V. F. Ochkov, *Mathcad 12 for Students and Engineers* (BHV Publisher House, Saint-Petersburg, 2005) (in Russian)
39. A. Solodov, V. F. Ochkov, *Differential Models. An Introduction with Mathcad* (Springer, Heidelberg, 2004)
40. V. F. Ochkov, *Units in Mathcad and Maple* (Finansy i Statistica, Moscow, 2002). in Russian
41. V. F. Ochkov, *Tips for Mathcad 2001 Users* (Publisher House of Moscow Power Engineering Institute, Moscow, 2001) (in Russian)
42. V. F. Ochkov, *Mathcad 8 Pro for Students and Engineers* (Computer, Moscow, 1999) (in Russian)
43. V. F. Ochkov, *Mathcad 7 Pro for Students and Engineers* (Computer, Moscow, 1998) (in Russian)
44. V. F. Ochkov, *MathCAD PLUS 6.0 for Students and Engineers* (Computer, Moscow, 1996)
45. V. D'jakonov, *Maple 9 in Mathematic, Physic and Education* (SOLON, Moscow, 2004) in Russian
46. G. Prokhorov et al., *Maple in Symbolic Calculations* (Petit Publisher House, Moscow, 1997). in Russian
47. K. Ebert, H. Ederer, *Computeranwendungen in der Chemie* (VCH Verlagsgeselschaft mbH, Weinheim, 1985)
48. N. Stepanov, M. Erlykina, G. Filippov, *Linear Algebra Methods in Physical Chemistry* (Moscow University Press, Moscow, 1976) (in Russian)
49. V. I. Korobov, Exponenta Pro. Math. Appl. **3–4**, 115–121 (2004) (in Russian)
50. V. Sabanin, N. Smirnov et al. Exponenta Pro. Math. Appl. **3–4**, 78–85 (2004) (in Russian)
51. V. D'jakonov, Exponenta Pro. Math. Appl. **3**, 10–19 (2003) (in Russian)
52. D. Kir'janov, Exponenta Pro. Math. Appl. **3**, 66–71 (2003) (in Russian)
53. V. Ditkin, A. Prudnikov, *Operational Calculus* (Vysshaja Shkola, Moscow, 1975) (in Russian)
54. R. Guter, A. Janpolsky, *Differential Equations* (Vysshaja Shkola, Moscow, 1976) (in Russian)

About the Authors

Viktor I. Korobov (korvik58@mail.ru) – Head of the Department of Physical and Inorganic Chemistry, Dnepropetrovsk National University (http://www.dsu.dp.ua). Research interests – the kinetics of electrochemical reactions, protection of metals against corrosion, methodological aspects of the use of mathematical packages in chemical education.

Valery F. Ochkov (ochkov@twt.mpei.ac.ru) – professor of Moscow Power Engineering Institute (technical university – http://www.mpei.ru). Specialist in the field of information technologies for the power engineering and chemical industry (see – http://www.trie.ru). Official beta tester Mathcad versions from 7 to 15 and Mathcad Prime. Active participant of the Mathcad Internet forum PlanetPTC (http://communities.ptc.com).

Index

MIX
Papier aus verantwortungsvollen Quellen
Paper from responsible sources
FSC® C105338

If you have any concerns about our products,
you can contact us on
ProductSafety@springernature.com

In case Publisher is established outside the EU,
the EU authorized representative is:
Springer Nature Customer Service Center GmbH
Europaplatz 3, 69115 Heidelberg, Germany

Printed by Libri Plureos GmbH
in Hamburg, Germany